3rd International Conference on Sustainable Construction Technologies & Advancements in Civil Engineering (ScTACE 2022)

IOP Conference Series: Materials Science and Engineering
Volume 1282

Bhimavaram, India
15-17 December 2022

Editors:

Pala Gireesh Kumar
S. Jayalakshmi

ISBN: 978-1-7138-8184-1
ISSN: 1757-8981

Printed from e-media with permission by:

Curran Associates, Inc.
57 Morehouse Lane
Red Hook, NY 12571

Some format issues inherent in the e-media version may also appear in this print version.

Institute of Physics
Dirac House, Temple Back
Bristol BS1 6BE UK

Phone: 44 1 17 929 7481
Fax: 44 1 17 920 0979

techtracking@iop.org

Additional copies of this publication are available from:

Curran Associates, Inc.
57 Morehouse Lane
Red Hook, NY 12571 USA
Phone: 845-758-0400
Fax: 845-758-2633
Email: curran@proceedings.com
Web: www.proceedings.com

TABLE OF CONTENTS

IOP Conf. Series: Materials Science and Engineering 1282 (2023) 011001 doi:10.1088/1757-899X/1282/1/011001

PREFACE

We were delighted with a high level of satisfaction and aspiration, indeed it's a great privilege to recapitulate the 3rd International Virtual Conference on Sustainable Construction Technologies & Advancement in Civil Engineering, ScTACE 2022. ScTACE 2020 & ScTACE 2021 extends its journey by appending more noteworthy and top notch accessories. The third edition of ScTACE in collaboration with Indian Concrete Institute – Student Chapter is reinvigorated in December 2022, aims to provide collaborative opportunities for researches, academicians, engineers, industry practitioners with one accord to discourse and share their grounding, indagation about most aspects of sustainable solutions and deliberate over the challenges encountered in the explication of the solutions.

The three day programme was excogitated with an even handed to the presenters as well as listeners with a bodacious array of 12 keynote and 5 invited guest speakers with prowess from IIT's, NIT, IISc, renowned Universities and industry from India and overseas. The Conference beheld a noteworthy representation with 350 entrants from variegated institutions, organizations across the country and overseas made this 3rd conference of its kind -a truly conference in spirit. The varied genres of trending technologies or materials in constituent to the sustainability and the series of interactive sessions promoted the indefatigable spirit of 350 learners and encouraged their determination in adding an extra skill to their forte. 57 papers were presented with divaricated 5 parallel panels, formed to be the heart of the conference and provided ample opportunity for discussion that brought new perspectives as a stepping stone in the enhancement of knowledge and skill of those working in the field of civil engineering. The proceedings provided a written record of the synergy with 173 paper submission with 21% of acceptance rate (44 accepted papers) publishing in the IOP Conference Series: Materials Science and Engineering owing to the Peer Review Policy that represented a solid framework which emerge a new interaction in the fullness of time. The event ended with the unveiling of the conference sovereign leaving everyone's experience with ScTACE-2022, a fruitful and long lasting one.

All in all the plenary lectures, progress and presentations bridged the gap making it possible for the people to stay productive and learn or to get au courant with newfangled technologies or skills. We divulge our unfeigned delight to all the external chairpersons for their admirable serenity and accorded stupendous time for colloquy, elucidation of doubts mooted by the partakers. We take the entitlement in acclaiming all the keynote and guest expertise for their credence and immense deliberation that aided the budding graduates and professionals to hone their skills in the art of civil engineering and pursue their long-procrastinated passions. We are also honor-bound to profess our acknowledgement to the management of Shri Vishnu Engineering College for Women for its verve in lending a hand and believing that there should be no reason for people to stop learning new skills in the competitive times we live in. We would like to express our gratitude and appreciation for all the reviewers who helped us maintain the capacity constraints and high quality of manuscripts included in the proceedings. Equally or exceedingly we bestow our commendation to the support team, committee members, and student coordinators for their stance and idiosyncrasy in handling the contenders and making this conference viable.

We promise to bring forth the same bright picture and charming landscape with the aforementioned support, participation and will continue to extend the success for a long time.

Conference Chair

Dr.Pala Gireesh Kumar
Associate Professor & HOD
Shri Vishnu Engineering College for Women

ANNEXURE - A

EDITORS OF SCTACE 2022

1. Dr.Pala Gireesh Kumar
Associate Professor & HOD,
Department of Civil Engineering,
Shri Vishnu Engineering College for Women (Autonomous), Bhimavaram.

2. Dr. S. Jayalakshmi,
Professor,
Department of Civil Engineering,
National Institute of Technology, Tiruchirappalli

ANNEXURE- B

CONFERENCE ADVISORY COMMITTEE

Dr. Ramez M. Hajj, University of Illinois, Champaign, USA

Prof. Krishna R. Reddy, University of Illinois, Chicago, USA

Prof. Rao S. Govindaraju, Purdue University, Indiana, USA

Prof. Srinivas S. Pulugurtha, University of North Carolina, Charlotte, USA

Prof. Venkatesh Merwade, Purdue University, Indiana, USA

Prof. Kumar, Praveen, IIT Roorkee

Prot. D. Rama Seshu, NIT Warangal

Prof. Moses Santhakumar, NIT Trichy

Prof. C.S.R.K. Prasad, NIT Warangal

Prof. K. Baskar, NIT Trichy

Prof. J. Murali Krishnan, IIT Madras

Dr. Christelle Abadie, University of Cambridge, UK

Prof. P. K. Agrawal, NIT Bhopal

Asst Prof. CHEN, Jiayu, City University of Hong Kong, Hong kong

Prof. C. H. Solanki, NIT Surat

Dr, Stavroula Kontoe, Imperial College London, UK

Prof. Anil Kumar Sachan, NIT Allahabad

Prof. Ashim Kanti dey, NIT Silchar

Prof. Cao Bin, NTU Singapore

Prof. G. J. Joshi, NIT Surat

Prof. B. B. Das, NIT Surathkal

Prof. Satyabrata Choudhry, NIT Silchar

Prof. Farhad Aslani, The University of Western Australia, Australia

Prof. Gitakrishnan Ramadurai, IIT Madras

Prof. Venkateswara Rao S, NIT Warangal

ANNEXURE - C

CONFERENCE TECHNICAL CORE COMMITTEE

Prof. A. Murali Krishna, IIT Tirupati

Dr. G. Bharath, Scientist, CSIR CRRI, Delhi

Prof. S. T. Ramesh, NIT Trichy

Prof. K Srinivasa Raju, BITS Pilani, Hyderabad

Prof. Basavaraju Manu, NIT Surathkal

Prof. Mathavakumar S, IIT Madras

Dr. Nisha Radhakrishnan, NIT Trichy

Prof. Devatha C P, NIT Surathkal

Prof. Radhikesh P. Nanda, NIT Durgapur

Dr. R. Gandhimathi, NIT Trichy

Prof. Dilip A. Patel, NIT Surat

Dr. V. Sunitha, NIT Trichy

Dr. S. Saravanan, NIT Trichy

Dr. Ramya Sri Mullapudi, IIT Hyderabad

Prof. Satyajit Patel, NIT Surat

Prof. Danish Ahmad, NIT Srinagar

Prof. Kishan Dharavath, NIT Bhopal

Dr. R Manjula, NIT Trichy

Dr. Ravi Shankar KVR, NIT Warangal

Dr. Marisamynathan S, NIT Trichy

Dr. Ambika S, IIT Hyderabad

Dr. B Raghuram Kadali, NIT Warangal

Prof. Hemant S.Chore, NIT Jalandhar

Prof. J. A. Naqash, NIT Srinagar

Dr. Sangeeta Kumari, NIT Jamshedpur

Dr. Kranthi Kumar Kuna, IIT Kharagpur

Dr. Somenath Mondal, NIT Jamshedpur

Dr. Anant Parghi, NIT Surat

Prof. I. V. Ramana Reddy, S. V. University, Tirupati

Prof. G. N. Pradeep Kumar, S. V. University, Tirupati

Dr. G. Sreedhar, Senior Associate Consultant, Infosys, Hyderabad

Prof. Ch. Sudha Rani, S. V. University, Tirupati

Prof. G Reddy Babu, GEC

Dr. P. Kesava Rao, C-GARD, NIRD, Hyderabad

Dr. K. Padma Kumari, JNTUK, Kakinada

Dr. M. V. S. S. Giridhar, JNTUH, Hyderabad

Dr. Deendayal Rathod, NIT Trichy

Dr. N. S. R. Prasad, C-GARD, NIRD, Hyderabad

Dr. Jeevan Joseph, NIT Trichy

Dr G George, NIT Trichy

Prof. M. Srimurali, S. V. University, Tirupati

Prof. D. V. Prasada Rao, S. V. University, Tirupati

Er. Rajesh Khanna, Regional Technical Services Manager, Andhra Pradesh

ANNEXURE - D

CONFERENCE PEER REVIEW COMMITTEE

IOP Conf. Series: Materials Science and Engineering 1282 (2023) 011001 doi:10.1088/1757-899X/1282/1/011001

S. No.	Name	Affiliation	Mail I.D	Specialization
1	Dr. Shaik Mahabu Subhani	Assistant Professor, Department of Civil Engineering, NIT Andhra Pradesh, Andhra Pradesh	subhani@nitandhra.ac.in	Structural Engineering-Concrete technology; Structural Health Monitoring; Smart materials and composites; Constitutive modelling;
2	Dr. Chalak Hanuman Devidas	Assistant Professor, Department of Civil Engineering, NIT Kurukshetra, Harayana	chalakhd@nitkkr.ac.in	Structural Engineering- Finite Element Analysis: Laminated Composite Structures, Sandwich Structures, Functionally graded material Application of FRP in Civil Engineering
3	Dr. Anant Parghi	Assistant Professor, Department of Civil Engineering, NIT Surat, Gujarat	amp@amd.svnit.ac.in	Structural Engineering- Seismic analysis and design smart materials and their structural application, seismic retrofitting of steel and masonry structures, recycle/reuse of industrial wastes for structural applications, finite element analysis
4	Dr. D. Ravi Prasad	Assistant Professor, Department of Civil Engineering, NIT Warangal, Telangana	drprasadravi@gmail.com	Structural Engineering- cement concrete composites, structural health monitoring
5	Dr. S. Anitha Priyadharshani	Assistant Professor, Department of Civil Engineering, NIT Warangal, Telangana	priyadharshanianitha@nitw.ac.in	Structural Engineering- Fiber Reinforced Polymer Composites; Stiffened panels; Finite Element Analysis
6	Dr. Nitesh A.	Assistant Professor, Department of Civil Engineering, NIT Silchar, Assam	nitesh@civil.nits.ac.in	Structural Engineering- Geotechnical Earthquake Engineering, Structural Earthquake Engineering
7	Dr. Prashanth J.	Assistant Professor, Department of Civil Engineering, NIT Silchar, Assam	prashanth@civil.nits.ac.in	Structural Engineering- Coastal Engineering, Marine Structures, Concrete Technology
8	Dr. Atanu Sahu	Assistant Professor, Department of Civil Engineering, NIT Silchar, Assam	atanu@civil.nits.ac.in	Structural Engineering- Vibroacoustics, Noise control, Structural dynamics and vibration control, Fluid-structure interaction, Non-linear hydrodynamics
9	Dr. Somnath Karmakar,	Assistant Professor, Department of Civil Engineering, NIT Durgapur,	somnath.karmakar@ce.nitdgp.ac.in	Structural Engineering- Blast impact (FEM-SPH Coupling) on bridges/structures, Wind impact (CFD) on bridges/structures,

		West Bengal		Retrofitting, Quality Control, and Constructional methodology of bridges/structures
10	Dr. Joy Pal	Assistant Professor, Department of Civil Engineering, NIT Hamirpur, Himachal Pradesh	jpal@nith.ac.in	Structural Engineering--Vibration Based Structural Health Monitoring: AE Based Identification of Source of Damage Location
11	Dr. Umesh Pandey	Assistant Professor, Department of Civil Engineering, NIT Hamirpur, Himachal Pradesh	ukp@nith.ac.in	Structural Engineering-
12	Dr. Dhiraj Raj	Assistant Professor, Department of Civil Engineering, NIT Jaipur, Rajasthan	dhiraj.ce@mnit.ac.in	Structural Engineering-Earthquake Engineering, Earthquake resistant Buildings and Construction Practices, Performance-Based Design of Structures, Seismic Vulnerability and Risk Evaluation of Structures
13	Dr. Mohit Jaiswal	Assistant Professor, Department of Civil Engineering, NIT Raipur, Chhattisgarh	mjaiswal.ce@nitrr.ac.in	Structural Engineering-Matrix analysis of structures, Wind Effect on high rise structures, Seismic Engineering
14	Dr. Meena Murmu	Assistant Professor, Department of Civil Engineering, NIT Raipur, Chhattisgarh	mmurmu.ce@nitrr.ac.in	Structural Engineering-Waste management in cement and concrete construction. Characterization of raw materials of cement and concrete. Optimization in cement and concrete
15	Dr. Govardhan Bhat	Assistant Professor, Department of Civil Engineering, NIT Raipur, Chhattisgarh	gov.ce@nitrr.ac.in	Structural Engineering-Structural Dynamics seismic Base Isolation Geotechnical Earthquake Engineering Earthquake protection of Buildings
16	Dr. Alfia Bano	Assistant Professor, Department of Civil Engineering, NIT Raipur, Chhattisgarh	alfia.ce@nitrr.ac.in	Structural Engineering-Structural Analysis, Composite Materials, Stability of Structures, Finite Element Analysis, Earthquake Engineering
17	Dr. M Abdul Akbar	Assistant Professor, Department of Civil Engineering, NIT Jalandhar, Punjab	akbarma@nitj.ac.in	Structural Engineering
18	Dr. Navdeep Singh	Assistant Professor, Department of Civil Engineering, NIT Jalandhar,	navdeeps@nitj.ac.in	Structural Engineering

		Punjab		
19	Dr. Neelam Rani	Assistant Professor, Department of Civil Engineering, NIT Jalandhar, Punjab	ranin@nitj.ac.in	Structural Engineering
20	Dr. Senthil Kasilingam	Assistant Professor, Department of Civil Engineering, NIT Jalandhar, Punjab	kasilingams@nitj.ac.in	Structural Engineering
21	Dr.. S. A. Waseem	Assistant Professor, Department of Civil Engineering, NIT Srinagar, Uttarakhand	shakeelahmad@nitsri.ac.in	Structural Engineering
22	Dr. Fayaz Ahmad Sofi	Assistant Professor, Department of Civil Engineering, NIT Srinagar, Uttarakhand	sofifayaz@nitsri.ac.in	Structural Engineering
23	Dr. P. N. Hiremath	Assistant Professor, Department of Civil Engineering, NIT Srinagar, Uttarakhand	parameshwar.n@nitsri.net	Structural Engineering
24	Dr. Sagadevan R	Assistant Professor, Department of Civil Engineering, NIT Srinagar, Uttarakhand	sagadevan@nitsri.net	Structural Engineering
25	Dr. Mitesh Surana	Assistant Professor, Department of Civil Engineering, IIT Ropar, Punjab	msurana@iitrpr.ac.in	Structural Engineering-Risk-based Seismic Design of Buildings, Seismic Response Evaluation of Structures, Nonlinear Modelling and Analysis of RC Buildings, Seismic Design of Acceleration Sensitive Non-structural Components
26	Dr. Putul Haldar	Assistant Professor, Department of Civil Engineering, IIT Ropar, Punjab	putul.haldar@iitrpr.ac.in	Structural Engineering-Seismic Vulnerability and Risk Evaluation of Structures, Seismic Evaluation and Retrofitting of Structures, Performance-Based Design of Structures, Nonlinear Modeling and Analysis of Structures
27	Dr. Jayanthi N.V.Prathyusha	Assistant Professor, Department of Civil Engineering, NITAP,	prathyusha@nitandhra.ac.in	Geotechnical Engineering-Geotechnical Engineering; Environmental Geotechnology; Advanced Geomaterial Characterization; Sustainable

IOP Conf. Series: Materials Science and Engineering 1282 (2023) 011001 doi:10.1088/1757-899X/1282/1/011001

		Andhra :Pradesh		geomaterials; Climate change and its impact on Geomaterials; Desiccation cracking in soils; Thermal properties of geomaterials;
28	Dr. (Mrs.) Shruti J. Shukla	Assistant Professor, Department of Civil Engineering, NIT surat, Gujarat	sdv@amd.svnit.ac.in	Geotechnical Engineering- Geo-technical engineering and soil improvement techniquesPursuing research in the field of piled raft foundation
29	Dr. Ankesh Kumar	Assistant Professor, Department of Civil Engineering, NIT surat, Gujarat	ankesh@amd.svnit.ac.in	Geotechnical Engineering- Machine Foundation, Soil Dynamics, Rock Mechanics, Analysis of Underground Structures
30	Dr. G. V. Ramana	Assistant Professor, Department of Civil Engineering, NIT Warangal, Telangana	gvramana@nitw.ac.in	Geotechnical Engineering-Rock Mechanics and Rock Engineering, In-Situ and Laboratory Rock Mechanics Investigations (Deformability characteristics, In-Situ Stress Measurements, Shear strength parameters
31	Dr. Sreevalsa Kolathayar	Assistant Professor, Department of Civil Engineering, NIT Surathkal, Karnataka	sreevalsa@nitk.edu.in	Geotechnical Engineering- Geotechnical Earthquake Engineering, Seismic Hazard and Risk, Landslides, Disaster Risk Reduction, Reinforced earth, Natural Geocells, Water Geotechnics
32	Dr. Manendra Singh	Assistant Professor, Department of Civil Engineering, NIT Hamirpur, Himachal Pradesh	manendra@nith.ac.in	Geotechnical Engineering- Soil Dynamics, Geotechnical Engineering, Soil-structure interaction, liquefaction, Underground tunnels
33	Dr. Jitendra Singh Yadav	Assistant Professor, Department of Civil Engineering, NIT Hamirpur, Himachal Pradesh	jyadav@nith.ac.in	Geotechnical Engineering- Soil stabilization using waste material, ground improvement techniques especially design of Geo-synthetic and fiber reinforced soil structures, earth-retaining structures
34	Dr. A. D. Prasad	Assistant Professor, Department of Civil Engineering, NIT Raipur, Chhattisgarh	adprasad.ce@nitrr.ac.in	Geotechnical Engineering- Geomatics Engineering
35	Dr. Anjana Bhasi	Assistant Professor, Department of Civil Engineering, NIT Calicut, Kerala	anjanabhasi@nitc.ac.in	Geotechnical Engineering-

36	Dr. Seethalakshmi P	Assistant Professor, Department of Civil Engineering, NIT Calicut, Kerala	seetha@nitc.ac.in	Geotechnical Engineering-
37	Dr. R.P. Shukla	Assistant Professor, Department of Civil Engineering, NIT Srinagar, Uttarakhand	rpshukla@nitsri.ac.in	Geotechnical Engineering-
38	Dr. Ritesh S. Ingale	Assistant Professor, Department of Civil Engineering, NIT Srinagar, Uttarakhand	ritesh.ingale@nitsri.ac.in	Geotechnical Engineering-
39	Dr. Supriya Mohanty	Assistant Professor, Department of Civil Engineering, IIT Varanasi, Uttar Pradesh	supriya.civ@iitbhu.ac.in, supriya.mohanty88@gmail.com	Geotechnical Engineering-Geotechnical Earthquake Engineering, Liquefaction Potential Evaluation, Nonlinear Dynamic Response Analysis, Soil Dynamics, Waste Utilization
40	Dr. B. Hanumantha Rao	Assistant Professor, Department of Civil Engineering, IIT Bhubaneswar, Odisha	bhrao@iitbbs.ac.in	Geotechnical Engineering-Ground Improvement: Expansive soils stabilization Suction measurements and suction-swelling
41	Dr. Lohitkumar Nainegali	Assistant Professor, Department of Civil Engineering, IIT Jharkand, Jharkand	lohitkumar@iitism.ac.in	Geotechnical Engineering-Soils Dynamics And Foundations; Numerical And Experimental Modeling; Stability Of Geotechnical Structures
42	Dr. Sita Rami Reddy	Assistant Professor, Department of Civil Engineering, NIT Nagpur, Maharasthra	dsramreddy@civ.vnit.ac.in	Transportation Engineering-Pavement Material Characterization, Pavement Analysis & Design, Pavement Evaluation
43	Dr. Ambika Kuity	Assistant Professor, Department of Civil Engineering, NIT Silchar, Assam	ambika@civil.nits.ac.in	Transportation Engineering-Asphalt mix characterization, Microscopic study on asphalt mix, Rheological characterization of asphalt binders
44	Dr. Sunny Deol G	Assistant Professor, Department of Civil Engineering, NIT Raipur, Chhattisgarh	sdguzzarlapudi.ce@nitrr.ac.in	Transportation Engineering-Mechanistic-Empirical Pavement Design, Performance evaluation of pavements using NDT methods,Highway traffic
45	Dr. M Sivakumar	Assistant Professor, Department of Civil Engineering, NIT Calicut,	sivakumarm@nitc.ac.in	Transportation Engineering-

		Kerala		
46	Dr. Shailja Bawa	Assistant Professor, Department of Civil Engineering, NIT Jalandhar, Punjab	bawas@nitj.ac.in	Transportation Engineering-
47	Dr. Jitendra Gurjar	Assistant Professor, Department of Civil Engineering, NIT Srinagar, Uttarakhand	jitendra.gurjar@nitsri.ac.in	Transportation Engineering-
48	Dr. Ankit Gupta	Assistant Professor, Department of Civil Engineering, IIT Varanasi, Uttar Pradesh	ankit.civ@iitbhu.ac.in; anki_ce11@yahoo.co.in	Transportation Engineering- Pavement Performance Modeling, Pavement Material Characterization, Pavement Management Systems, SuperPave Design, Sustainable Pavements, Traffic Flow Modeling, Intelligent Transport Systems, Pedestrian Behavior Modeling, Road Safety.
49	Dr. Priyansh Singh	Assistant Professor, Department of Civil Engineering, IIT Indore, Madhya Pradesh	priyansh@iiti.ac.in	Transportation Engineering- Pavement Materials Characterization and Modeling, Pavement Design, Construction and Evaluation, Pavement Recycling, Innovative Materials and Technologies in Pavement Engineering
50	Dr. Anush Konayakanahalli Chandrappa	Assistant Professor, Department of Civil Engineering, IIT Bhubaneswar, Odisha	akc@iitbbs.ac.in	Transportation Engineering- Sustainable Pavement Materials and Technologies; Pavement Evaluation; Urban Heat Island Effect
51	Dr. Baranidharan Sundaram	Assistant Professor, Department of Civil Engineering, NIT AP, Andhra Pradesh	baranis@nitandhra.ac.in	Environmental Engineering- Water & Wastewater Treatment Process Fate and Transport of Emerging Contaminants & Monitoring Emerging Contaminants & Toxicity Studies Risk and Life Cycle Assessment
52	Dr. Rakhee Das	Assistant Professor, Department of Civil Engineering, NIT Rourkela, Odisha	dasr@nitrkl.ac.in	Environmental Engineering- Physicochemical water treatment | Urban water supply management | Water policies, law and legislations | PPP in water supply project | Stakeholder management
53	Dr. Namrata D. Jariwala	Assistant Professor, Department of Civil Engineering, NIT surat, Gujarat	ndj@ced.svnit.ac.in	Environmental Engineering- Health related issues in solid waste mangement practice, Environmetal Education, Environmet and health

54	Dr. George. K. Varghese	Assistant Professor, Department of Civil Engineering, NIT Calicut, Kerala	gkv@nitc.ac.in	Environmental Engineering-
55	Dr. Ashootosh Mandpe	Assistant Professor, Department of Civil Engineering, IIT Indore, Madhya Pradesh	asmandpe@iiti.ac.in	Environmental Engineering- Bio-valorization of solid wastes, Additive-aided-Thermophilic composting, Contaminants of Emerging Concern, Lifecycle Assessment
56	Dr. Mayur Shirish Jain	Assistant Professor, Department of Civil Engineering, IIT Indore, Madhya Pradesh	mayur.jain@iiti.ac.in	Environmental Engineering- Rapid Composting Techniques, Kinetic modelling of Bio-waste degradation, Circular economy in environmental engineering, Soil Revitalization via waste utilization
57	Dr. Indramani Dhada	Assistant Professor, Department of Civil Engineering, IIT ropar, Punjab	idhada@iitrpr.ac.in	Environmental Engineering- Environmental Engineering: Indoor and ambient air quality assessment; Source apportionment studies and modelling, Health Risk Assessment of pollutants, Fate processes of organic pollutants
58	Dr. Anurag Sharma	Assistant Professor, Department of Civil Engineering, NIT Rourkela, Odisha	sharmaan@nitrkl.ac.in	Water Resources Engineering- Turbulence, Sediment Transport, Fluid Mechanics, Open Channel Flow, Eco Fluvial Hydraulics
59	Dr. Litan Kumar Ray	Assistant Professor, Department of Civil Engineering, NIT Warangal, Telanagana	litan@nitw.ac.in	Water Resources Engineering- Stochastic hydrology, Flood & drought estimation, Real-time flood forecasting modelling, Hydrologic Modelling, climate change and climate variability study
60	Dr. Susmita Ghosh	Assistant Professor, Department of Civil Engineering, NIT silchar, Assam	susmita@civil.nits.ac.in	Water Resources Engineering - Groundwater Engineering; Groundwater modelling and Management; Hydrological modelling; Hydrodynamic modelling; Hydraulics
61	Dr. Ray Singh Meena	Assistant Professor, Department of Civil Engineering, NIT hamirpur, Himachal Pradesh	rsmeena@nith.ac.in	Water Resources Engineering- Hydrology Hydraulics River Engineering Sedimentation Groundwater Hydrology Remote Sensing and GIS
62	Dr. Rutuja M. Chavan	Assistant Professor, Department of Civil Engineering, NIT Bhopal,	rutujachavan297@gmail.com	Water Resources Engineering- Water Resources Engineering and Management

IOP Conf. Series: Materials Science and Engineering 1282 (2023) 011001 doi:10.1088/1757-899X/1282/1/011001

		Madhya Pradesh		
63	Dr. Maheswaran Rathinasamy	Assistant Professor, Department of Civil Engineering, IIT Hyderabad, Telangana	rmaheswaran@ce.iith.ac.in	Water Resources Engineering
64	Dr. Priyank J. Sharma	Assistant Professor, Department of Civil Engineering, IIT Indore, Madhya Pradesh	priyanksharma@iiti.ac.in	Water Resources Engineering-Statistical Hydroclimatology, Climate Change Impact on Basin Hydrology, Streamflow Analysis and Modelling, Reservoir Operation and Management
65	Dr. Kaustav Bakshi	Assistant Professor, Department of Civil Engineering, IIT Indore, Madhya Pradesh	kaustav.bakshi @ iiti.ac.in	Structural Engineering, Static and dynamic studies on laminated composite shell roofs, First and progressive ply failure studies, Finite element method; Geometric nonlinearity
66	Dr. Deendayal Rathod	Assistant Professor, Department of Civil Engineering, NIT Trichy, Tamil Nadu	deendayal@nitt.edu, deendayalrathod123@gmail.com	Geotechnical Engineeering-Migration of Sulphate under diffrent conditions by using Electro-kinetic Method, Geosynthetics,Removal of Fluoride by using IISC method
67	Dr. K. V. R. Ravi Shankar	Assistant Professor, Department of Civil Engineering, NIT Warangal, Telangana	ravikvr@nitw.ac.in	Transporatation Engineering-Mixed traffic flow modeling and simulation model development; Non-lane based behavior in mixed traffic conditions; Pedestrian behavioural analysis, modelling, and simulation;
68	Dr. Sangeeta Kumari	Assistant Professor, Department of Civil Engineering, NIT Jhamsedpur,		
69	Dr. Greegar George	Assistant Professor, Department of Civil Engineering, NIT Trichy, Tamil Nadu	greegar@nitt.edu	Structural Engineering, Structural Dynamics, Probabilistic modelling, Uncertainty Quantification, Global Sensitivity Analysis
70	Dr. R. Manjula	Assistant Professor, Department of Civil Engineering, NIT Trichy, Tamil Nadu		
71	Dr. Somenath mondal	Assistant Professor, Department of Civil Engineering, NIT Jhamsedpur,		

72	Dr. P. Sridhar	Associate Professor, Department of Civil Engineering, SVECW (A), Bhimavaram		Geoinformatics Engineering, Structural Mechanics
73	Mr. N. Pavan Kumar	Assistant Professor, Department of Civil Engineering, SVECW (A), Bhimavaram		Remote Sensing, Water Resources Engineering, Advance Surveying
74	Mr. Ramgopal. L	Assistant Professor, Department of Civil Engineering, SVECW (A), Bhimavaram		Structural Engineering, Concrete Technology, Repair and Rehabilitation

IOP Conf. Series: Materials Science and Engineering 1282 (2023) 011002 doi:10.1088/1757-899X/1282/1/011002

Peer Review Statement

All papers published in this volume have been reviewed through processes administered by the Editors. Reviews were conducted by expert referees to the professional and scientific standards expected of a proceedings journal published by IOP Publishing.

- **Type of peer review:** Double Anonymous
- **Conference submission management system:** Procedia Conference Managing System
- **Number of submissions received:** 210
- **Number of submissions sent for review:** 173
- **Number of submissions accepted:** 44
- **Acceptance Rate (Submissions Accepted / Submissions Received × 100):** 21
- **Average number of reviews per paper:** 2
- **Total number of reviewers involved:** 74
- **Contact person for queries:**
  Name: Dr. PALA GIREESH KUMAR
  Email: hodcivil@svecw.edu.in
  Affiliation: ASSOCIATE PROFESSOR & HEAD OF THE DEPARTMENT, DEPARTMENT OF CIVIL ENGINEERING, SHRI VISHNU ENGINEERING COLLEGE FOR WOMEN (AUTONOMOUS), BHIMAVARAM, ANDHRA PRADESH, INDIA

Parametric Study on the Stability of Slopes Subjected to Earthquake Forces

Partha Pratim Boruah[1], **Arunav Chakraborty**[2]
[1] M. Tech, Department of Civil Engineering, Tezpur University, Tezpur, Assam
[2] Assistant Professor, Department of Civil Engineering, Tezpur University, Tezpur, Assam
E-mail: parthaboruah101@gmail.com

Abstract. For thousands of years, engineers have been grappling with slope instability as an issue in geotechnical and geological engineering. Earthquake loading has a great impact in slope stability. The present study is to investigate the impact of various soil properties and slope geometry on FOS in the presence as well as in the absence of seismic loading. Here, pseudo-static methods are used to analyses seismic slope stability under earthquake loading. From parametric studies, present work concluded that the FOS values start to decline drastically with earthquake impact. Moreover, various prediction models have been developed using Multiple Linear Regression (MLR) & Multiple Non-Linear Regression (MNLR) considering all the data. The input data of slope stability estimation consists of the values of soil parameter, slope parameter and earthquake loading parameters. As an output, MLR and MNLR predict FOS values as a function of these parameters. The model is then checked by comparing the results to another ten actual situations. It was found that the predicted FOS using MNLR gives satisfactory results.

Keywords: Limit equilibrium methods, Factor of safety, Linear Regression, Non-Linear Regression, Dynamic slope stability, Seismic Slope stability.

1. Introduction

Geotechnical and Geological engineers have grappled with slope instability for thousands of years. It has resulted loss of life as well as property for the human race, and lakhs of crores have been spent on protecting and restoring the environment. Slope instability can be caused by a variety of factors such as soft or weak soil, erosion from wind and water, rainfall, external loading, deforestation, earthquakes, etc.Over the past few decades several slopes have been severely damaged by earthquakes. Hence there is a need of analyzing slope stability under seismic loading.Marrapu and Jakka [1] created two distinct multiple regression models, verified their precision in the estimation using actual field data, and were satisfied with the findings. Wang and Khan [2] found correlations between various soil and slope parameters using the SLIDE software to forecast slope stability assessment. Zhou et al [3] investigate slope prediction using logistic regression and conclude that it is the best predictive approach for slope stability prediction.Huang and Peng [4] conducted a seismic slope stability analysis for saturated and unsaturated soil and found that dynamic pore water pressure influence is significant towards stability. A soil slope was examined using limit equilibrium techniques by Choudhury et al. [5]. They carried out in-depth analysis, which showed that the dynamic FOS depends on friction angle of the soil. With increasing horizontal and vertical acceleration, the dynamic FOS significantly decreases. According to some researchers, limit equilibrium analysis might be a little bit non-conservative. Using pseudo-static and upper-bound limit analysis, Zhao et al. [6] developed many stability charts and these charts were very helpful in determining the critical slope height and critical fracture depth. Limit analysis was used by Gong et al. [7] to access the seismic slope stability for non-homogeneous anisotropic slopes. The problem is made simpler by using the pseudo-static approach. They got to the conclusion that the anti-slide piles have a significant impact on the soil. Additionally, they observed that the anti-slide pile required a greater stabilizing force higher the anisotropic coefficient. The outcomes of an ANN model were shown to be trustworthy when compared to conventional limit equilibrium methods. Lin et al. [8] used a neural network to assess the likelihood that a roadway slope will fail. They constructed models to look at slope failure characteristics and identified the influence of various elements (variables) on slope stability. Abdalla et al. [9] use of ANN to forecast the FOS for slope failure in clay . Two multilayer perceptions ANN models based on various slope stability strategies were utilized to calculate the minimum FOS utilizing different sets of data related to geometry and shear strength properties.

The study's objectives include:

•The effect of various parameters of soil properties and slope geometrics with respect to FOS of a slope both in presence and absence of seismic loading.

•To develop a correlation between soil parameter as well as geometrical parameters of a slope using statistical approach such as MLR and MNLR.

•To verify the prediction models using some of the case studies along National Highway of Assam Meghalaya boarder (NH-6) and also to carry out a comparison between the results obtained from MLR and MNLR.

2. Methodology

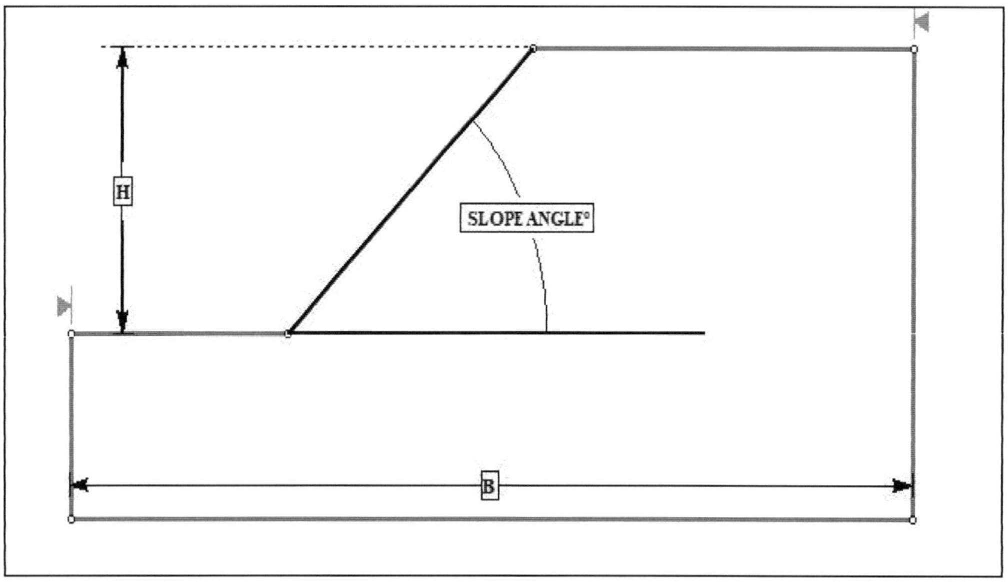

Fig 1: The slope model.

The methodology involves calculating the FOS using ordinary slice methods using a number of artificially simulated slopes with changed slopes, soils, and earthquake parameters. The slope model is shown in Fig 1.Parametric study of seismic slope stability has been carried out at first. For parametric study, cohesion is varying and keeping other parameters constant both in presence and absence of seismic load and Several FOS values by ordinary slices methods are obtained. Similarly, for parametric study slope angles, height of the slope, internal angle of friction of the model are varying keeping other parameter constant both in presence and absence of seismic load and several

FOS values are obtained. For plotting of variation graph considering the FOS with respect to internal angle of friction at different values of cohesion the following parameters are kept constant i.e., height of slope (H) = 20 m, Slope angle (β) = 34◦ and unit weight of the soil(γ)=19 KN/m^2.For plotting of variation graph considering the FOS with respect to angle of slope at different values of cohesion the following parameters are kept constant i.e., Slope Height (H)= 20 m, Friction angle(ø) =30°, unit weight of soil(γ)=19 KN/m^2.

To calculate the FOS, 480 artificial slope cases with wide range of geometry, soil properties, and earthquake properties were analyses using ordinary methods of slice and FOS values were calculated. These values were used in the development of the regression models using MLR and MNLR. Here, height of the slope (H), cohesion (c), angle of internal friction (φ), slope inclination (β), unit weight of soil (γ), and Horizontal Seismic coefficient (K_h) were used as input parameters. FOS served as a parameter for the output.

3. Results and Discussion:

Parametric study of seismic slope stability has been carried out. All the results obtained from LEM approach has been complied together and FOS variation with respect to other parameter has been represented in a feasible and comprehensible manner. By examining the projected and analytical outcomes for 10 susceptible slope scenarios along NH6, the model's accuracy was determined.

3.1 Parametric study findings:

a) Variation of FOS with respect to internal angle of friction at different values of cohesion:
i) Without seismic loading conditions: A graphical representation of the data has been attempted here with respect to FOS and internal angle of friction at different values of cohesion.

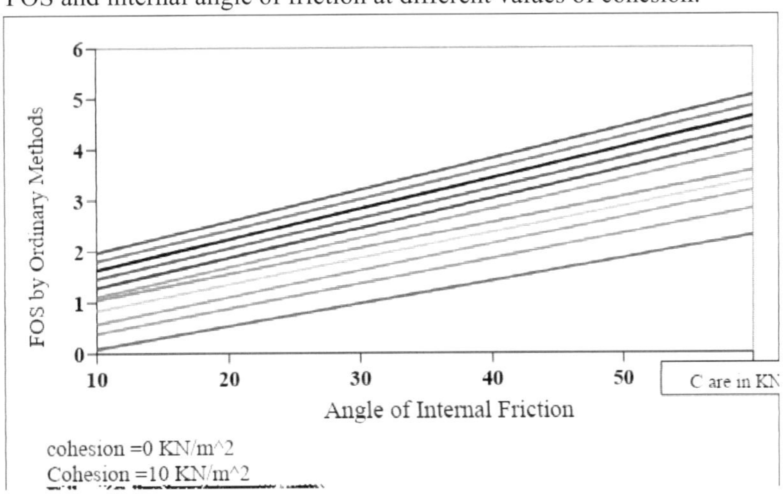

Fig 2. FOS vs Angle of internal friction at different values of cohesion in absence ofseismic loading.

It can be clearly concluded from the graphical representation that as friction angle is increasing, the FOS values are also showing increasing trends. The trend is parabolic nature. This is due to the fact that higher friction angle of soil, better is the load handling features along with higher adhesion. For high value of cohesion and internal angle of friction it is observed that FOS is also large.

ii) With seismic loading conditions

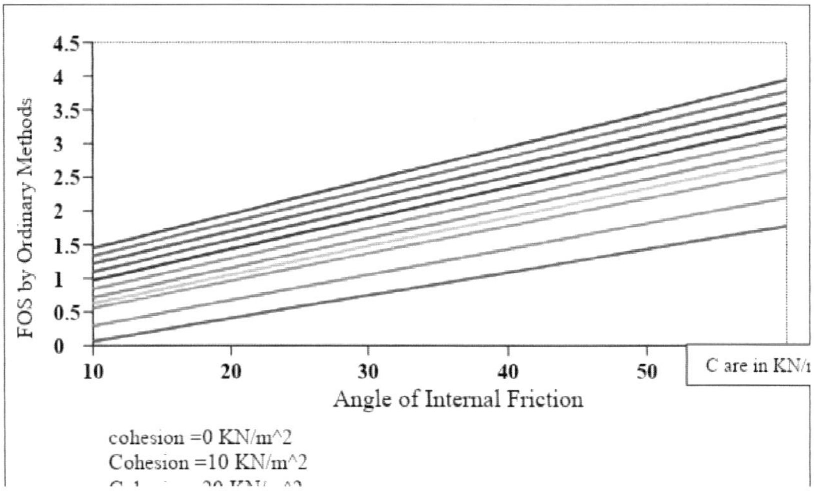

Fig 3. FOS vs Angle of internal friction at different values of cohesion in presence of Seismic loading

It can be clearly concluded from the graphical representation that as friction angle is increasing, the FOS values are also showing increasing trends in presence of seismic loading (considering the earthquake data by using PGAH as 0.267 m/s^2 and peak ground vertical acceleration as 0.139 m/s^2).For slope angle 60 degree and cohesion value 100 KN/m^2 the FOS of the model without considering seismic loading is found to be 5.412 and with considering seismic loading it is found to be 4.193. Hence dynamic load has a large impact on FOS of a slope.

b) Variation of FOS with respect to angle of slope at different values of cohesion:

i) Without seismic loading conditions: A graphical representation of the data has been attempted here with respect to FOS and angle of slope at different values of cohesion.

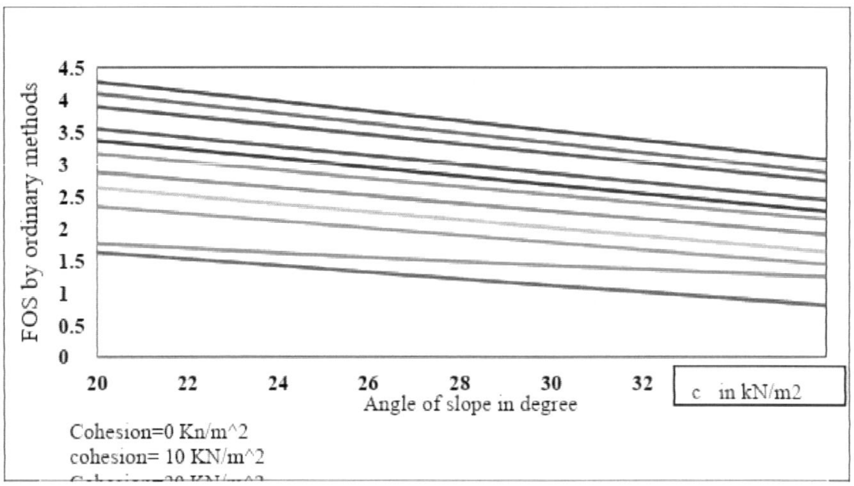

Fig 4. FOS vs Angle of Slope at different values of cohesion in absence of seismic loading.

It can be clearly concluded from the graphical representation that as slope angle is increasing, the FOS values are showing decreasing trends. That means slope angle is inversely proportional to FOS. The trend is linear. For high value of cohesion, it is observed that FOS is also large.

ii) With seismic loading conditions: Considering the earthquake data by using PGAH as 0.267 m/s^2

and peak ground vertical acceleration as 0.139 m/s^2. It is found that with the impact of seismic loading the FOS starts decreasing drastically as compared to without seismic loading condition case.
It can be clearly concluded from the graphical representation that as slope angle is increasing, the FOS values are showing decreasing trends. That means slope angle is inversely proportional to FOS. The trend is linear. For high value of cohesion, it is observed that FOS is also large.
ii) With seismic loading conditions: Considering the earthquake data by using PGAH as 0.267 m/s^2 and peak ground vertical acceleration as 0.139 m/s^2. It is found that with the impact of seismic loading the FOS starts decreasing drastically as compared to without seismic loading condition case.

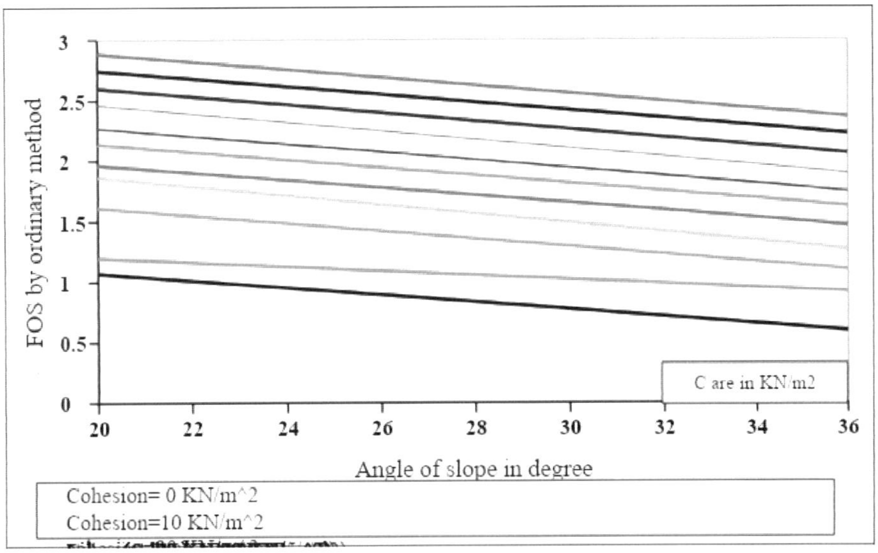

Fig.5. FOS vs Angle of Slope at different values of cohesion in presence of seismic loading.

For slope angle 36 degree and cohesion value 100 KN/m^{2}the FOS of the model without considering seismic loading is found to be 2.985 and with considering seismic loading it is found to be 2.334. Hence dynamic load has a great impact on FOS of a slope.

3.2 Multiple Linear Regressions Findings

The following predicted equation have been developed using MLR.

Table 1. Summary output of MNLR analysis

Regression Statistics				
R^2			0.82	
MSE			0.08	
Stability Parameter	Coefficients	Standard Error	t Stat	P value
Intercept	4.067	0.291	13.968	< 0.0001
Height of Slope (H)	-0.129	0.005	25.059	< 0.0001
Cohesion (c)	0.031	0.001	16.978	< 0.0001
Slope angle (β)	-0.013	0.001	17.613	< 0.0001

Unit Weight (γ)	-0.109	0.017	-6.217	< 0.0001
Internal angle of friction (φ)	0.034	0.002	16.804	< 0.0001
Horizontal Seismic coefficient (K_h)	-0.560	0.078	33.914	< 0.0001

$$Predicted\ Factor\ of\ safety\ MLR = 4.07 - 0.13\times H + 0.03\times c - 0.01\times\beta - \quad (1)$$

3.3 Multiple Non-Linear Regressions Findings

From the table it has been found that the value of R^2 has found to be 0.88. Further a correlation has been established between soil and slope parameter to predict FOS value. The following equation has been from using MNLR technique.

Table 2. Summary output of MNLR analysis

Regression Statistics	
R^2	0.88
MSE	0.05

$$Predicted\ Factor\ of\ safety\ using\ MNLR = \frac{10.75\times H^{-0.65}\times c^{0.54}\times(tan\beta)^{-0.12}}{\gamma^{0.98}\times(tan\varphi)^{-0.26}\times K_h^{0.39}} \quad (2)$$

3.4 Case Study

A few case studies are conducted to test the validity of prediction models independent of the data used to build them. Furthermore, error analysis is used to test the model's suitability.

For case studies the national highway from Jorabat to Shillong is taking into consideration as it includes lots of hillslope, slope collapse is prevalent along the Jorabat-Shillong roads. A survey work has been performed along Jorabat and Umling over a distance of 25 km to determine different slope geometry. Undisturbed soil samples are obtained usingShelby tubes having 20 cm length and 3.8 cm as diameter with an area ratio of 10%. Consolidated drained test is carried out to obtain the shear parameters. To incorporate seismic loading parameter the earthquake of magnitude of 6 which is occurs in Assam in the year 2021 is taking into consideration. The PGAH of that particular earthquake at epicenter (K_h=0.5) and at Jorabat area (K_h=0.25) is considered which is collected from USGS data center website. Finally, the FOS values are obtained for all the 10 vulnerable sites using LEM approach. The test results for 10 real slopes case studies are given in the Table 3.

Fig 6 shows the results of MSE values for MLR and MNLR at different $K_h$values. It has been found that MSE values for MNLR found to be lowest as compared to MSE values of MLR in both cases. Similarly in Fig 7 shows the results of MAE values for MLR, and MNLR at different $K_h$values. It has been also found that MAE values for MNLR found to be lowest as compared to MAE values of MLR in both cases. Thus, FOS values obtained by MNLR have higher accuracy .

This study indicates that earthquakes are also responsible for land sliding along NH 6. Almost all the hill slopes along NH 6 are wet or moist, which becomes a serious issue during earthquakes. Earthquake having long duration weaken and destroy mountainous slopes, leading to instability. Engineers will benefit greatly from the results of this study and can be used to evaluate slope stability and produce sustainable hazard assessment maps that consider earthquake effects.

Table 3. Case studies for 10 vulnerable slopes along NH 6.

No	Latitude & Longitude	H (m)	c (KN/m^2)	φ (°)	β (°)	γ (KN/m^2)	FOS without seismic loading	FOS considering K_h=0.5			FOS considering K_h =0.25		
							LEM	LEM	MLR	MNLR	LEM	MLR	MNLR
1	26°05'56" N 91°51'44" E	20	39.5	30.2	50	17.6	1.514	0.799	0.845	0.749	1.086	0.985	0.981
2	26°05'30" N 91°51'43" E	23	39	30	50	17.3	1.381	0.727	0.467	0.689	1.001	0.607	0.903
3	26°05'20.06N 91°51'43.08E	18	38.7	30.5	60	17.8	1.374	0.755	0.968	0.752	1.019	1.108	0.986
4	26°04'31" N 91°51'42" E	25	39	31.2	55	17.9	1.23	0.653	0.127	0.626	0.893	0.267	0.82
5	26°04'25" N 91°51'41.48E	18	39	30	50	17.3	1.584	0.833	1.117	0.808	1.139	1.257	1.059
6	26°03'44" N 91°51'41.34E	22	39	30	50	17.3	1.43	0.75	0.597	0.71	1.02	0.737	0.93
7	26°03'37" N 91°51'41.20E	18	38.5	30.7	48	17.5	0.964	0.688	1.121	0.806	1.262	1.261	1.057
8	26°03'21" N 91°51'41.08E	26	37.9	30	45	17.3	1.428	0.735	0.094	0.64	1.007	0.234	0.839
9	26°02'55" N 91°51'41" E	20	38.5	29	50	17.5	1.464	0.774	0.79	0.733	1.05	0.93	0.961
10	26°02'48" N 91°51'40.89E	20	39.2	29.7	55	17.5	1.375	0.739	0.782	0.73	1.004	0.922	0.957

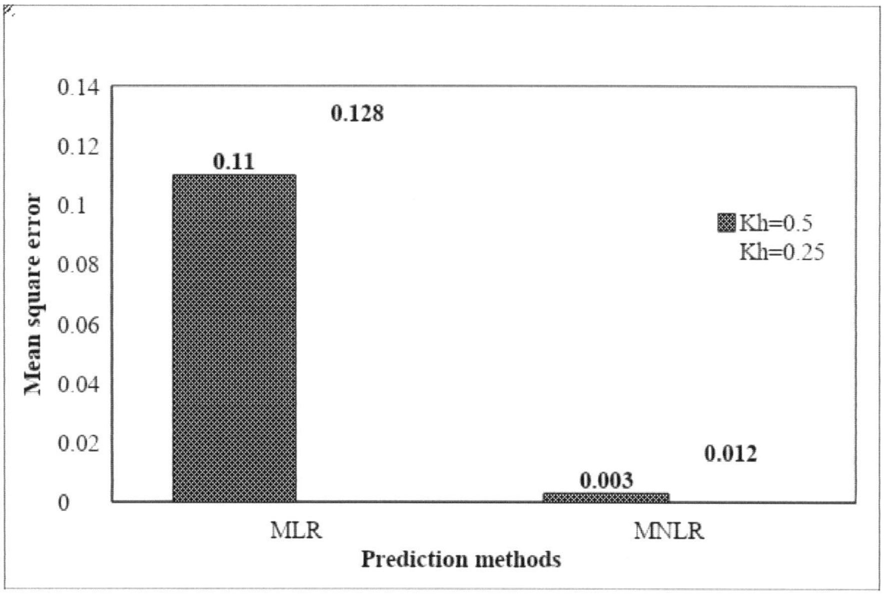

Fig 6. Mean Square error V/S Prediction methods

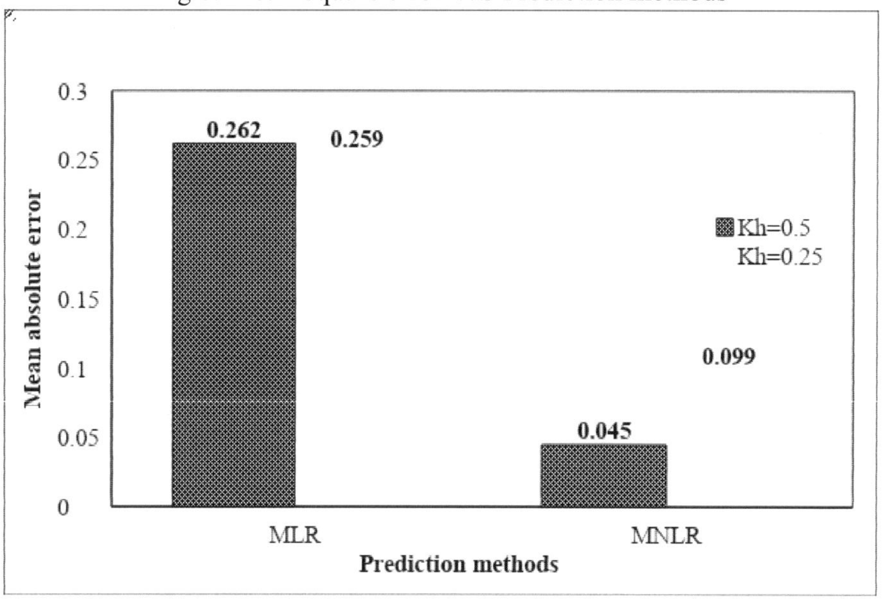

Fig 7. Mean Absolute error V/S Prediction methods

4. Conclusions

A parametric study of seismic slope stability is conducted. Additionally, the FOS values are used for making prediction models using MLR, MNLR. The parameters such as slope height, cohesion, angle of internal friction, angle of slope, unit weight of the soil, and horizontal seismic coefficient are act as input variable and output parameter is FOS. Some case studies along NH-6 are used to test the accuracy of prediction models. Based on the above presented results, the following inference can be made.

a. In parametric studies, FOS values increase as friction angle increases at different values of cohesion, both in the presence and absence of seismic loading following the parabolic trends. The FOS values decrease when earthquake loads are applied. As friction angle increases, soil strength increases, therefore FOS values increase. Whereas Earthquakes cause failure in two different ways. Because of loss of effective stress, vibration from an earthquake can lead to liquefaction of uniformly graded, fine-grained sediments. As a result of earthquakes, factor of safety can also decrease due to increase in shear stress.

b. From parametric studies it has been found that FOS values starts decreasing as the angle of slope increases at different values of cohesion both in presence and absence of seismic loading following the linear trends. In presence of earthquake loading FOS values starts decreasing drastically. Slopes with steeper gradients are more likely to fail.

c. Regression analysis can be quite useful for forecasting slope stability. The results of the MLR and MNLR models generally agree with the results of LEM studies.

d. The MNLR model is shown to have lower values of MSE of 0.003 and 0.012 for K_h =0.5 and K_h=0.25 respectively as against 0.11 and 0. 128.Therfore predicted results of MNLR model are found to be more closer as compared to MLR model.

e. The MAE values obtained from MNLR model are 0.045 at K_h=0.5 and 0.099 at K_h=0.25 against 0.262 and 0.259 from MLR model respectively. Hence MNLR prediction is more accurate as compared to MLR.

f. Accordingly, the present study is extremely valuable for engineers to develop a landslide hazard zone map taking earthquake impact into consideration.

5. References

[1] Marrapu, B.M. and Jakka, R.S., 2017. Assessment of slope stability using multiple regression analysis. Geomech Eng, 13, pp.237-254.
https://doi.org/10.1061/12321

[2] Khan, M.I. and Wang, S., 2021. Slope Stability Analysis to Develop Correlations between Different Soil Parameters and Factor of Safety Using Regression Analysis. Polish Journal of Environmental Studies, 30(5).
https://doi.org/10.15244/pjoes/131203

[3] Zhou, L.Y., Shan, F.P., Shimizu, K., Imoto, T., Lateh, H. and Peng, K.S., 2017, August. A comparative study of slope failure prediction using logistic regression, support vector machine and least square support vector machine models. In AIP Conference Proceedings (Vol. 1870, No. 1, p. 060012). AIP Publishing LLC.
https://doi.org/10.1063/1.4995939

[4] Huang, S. and Peng, Y., 2018. Seismic stability analysis of saturated and unsaturated soil slopes using permanent displacement. Advances in Civil Engineering, 2018, pp.1-9.
https://doi.org/10.1155/2018/1786392

[5] Choudhury, D., Basu, S., and Bray, D. J. 2007 Behaviors of slopes under static and seismic conditions by limit equilibrium method. Embankments, Dams, and Slopes: Lessons from the New Orleans Levee Failures and Other Current Issues, pp. 1-10.
https://doi.org/10.1061/40905

[6] Zhao, L., Cheng, X., Zhang, Y., Li, L. and Li, D. 2016 Stability analysis of seismic slopes with cracks. Computers and Geotechnics, vol. 77, pp. 77-90.
https://doi.org/10.1016/j.compgeo.2016.04.007

[7] Gong, W. B., Li, J. P. and Li, L. 2018 Limit analysis on seismic stability of anisotropic and nonhomogeneous slopes with anti-slide piles. Sci. China Technol. Sci., vol. 61, pp. 140-146.
https://doi.org/10.1007/s11431-017-9147-8

[8] Lin HM, Chang SK, Wu JH, Juang CH 2009 Neural network-based model for assessing failure potential of highway slopes in the Alishan, Taiwan area: pre- and post-earthquake investigation. EngGeol 104(3–4):280–289

[9] Abdalla, J.A., Attom, M.F. &Hawileh, R 2015. Prediction of minimum factor of safety against slope failure in clayey soils using artificial neural network. Environ Earth Sci 73, 5463–5477.
https://doi.org/10.1007/s12665-014-3800-x

Analysis of Strength Parameters of Paving Blocks with inclusion Sea Sand

Ashwini R. Patil[1], Dr. Ashok B More[2], Dr. Arun Kumar Dwivedi[3], Pratik B. Abhang[4]
[1]Assistant Professor, D.Y. Patil College of Engineering, Pune & Research Scholar Sandip University Nashik, India
[2]Associate Professor Department of Civil Engineering, D.Y. Patil College of Engineering, Pune 411044, Maharashtra, India
[3]Professor, Civil Department, Sandip University, Nashik 422001
[4]Departmentof Civil Engineering, D. Y. Patil College of Engineering, Pune 411044, Maharashtra, India
Email id: arpatil@dypcoeakurdi.ac.in

Abstract: Start with the Scarcity of materials in concrete so sand has steadily gained prominence as the most frequent building material used in the construction industry. As the population increases, so does the need for residential and commercial space. The most essential resource for mankind, river sand, is being mined out of riverbeds more frequently. One ton of regular Portland (OPC) cement is produced in India for every tonne of CO2 that is released into the environment. By substituting cementitious materials like fly ash and ground-granulated blast furnace slag for ordinary cement, we can lower the number of greenhouse gases that are emitted into the atmosphere. To create sustainable concrete pavement blocks, this experimental work has been carried out with the use of cementitious elements including fly ash and ground granulated furnace slag (GGBS) as a replacement for cement, and to enhance public transportation, this research is focused on creating paver blocks that contain sea sand as fine aggregate. This experimental investigation was conducted to investigate different sustainable concrete strength properties with 0% sea sand, 50% Sea Sand and 50% M sand, and 100% sea sane used to partially replace the mix proportion. Results of mechanical parameters such as Compressive strength, water absorption, abrasion resistance, and carbonation of these sustainable concrete pavement blocks are presented also a cost analysis of paving blocks was evaluated.

Keywords: Carbonation, Sea sand, Fly ash, Ground Granulated Blast Furnace, Water absorption, Paver blocks.

1. INTRODUCTION

Making concrete paver blocks with Geopolymer concrete is an environmentally friendly process called "Green Paver Block."[1],[2].Cement concrete is currently the second most consumed commodity on earth due to the rapid growth of infrastructure. Because of their quick and simple installation, superior appearance, and finish, cement concrete paver blocks have become the preferred option. Precast concrete paver blocks are the most popular option for paving footpaths, parking lots, bus stops, industries, etc. The majority of constructions use river sand as fine aggregates, and as demand rises, sand availability decreases. This leads to illegal sand grabbing from river beds, and the demand for river sand in construction is currently very high. Therefore, it has been very difficult to discover a substitute for river sand that has the same benefits as river sand, such as ease of availability, eco-friendliness, and low cost [4]-[6]. Sea sand, a material that may be utilized for building as well as concrete components like paver blocks, is an alternative to river sand[7]-[9] Because sea sand contains salt, replacing it with concrete affects the building and concrete products in numerous ways. Further research has been done to establish how the chemical and physical differences between river and sea sand affect the mechanical properties of concrete [11]. To improve concrete durability, sea sand was added to the mix when creating ultra-high-performance concrete. The development of non-corrosive fiber-reinforced polymers and their incorporation into concrete were part of an experimental program to lessen the impact of corrosion, which is one of the most important factors in the deterioration of concrete. 15 different mix proportions were made and evaluated using equal amounts of cement, silica fume, extra fine materials, and water [10].

To reduce the salt content in sea sand and improve the construction of concrete goods, researchers have been conducting several studies. On the other hand, M-sand, which is widely used in concrete and has the same qualities as river sand because it is made by crushing rocks, is more expensive and eventually becomes unprofitable. Taking into account all the variables, it has been tried to show how effectively sea sand may be used as an alternative in full or partial replacement of river sand through this experimental investigation. In this study, the impact of sea sand on concrete's compressive strength was investigated. This study attempts to bring all of the efforts together and give a comprehensive explanation of how sea sand may be used effectively in concrete. It also displays the results of using sea sand in concrete. When sea sand was substituted for 20% of the river sand in concrete cubes during testing after 7, 14, and 28 days, the compressive strength increased from 39.676 N/mm2 to 44.908 N/mm2 [1]. An accelerated test with a simulated marine environment was designed for the degradation of sea-sand concrete that was partially submerged in seawater. The mass, water absorption, and mechanical properties (compressive strength and dynamic modulus) were tested, and the impact of seawater on cement paste hydrates was thoroughly examined using a thermodynamic model [2]. The compressive, split tensile, and flexural strengths of concrete made from sea sand are gathered from Thalikulam in the Thrissur district of Kerala at Snehatheeram beach or Love shore as a complete utilization of percent M-sand, percent sea sand, and partial utilization of M-sand and sea sand for fine aggregate [3]. Because they are simpler to install and have a superior appearance and finish, concrete paving blocks make suitable materials for footpaths and roadways.

Paver blocks may easily be replaced when they break and need no maintenance industrial waste can take several shapes, some of which can be utilized in the construction industry to recycle waste and foster a greener environment [12]. In order to determine the most affordable material that can replace fine aggregate in construction, the salt content of sea sand was removed using hot water and chemical treatment, and the performance of concrete was evaluated by utilizing sea sand as fine aggregate. The compressive strength of sea sand, which was substituted for fine aggregate in concrete after being treated with acetic acid, was then evaluated using the conductivity test. River sand was used in place of some of the sea sand in the fine aggregate as part of an experimental study of the strength characteristics of cement concrete.

2. Experimental program

2.1 Materials used

2.1.1 Cement

The cement used complies with IS: 8112:1989. It is Ordinary Portland Cement (OPC) of Grade 43. The building industry's most used raw material is concrete. According to the Bureau of Indian Standards, concrete's assessment number reveals the minimal compressive quality that concrete must attain in just 28 days (BIS). When Portland cement and water are combined, chemical processes take place to release energy and result in the cement paste event, which allows the substance to solidify. For OPC cement 43 grade, the concrete's minimum compressive quality at the end of the 28th day must be less than 43 Mpa or 430 kg/sq.cm. Numerous types of concrete can employ OPC 43.

Table 1 Physical characteristics of cement

Properties	Unit	Element
8 days compressive strength	Mpa	1.5
Initial setting time	Minute	4
Specific surface area	m^2/gm	907
Final setting time	Minute	21

2.1.2 Aggregates

The 10 mm coarse aggregate used complied with Table 2 of IS 383 and was used. The most fundamental components are coarse aggregates, which make up 70 to 75 percent of the volume in concrete formulations. Coarse particles are used in concrete, railroad ballast, and other purposes. It uses the gross total size, which is rated in accordance with IS 383:1970. The term "coarse totals" refers to particles larger than 4.75 mm. The typical measurement range for width is 9.5 to 37.5 mm. To achieve adequate toughness, the aggregate used in paver construction must be sound and free of brittle or honeycombed particles. The largest coarse totals utilized in the manufacture of paver blocks will be 12 mm. Because they are so heavy, aggregates have a significant impact on a structure's ability to transfer loads.

2.1.3 Sea Sand

This sand is obtained from distant seabed excavations or ocean beaches. Similar to stream sand, ocean sand has finely tuned grains and a little deeper color. Sea sand from Zone II that complied with IS 38sieve had an IS sieve classification of 1.18mm was used.

2.1.4 Fly ash

Fly ash is the material that is used the most on the world (Class C). Fly ash was supplied by a ready-mix concrete factory located in Kivale, Pune (Maharashtra). If homogenous blending with cement is achieved, fly ash that complies with Grade 1 of IS 3812 (Part 1) can be used as a partial replacement for OPC.

2.1.5 Granulated Ground Blast Furnace (GGBS)

GGBS can be produced by drying and pressing molten iron slag from blast furnaces. It is one of the primary basic ingredients used to create a cement less binder with pozzolanic and cementitious characteristics. A blend of calcium, silicates, and alumina that contains around 90% GGBS satisfies the pozzolanic material criterion. If homogeneous blending with cement is accomplished, GGBS in accordance with IS 12089 may be used as a partial replacement for OPC cement. In the furnace, GBS is created. The intensity, hardness, and look of concrete are enhanced by the addition of CSH (calcium silicate hydrates) to ground-granulat ed slag from a burning furnace. GGBS was obtained from BG Shrike's Concrete Testing Lab located in Pune.

2.1.6 Superplasticizer

Suryotthan Melamine superplasticizer was utilized in this experiment to accelerate quick hardening and make free-floating concrete for precast applications.

2.1.7 Water

Water is frequently recognized as a crucial liquid component for all types of labor. In this investigation, drinking water is used for casting and curing. A paste is created when water and cement are combined, which holds all of the aggregate together. Due to the water-cement ratio (w/c %), water plays a crucial part in concrete. In this study, the w/c ratio was reduced to 0.45 in compliance with IS 456:2000.

2.1.8 Tests

2.1.8.1 Slump Cone Test

As the project progresses, this test is carried out in the lab or on the construction site to ascertain the usefulness or consistency of a concrete mix. To ensure that the cement is constant throughout its development, it is done clump by clump. The slump value, which represents the water-to-concrete ratio, is used to gauge workability, or how readily concrete can be poured. On the other hand, various factors like as material characteristics, blending techniques, measures, and admixtures all affect the slump value.

2.1.8.2 Compression test

On cube samples, compression strength tests were conducted using compression testing equipment. Three samples from each batch were used to compute the average strength values in this study. The testing apparatus bearing plates need to be cleaned. The dimensions of the bearing plates match those in the examples. Up until the specimen could no longer support it or delimitation took place, the load would be steadily increased at a rate of 153 $N/mm^2/min$. Recorded were the loads that were applied to the specimen at the highest intensities.

Fig. 1. Compression testing machine used to measure the compressive strength of samples

2.1.8.3 Abrasion resistance test

Cast rubber molds in the form of Milano's were used to create the test specimens for the abrasion test. To determine crushing, deterioration, and disintegration, the aggregate hardness and abrasion resistance of the 12 specimens are measured at 7, 14, and 28 days. The test specimens were manufactured in the same shape for each test. 12 blocks were cast for each mix percentage, and they were assessed 7, 14, and 28 days later.

2.1.8.4 Water absorption

After casting, the cubes were submerged in water for 28 days to cure. The paver blocks were then weighed, and the outcome was noted as the paver block's wet weight. The specimens were then uniformly bulked up by oven drying than at 1100 C before being weighed once more. Paver blocks must be able to absorb some water to effectively bond with mortar because they are not water resistant. The initial rate of water absorption in milligrams is used to assess if pavers need to be wet before laying.

2.1.8.5 Carbonation test

The phenolphthalein indicator is sprayed on freshly exposed concrete that has been cracked away from the building or on split cores to conduct carbonation testing. A solution of the phenolphthalein indicator, which glows pink when in contact with alkaline concrete with pH values higher than 9, and goes colorless at lower pH levels, is used to measure the depth of carbonation.

2.2 Casting and Curing of specimens

After the mix design was determined, the ingredients GGBS and Fly Ash were added in the ratios of 1:1, 1:1.2, and 1:1.3. Fly ash is used in percentages of 25 percent, 30 percent, and 40 percent. The GGBS proportion is held constant at 25%. Each component was completely dried before being mixed. This dry mixture was then homogeneously mixed once more after the necessary quantity of superplasticizer of the suitable grade was added. This mixture was homogeneously blended once again after being given additional water and superplasticizer. The molds were filled with wet concrete, which was hand-compacted in three layers before being vibrated to achieve the desired level of compaction.

Blocks made of Geopolymer that measure 250 x 120 x 80 mm. As required, these specimens were cast, and they were kept at room temperature for curing. After casting for a day, the specimens were removed from the mold.

Table 2

Mix proportions used for casting of paver blocks.

Details of replacement	Cement (kg)	Fly ash (kg)	Coarse aggregate (kg)	Water (kg)	Superplasticizer (ml)	W/C ratio	GGBS (kg)	Fine aggregate (kg)	
								M-sand	Sea sand
Conventional paver blocks	2.5	-	11	1.5	18	0.45	-	6	-
Batch 1 25% GGBS+25% Fly ash	10	3.3	40	6	70	0.45	3.3	-	24
Batch 2 25% GGBS+30% Fly ash	10	4	40	6	70	0.45	3.3	12	12
Batch 3 25% GGBS+ 40% Fly ash	10	4.5	40	6	70	0.45	3.3	12	12

The casted specimens are demolded after the concrete haves completely hardened, typically after 24 hours. After being removed from the mold, the specimens are stored outside to cure. In the study, cure times of 7, 14, and 28 days were examined. After being cured, the specimens are checked for compression, abrasion, water absorption, and carbonation.

2.2.1 Casting procedure

Fig. 2. Weighing of materials

Fig. 3. Casting of paver blocks of three mixes

Fig. 4. Vibration of paver molds on vibration belt

Fig. 5. Demolding of paver blocks

3 Results

3.1. Compression Test Result

Table 3 presents the observation on the Compressive strength test for Geopolymer concrete paver blocks. The tested results of compression strength of paving blocks which are cured for 7, 28, and 56 days are tabulated in the table.

Table 3

Compressive strength of paving blocks on different curing days

Mix Proportion	Sea Sand (%)	M-Sand (%)	Compressive Strength(N/mm^2)		
			7th Day	28th Day	56th Day
Controlled Concrete	0	100	32.24	38.35	58.25
25% Fly ash +25%GGBS	0	100	35.50	42.37	59.97
30% Fly ash +25%GGBS	50	50	35.16	45.93	59.98
40% Fly ash +25%GGBS	100	0	35.10	49.07	61.74

3.2 Water Absorption Test Results

Geopolymer concrete paver blocks' ability to absorb water is assessed after curing; elements affecting water absorption include the kinds of additives used, temperature, and exposure duration. The data clarifies how the material performs in damp or humid conditions. According to IS 15658:2006, water absorption should be less than 7% after 24 hours. The results of the experiments are shown in Table 8, and Figure 9 displays the percentage of water absorption at various mix ratios.

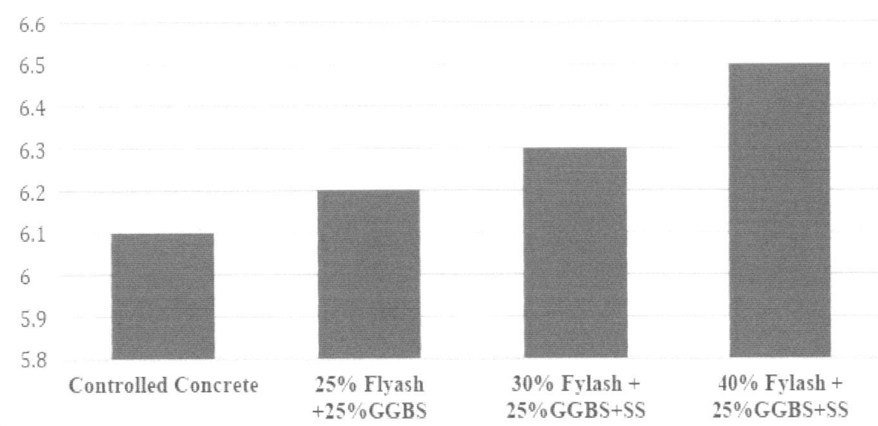

Fig. 6. Water absorption at different mix proportion

3.3 Abrasion Resistance Test Results

The abrasion test was conducted both IS 15658:2006, which calls for rubbing a fast-moving ball race on the paver surface to leave a circular groove on the paving block.

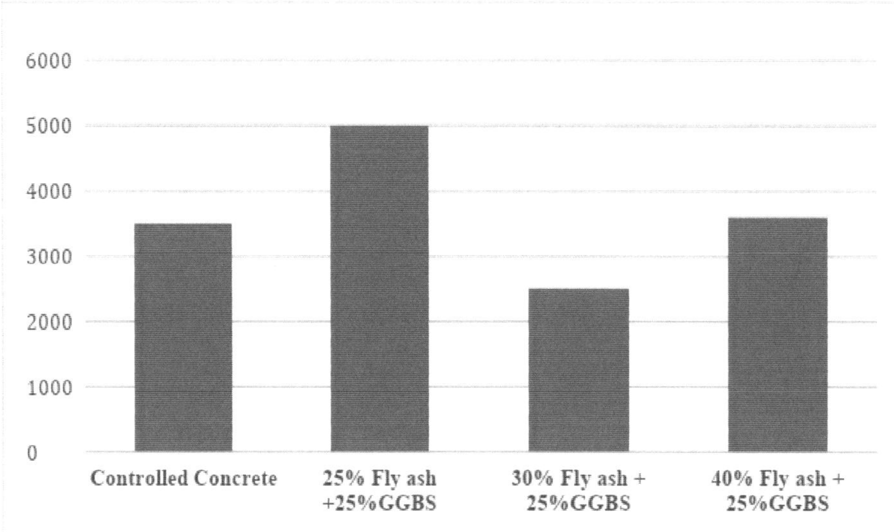

Fig. 7. Abrasion resistance in mm^3 per 5000 mm^2

3.4 Carbonation Test Results

Concrete is carbonated by phenolphthalein indicator on newly exposed concrete surfaces that have been torn away from the building or on split cores, which causes a gradual neutralization of the alkalinity induced by atmospheric carbon dioxide and Sulphur dioxide from the surface inwards. To evaluate the depth of carbonation, a phenolphthalein indicator solution is employed, which turns pink when in contact with alkaline concrete that has a pH higher than 9, and colorless when the pH is lower.

Table 4
Presence of carbonation

Sr. No	RCC Member	Carbonation Present		
		50 mm	100 mm	150 mm
1	Paving Block	No Carbonation	No Carbonation	No Carbonation

3.5 Cost Analysis

Table 5
Cost for conventional paver block for 12 blocks

Materials	Quantity (kg)	Price/kg	Total price (Rs)
Superplasticizer	0.07	33	2.31
Coarse aggregate	40	0.7	28
Cement	20	7	140
Water	6	1	6
Fine aggregate	24	3.6	86.3
Total cost			263

Table 6
Cost for Mix type I25% GGBS+25% ash for 12 blocks

Materials	Quantity (kg)	Price/kg	Total price (Rs)
M-sand	-	-	-
Superplasticizer	0.07	33	2.31
GGBS	3.3	2.5	8.25
Fly ash	3.3	3	9.9
Coarse aggregate	40	0.7	28
Sea sand	24	3	72
Water	6	1	6
Cement	10	7	70
Total cost			196.46

Table 7
Cost for Mix type II 25% GGBS+30% Fly ash for 12 blocks

Materials	Quantity (kg)	Price/kg	Total price (Rs)
Cement	10	7	70
Water	6	1	6
Fly ash	4	3	12
Coarse aggregate	40	0.7	28
Sea sand	12	3	36
M-sand	12	3.6	43.2
GGBS	3.3	2.5	8.25
Superplasticizer	0.07	33	2.31
Total cost			205.76

Table 8

Cost for Mix type III 25% GGBS+40% Fly ash for 12 blocks

Materials	Quantity (kg)	Price/kg	Total price (Rs)
Superplasticizer	0.07	33	2.31
Fly ash	4.5	3	13.5
GGBS	3.3	2.5	8.25
Coarse aggregate	40	0.7	28
Water	6	1	6
M-sand	12	3.6	43.2
Sea sand	12	3	36
Cement	10	7	70
Total cost			207.26

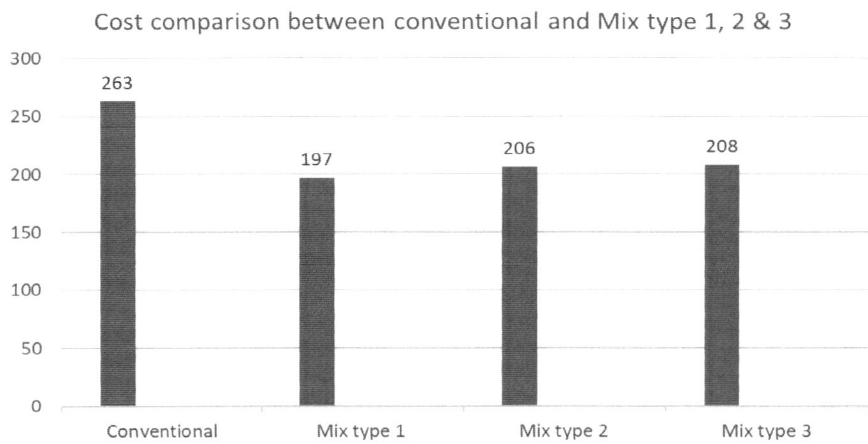

Fig.8 . Cost comparison between conventional and Mix types 1, 2 & 3

4 Conclusion

The following conclusions about various sustainable materials used in concrete paver blocks in the context of achieving sustainable development and effective waste resource utilization were drawn from the literature that was compiled and examined.

1) At day 56, 100 percent S-sand, 40 percent FA, and 25 percent GGBS exhibited a compressive strength up to 61.74 MPa. This reduces building costs and enables the safe disposal of waste slag produced in vast numbers across various industries. The ultimate goal is for Geopolymer concrete paver blocks to completely displace conventional concrete paver blocks as the most popular artificial material used in construction.

2) Fly ash addition resulted in a rise in water absorption values of 5.67, 6.40, and 6.71 respectively.
3) The resistance to abrasion remained constant at 1875 mm3 per 5000 mm2 for replacement of sea sand.
4) Using Fly ash and GGBS instead of cement enables environmentally beneficial use as a paving unit while also lowering building expenses.
5) Ideally, sea sand-infused paver blocks will replace conventional concrete as the most widely used man-made material in the construction industry.
6) According to the findings, there was no carbonation present at any depths of 50 mm, 100 mm, or 150 mm, preventing the specimens from experiencing an increase in porosity brought on by the effects of carbonation.

Reference

[1] S. Sidhardhan, S. Jansi Sheela, Jesus S Meylin, Study on sea sand as a partial replacement for fine aggregate (2017)
[2] Xu He, Jikai Zhou, Mechanical characteristics of sea-sand concrete in the simulated marine environment (2020)
[3] K. Yamazaki, Viewpoints on using sea sand in concrete as fine aggregates in Japan, Concr. J. 12 (10) (1974) 72–76 (in Japanese).
[4] Ashwini Manjunath B T, Lakshmi K, Kiran M S, Sahana C K, Use of Sea Sand in Partial Replacement of River Sand in Concrete for Construction (2018)
[5] Jianzhuang Xiao, Chengbing Qiang, Antonio Nanni, Kaijian Zhang 'Use of sea sand and seawater in concrete construction: Current status and future opportunities
[6] Jonionbi Jonbi, Mohamad Ali Fulazzaky 'Modeling the water absorption and compressive strength of Geopolymer paving block'
[7] Kartheek Thunga, T. Venkat Das an experimental investigation on concrete with replacement of treated sea sand as fine aggregate
[8] K. Ashok Kumar and Dr. P. Partheeban Experimental Investigation on Compressive Strength of Geopolymer Concrete Paver Block
[9] C. Banupriya, Sharon John, R. Suresh, E. Divya and D. Vinitha Experimental Investigations on Geopolymer Bricks/Paver Blocks
[10] Jin-Guang Teng, Yu Xiang, Tao Yu, and Zhi Fang, Development and mechanical behavior of ultra-high-performance seawater sea-sand concrete (2019).
[11] Falah M. Wegian Effect of seawater for mixing and curing on structural concrete.
[12] Manikandan, R., and S. Revathi. 2018. "Behavioral Study on Treated Sea Sand as a Fine Aggregate in Concrete." Journal of Advances in Civil Engineering.

IOP Conf. Series: Materials Science and Engineering 1282 (2023) 012003 doi:10.1088/1757-899X/1282/1/012003

Influence of silane treated nano eggshell powder on mechanical and durability properties of concrete

[1]M R Lavanya , [2]V Johnpaul , [3]N Balasundaram and [4]G.Venkatesan

[1]Research Scholar, Department of Civil Engineering Karpagam Academy of Higher Education, Coimbatore-641 021, TamilNadu, India

[2]Assistant Professor, Department of Civil Engineering, Karpagam Academy of Higher Education, Coimbatore - 641 021, TamilNadu, India

[3]Professor, Department of Civil Engineering, Karpagam Academy of Higher Education, Coimbatore - 641 021, TamilNadu, India

[4]professor, Department of Civil Engineering,University college of Engineering,BIT Campus, Amma Uninersity, Tiruchy –620024.

E-mail: Johnpaul.v@kahedu.edu.in

Abstract.: In order to test concrete's sustainability, this study substitutes nano eggshell powder (nESP) for cement in a silane-treated environment. The results showed that the silane-treated concrete mixtures outperformed the untreated ones in terms of performance. nESP was replaced by 5 to 20% with in cement of 5% along with constant replacement of 30% fly ash by weight of cement. It was found that partial cement substitution with nESP up to 10% produced a sample with greater strength than the control sample. The filling and reinforcing properties of the nESP and the pozzolanic effect of flyash after silane treatment produced favorable results when mechanical strength was evaluated. The increased electrical resistance with age may be caused by the increased hydration products and excess CSH gel formation induced by the pozzolanic action of the fly ash in concrete.

Keywords: Nano eggshell powder, flyash, silane treatment, mechanical behaviour.

1. INTRODUCTION

Concrete and cement-based building materials are currently the most popular and adaptable building materials worldwide [24]. However in challenging environments, water and toxic ions always find their way into concrete structures, causing corrosion and ultimately damaging the reinforced concrete structure and shortening its useful life [25]. Reducing the use of energy, natural resources, and greenhouse gas emissions has become essential because it is anticipated that the amount of cement and concrete production will increase overtime [11]. Fly ash has been thoroughly investigated as a cement-free option in the building industry for many years to produce structural concrete. The heating process and the early-age cracking potential of concrete have been discovered to make fly ash more workable and reduce the rate of hydration [1]. An efficient way to increase compressive strength at a young age is to use enhanced fine fly ash in cement-based products. The cement-fly ash system features dilution, improved pozzolanic reaction, nucleation, and physical filling effects due to the inclusion of fine fly ash [32]. Saha 2018 investigated how fly ash affected the quality of concrete's durability. In conclusion, concrete containing fly ash showed lower compression strength at young ages than conventional concrete, but the strength gradually increased as the curing of the concrete advanced. To increase the mechanical strength development of standard concrete, fly ash is utilized as a water reducer [23]. The performance of concrete is enhanced in terms of density and strength by adding $SiO2$ and $Al2O3$ to the fly ash [8] .Continuous landfill expansion and significant environmental problems are brought on by the disposal of waste products like eggshells. Due to its availability and chemical makeup, this solid waste has significant environmental impact when not adequately treated.

Researchers have sought to use waste materials as a replacement for cement to make the building sector more sustainable and at the same time reduce the waste that is produced [29].There are now two major groups of popular techniques for preventing moisture transport. The first is enhancing the concrete's compactness by adjusting the water-cement ratio, adding water reduction agents, and mixing in mineral admixtures [36]. Khan et al. 2014 discovered that denser and less permeable concrete could be developed with the inclusion of fine particle sized mineral admixtures with greater surface area. Also the concrete fluidity can be reduced significantly due to the ultra- fine particles and helps to create a new way for maximum workability [6]. Implementing polycarboxylate based admixtures into the concrete can enhance workability and the ability to counteract the side effects of nano- particles[7]. The second strategy involves usage of coatings such as osmotic, surface-sealing and surface- enhanced to prevent water from coming into contact with concrete [20]. Their primary method of protecting concrete is through the physical or chemical reaction of the coating, which creates a shield-like layer over the surface that can prevent dangerous chemicals from penetrating inside, increase durability and service life. Cement-based material's anti-permeability and anti-icing capabilities have been improved by super hydrophobic surfaces with more than 150° water contact angles and within 100 roll-off angles [5][31]. Materials from water infiltration for the anti- permeability performance were protected by excellent water repellency, allowing concrete to be more durable [17]. Super hydrophobic surfaces can be applied to concrete in two ways: externally and internally. A hydrophobic coating was used to treat the surface [13].By treating Nano materials of TiO2 and SiO2 with perfluorodecyltriethoxysilane, created super hydrophobic coatings that they sprayed on the cement surface [11]. To lessen the amount of ice depositing on the pavement, [2] developed a super hydrophobic coating using Nano-materials. In order to duplicate and transfer the micro-textured structure to the concrete surface, [11][15] employed polydimethylsiloxaneas a template and created a uniform surface coating through the mechanism of spraying siloxane based compounds. However, the long evity of super Hydrophobic surfaces created using this technology frequently run into issues because the water proofing capacity of the coating degrades with time and peeling[16]. Super hydrophobic particles were added to cement pastes to address internal concerns. [3] incorporated super hydrophobic Nano-particles to cement pastes to boost the hydrophilic nature of calcium silicate hydrate (C-S-H) gels and avoid excessive surface energy there by reducing the permeability. Scott et al. Mixed super hydrophobic admixtures with tailored cementitious composites increased the water-repellence and extend the service life of concrete [19]. In [28] modified concrete using fluoro alkysilane during the hydration process and covered it with metal mesh to produce super-durable super hydrophobic concrete. The application of silanes, according to evidence from various studies, dramatically reduces water up take, which minimizes the entrance of chlorides and consequently lowers the risk of corrosion to the reinforcement [25]. However, factors like surface flaws, cyclic wetting and drying, applicator expertise, surface preparation, application rates, and the immediate environment can all impact on how well they operate. Numerous techniques have been used to increase the permeability of concrete, including its compactness and is made possible by the employment of water-repellent admixtures, silane impregnation, isolation and super hydrophobic coatings [18][4]. However, each of these techniques has drawbacks of its own. For instance, increasing the concrete compactness would not alter its hydrophilic properties, water continues to be absorbed by concrete. Integral hydrophobic additive significantly reduces the concrete strength [32].The surface barrier coating process is susceptible to failure due to cracking and peeling resulting in poor weather resilience[10]. The surface silane impregnation technology is too expensive to be applied extensively in the building materials industry. Hydrophobic agents and potential fillers are the critical raw ingredients for super hydrophobic coatings [35]. The super hydrophobic covering makes it harder for water to be absorbed. Therefore, the super hydrophobic coating technique is the best for lowering concrete permeability. The main barriers to the practical deployment of super hydrophobic coatings are their high cost, poor durability, and challenging manufacturing processes [14]. Therefore, it is crucial to create super hydrophobic coatings for concrete surfaces that are in expensive, simple to prepare, and have the suitable durability. The current study aims to evaluate the effects of silane additive on concrete incorporating waste Nano egg shell particles as an integral addition and surface treatment of aggregate. The primary goal of the current study is to investigate the effects of eggshell powders at the Nano scale on the strength and durability behaviour of concrete under silane-treated and untreated circumstances.

2. MATERIALS

OPC53 grade is chosen as a primary binder and table1 represents their oxide composition. It was determined that the cement's physical characteristics met the standards outlined in IS: 12269-1987 for 53grade OPC. At a consistent weight ratio of 30%, Class F fly ash from Astraa Chemicals in Tamil Nadu was used as a substitute. The fine aggregate is naturally derived river sand and 10mm crushed stone as coarse aggregate is employed in this study. The physical properties and particle size distribution curves are listed in table 2 and figure 1 respectively. The data presented for cement, fly ash, river sand and aggregate are provided from their collected sources. Raw eggshells (ES) were gathered from nearby bakeries and eateries. They were cleaned with tap water to eliminate the organic debris and other surface dusts. After that, they were dried in an electric oven at 110 ± 50C for a day. The dried eggshells were heated to 800^0C in an electric furnace it is then immediately crushed and milled to minimize their particle size. In order to lower the carbon content, the 150mm sieved shell powders are allowed to expose in the electric furnace at 800^0C for 6 hours. The final Product nESP, contained particles with a diameter of 524nm and the XRD pattern is given in figure 2. After being sieved the Nano eggshell powders were stored in sealed containers and used. Sika superplasticizer (SP) was applied at constant ratio of 1% to support the concrete fluidity.

Table 1 Oxide composition of materials used

Oxides (%)	Cement	Fly ash	ESP
SiO2	26.01	49.90	5.772
CaO	53.60	3.23	92.818
SO3	2.77	0.88	1.257
Al2O3	8.54	24.01	-
Fe2O3	4.09	14.40	-
MgO	0.35	0.98	-
MnO	0.13	-	-
K2O	0.90	2.46	-
TiO2	0.42	-	-
P2O5	0.17	-	-
Na2O	0.18	-	-
PdO	-	-	0.152
LOI (%)	2.70	3.50	48.1

Table 2 Physical parameters of material

Parameters	Value
Cement	
Specific gravity	3.14
Blaine's surface area	329 m^2/kg
Initial time	59min
Final time	137min
River sand	
Specific gravity	2.7
Fineness modulus	3.1
Water absorption (%)	0.67
Coarse aggregate	
Specific gravity	2.68
Fineness modulus	6.7
Water absorption (%)	0.55

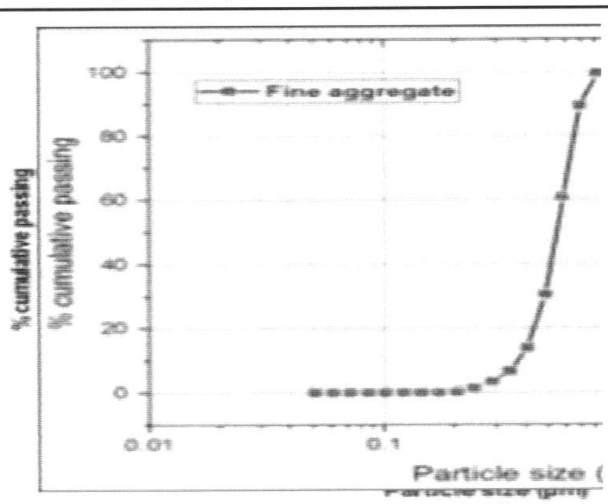

Figure 1 Particle size distribution curves of raw materials

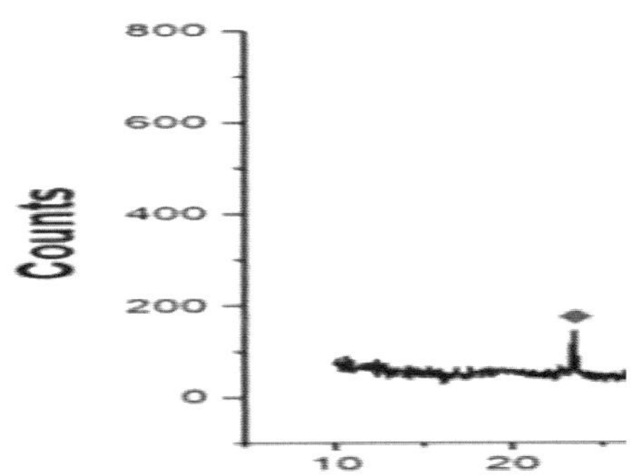

Figure 2 XRD of eggshell powder employed

2.1 MIX PROPORTION

Three different series of concrete mixtures and a control concrete mix with no replacement were made by making four mixes for each series with cement substitutions of 5%, 10%, 15%, and 20% nESP. Each mix undergoes three trial mixes upon which the average will be considered for obtaining the result and hence 12 mixes are casted for each series in addition to three trial mixes casted for reference mix. The fly ash replacement ratio is fixed as 30% for all the concrete mixes except the reference mix. The water to binder ratio was 0.37, and the total binder content was 500 kg/m3. The first stage involved combining the solid components with coarse and fine particles for three minutes. After the mixed water was absorbed, the binder components such as cement, nESP were added deliberately and the mixture was mixed for five minutes. The second half of the leftover water and SP were added to the concrete mixture after it had been mixed for 5 minutes.

Table 3 Mix details of developed concrete specimens

Mix Id	Cement	Flyash	nESP	Fine	Coarse	Silane	Treatment
	Kg/m^3						
REF	500	-	-	760	910	-	Untreated
nES5	325	150	25	760	910	-	
nES10	300	150	50	760	910	-	
nES15	275	150	75	760	910	-	
nES20	250	150	100	760	910	-	
nES5	325	150	25	760	910	2.5	Silane treated mixes
nES10	300	150	50	760	910	5	
nES15	275	150	75	760	910	7.5	
nES20	250	150	100	760	910	10	
nES5	322.5	150	25	760	910	2.5	Silane addition mixes
nES10	295	150	50	760	910	5	
nES15	267.5	150	75	760	910	7.5	
nES20	240	150	100	760	910	10	

Two different series are created to test the effects of silane surface treatment and integral silane addition on the characteristics of concrete. One contained reactive aggregates that had been silane-treated, while the other had silane agents in varying amounts. The silane-treated reactive aggregates and silane agent addition range from 0.5 to 2% by weight of cement represents each series as shown in Table 3.The concrete mixtures were pre-mixed with water and the liquid ingredient before adding the silane.

3. METHODOLOGY

According to IS: 7320:1974, the fresh state behaviour of created concrete mixtures was assessed using slump flow. According to IS:456:2000, compressive and flexural strength tests were employed on cubic and prismatic specimens of 100 mm and 160x40x40mm3 respectively. On cylindrical specimens measuring 100mm x 200mm, the split tensile strength test was conducted following IS: 5816-1959. After that, the100 mm cube samples underwent a UPV test following IS: 13311 Part-1:1992.According to ASTM C 642, absorption research was carried out to determine the concretes' relative porosity permeable void space. The ability of concrete to withstand the flow of an electric current while being impacted by chloride ions is known as electrical resistivity, and measurements of electrical resistivity were made using the Wenner probe method over cylindrical specimens. The test specimens employed for testing are presented in figure 3.

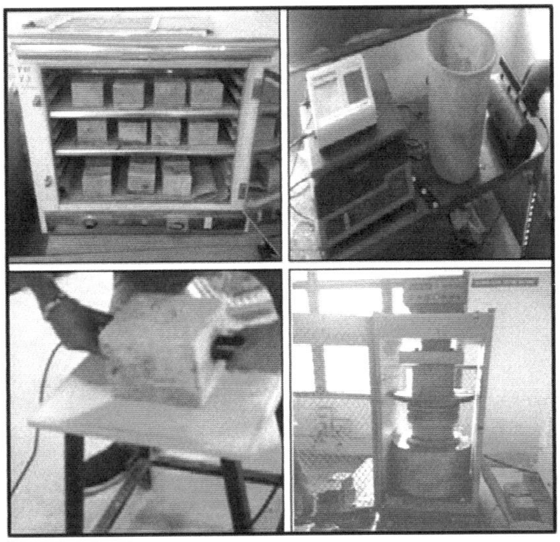

Figure 3 Test specimens used in research

4. RESULTS AND DISCUSSIONS

4.1 Workability measurements

The workability of Nano sized egg shell concrete is less compared to regular Portland cement concrete. An increase in cement content caused by nESP also makes concrete less workable because of its smooth and spherical surface, which in turn increases the flow ability and fluidity in the concrete mixtures. With increasing eggshell powder content, concrete took longer to set, which might be explained by the hydrophobic properties of eggshell powder, which made more free water available as a result the setting time was lengthened (Sophia et al. 2019b). It was determined that silane was a successful treatment for the concrete since the silane treated concrete mixes had superior workability to the untreated and silane added mixes which can be evident in fig 4. Although untreated concrete specimens containing nESP have a poorer workability and slump, they are also compactable, and the addition of nano eggshells does not affect the qualities of fresh concrete. This is because of ESP's high water absorption rate, which depletes the water supply and limits flow ability. The slump falls within the permissible range of 75±25 mm, for all proportions of nESP and this is mainly due to the combined effect of fly ash and nESP. Due to its low water absorption rate, eggshell powder is excellent as filler in concrete mixtures.

Figure 4 Slump values of developed concrete mixes

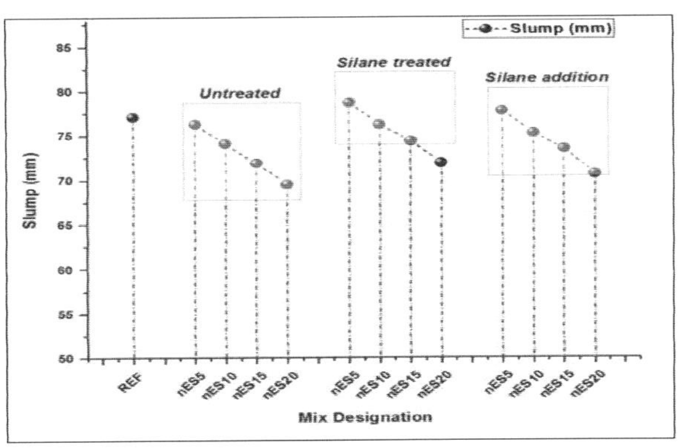

4.2 Mechanical strength

Figure 5 represents the mechanical strength results of the silane treated and untreated Nano eggshell substituted concrete. The compressive strength of the nESP-substituted concrete mixtures increased up to 10%. The filling effect that the nESP demonstrated up to a certain threshold level, which sealed the interstitial holes in the concrete matrix thus functioning as an active bio-filter, may be used to explain the gain in compressive strength. However once nESP exceeded a 10% cement replacement rate, compressive strength values declined. Beyond the threshold addition level, a drop in compressive strength maybe caused by a rise in nESP, which took up spaces that could have been used to make concrete, weakening the crystal development of concrete. The pozzolanic character of fly ash makes it feasible for the calcium hydroxide (C- H) to react with the silica produced during the hydration of cement, resulting in the creation of calcium silicates responsible for the strength of the cement mix. By filling up the Voids already present, nESP served as filler, resulting in a more compact concrete mix and the creation of higher compressive strength. For nESP10 under the untreated, silane treated, and silane supplemented conditions, the rise in percentage was roughly 9%, 17%, and 13%, respectively at 28 days. When nESP is used with high cement content, calcium hydroxide and silica in the mixture are reduced due to the lower cement content. Because of this, there is not enough silica to react with C-H to create C-S-H gel, which leads to a shortage of compressive strength. Along with the pozzolanic flyash, then nESP also served as a filler by filling in the gaps in the internal structure of the concrete, which increased the materials density and compressive strength. The high CaCO3 concentration of eggshells, aids in the hydration of cement paste, nESP inclusion improves concrete compressive strength at all ages.CaCO3 combines with C3A from the cement during hydration creating a ucleation site or an enancedhydration process. Due to the fly ash content, the concrete hydrates more completely and forms more CSH gel, improving the internal micro structure and reducing voids and porosity. Beyond a 10% replacement rate, nESP raises the water demand, which raises the water-cement ratio and lowers the strength. The flexural results thus show that the nESP effectively improved the flexural strength of the concrete mixes due to their surface properties that increased the cohesion between the cement matrix and the filler surface there by improving the flexural strength of concrete. Generally, the flexural strength of concrete increases with the increase of nESP content up to a certain point. The optimal flexural strength attained for nESP10 is about 27.02% for silane treated concrete mixes whereas the silane addition showed 23.67% strength increment. The filling effect of nano eggshells can increase the flexural strength of concrete, so the flexural strength is higher than the control mix. The activation of Nano eggshell powder in combination with fly ash and their finer nature improved flexural performance of the concrete when compared to the reference mix.

Figure 5 Mechanical behaviour of developed concrete mixes

The improved split tensile strength of concrete up to 10% nESP is mainly due to the filling capacity of the Nano sized ESP. The fine nature of ESP and fly ash enhanced the interlocking mechanism between the cement and aggregates is more evident in the silane treated concrete mixes. The reduced splitting tensile strength beyond 10%may be due to the loss in the moisture of the concrete thereby increasing the pores in the concrete. The silane treated mixes showed enhanced tensile strength of 25.22MPa followed by silane added mixes proving that silanes super hydrophobic nature is more effective when treated rather than their integral addition.

5. Ultrasound Pulse Velocity

As the substitution ratio of the nESP grew, ultrasonic velocities decreased which can be evident from figure 6. This decrease in UPV could be brought by the concrete specimens having pores. The increased capacity for absorption and the existence of micro pores in the egg shell powders also decreased the ultrasonic velocities. Due to the inside structure's numerous voids and pores, the quality of the concrete drastically decreased at10% of nESP. Pulse velocity was utilized to gauge the compressive strength of concrete inner cracks and voids that undermined the material's structural integrity decreased the pulse velocity. The increased UPV value of nESP-replaced concrete mixes compared to conventional concrete mix served as evidence of the beneficial effects of Nano eggshell powder. The porosity of the concrete mixes was reduced due to then ESP's enhanced specific surface and faster rate of hydration due to fly ash, which impacted the ultrasonic pulse velocity at all ages. The denser internal structure of the concrete caused due to combined action of fly ash and nESP enhanced the pulse velocity, which in turn represents the filling capacity of the nESP. The concrete composites' stiffness was increased by the nESP's high particle packing density, which led to the greatest improvement. The interaction between the surface of the bio-filler and the cement matrix, in particular, has a positive impact on flexural strength as a result of replacing shell powder. The UPV value evaluated at 28 days for the nESP10 mix at the untreated, silane treated, and silane supplemented conditions, respectively, improved by 6%, 11.7 %, and 9.6 % when compared to the reference concrete illustrating the good synergistic effect of cement and nano eggshell powder. The decrease in ultrasonic pulse velocity is due to the enhanced pore structure and micro cracking of the concrete with nESP and fly ash replacement. The residual ultrasonic pulse velocity significantly differed from the control mix's velocity due to nESP's interlocking behaviour. All of the nESP replaced concrete mixes showed improved ultra sound velocity than the control mix at all ages and were rated as having "good and medium" quality by IS 13311-1. The improvement in mechanical strength of the concrete mixes was further substantiated by the higher Ultra-sonic pulse velocities. The production of CSH mainly due to pozzolanic activity of fly ash, which can self-heal and close opened pores, may have increased, which is another explanation for the increase in ultrasonic pulse velocity.

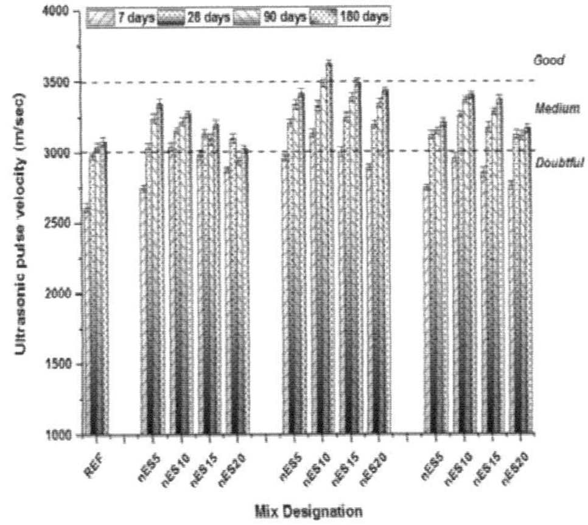

Figure 6 UPV results showing the quality of developed concrete mixes

5.1 Water absorption

The water absorption values of the nESP replaced at 7, 28, 90, and 180 days are shown in figure 7. The use of nano egg shell powder in combination with the fly ash reduced the water absorption up to 10% nESP due to their filling ability and hydrophobic inclination. However, it was shown that the water resistance qualities were reduced when compared to the reference mix after a nESP substitution of more than 10%. Increased Heterogeneity and voids is due nESP's inadequate dispersion in the concrete matrix. The 10% nESP that had been silane treated out performed all roughly by 32% among all specimens at 28days. This might be because of the tightly packed concrete structure that the nESP exhibits. Because there were less pores in the concrete, the water absorption values decreased. Additionally, the nESP's filling properties and pozzolanic nature of fly ash significantly lowered the water absorption values, which closed pores and permeable voids. The pores in the concrete's surface were refined by the nESP's finer texture, which closed any porous voids.

5.2 Porosity

The fly ash as binder replacement produced more secondary C-S-H gel while serving as excellent filler for the concrete mixture and adding additional calcium. As a result, the gel can fill up any gaps in the concrete, reducing water absorption rate. Compressive and flexural strengths demonstrate the importance of C-S-H gel as a main strength-producing reaction product of cement hydration. It also acts as a porosity reducer, producing dense concrete. Additionally, including nano-ESP in the concrete significantly contributed to the reduction of porosity, which raised the concrete's compressive strength. Porosity studies from figure 8 shows that the specimen's porosity values reduced after the nESP substitution. However, for the specimens treated with nESP, especially at more significant proportions, the increases in the porosity values were minor. The porosity reduction of about 33.4%, 36.5%, 36.9%, 40.6% is attained for nESP10 mix at 7days, 28 days, 90 days and 180 days respectively. The leading causes of the reduction in porosity are the filling effect brought on by the replacement of cement with nESP and the decrease in the micro cavities is due to high specific surface area of the nESP. The nESP and fly ash reduces the number of accessible pores in the concrete by successfully bonding with the cement paste due to its spherical surface features. The pozzolanic activity of fly ash mainly encourages the development of C-S-H gel, which fills the pores in the concrete and reduces their volume. Nano particles with small particle sizes help to improve the microstructure of concrete by suffusing the pores in the cement paste and thereby improving the packing level. Nanoparticles further boost the strength of cement-based products by accelerating the hydration of cement, which leads to the creation of calcium-silica-hydrate and helps make concrete even denser.

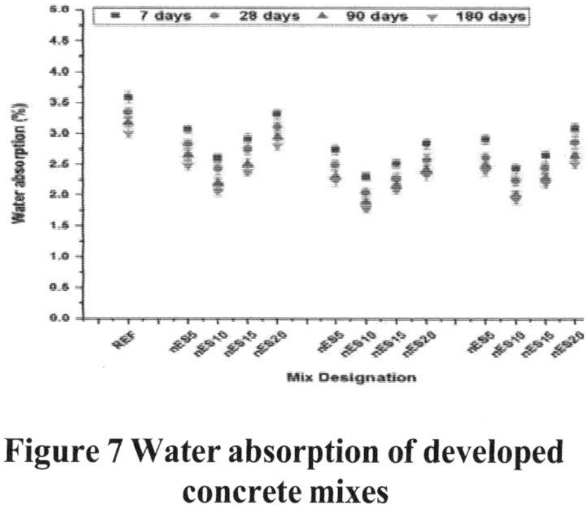

Figure 7 Water absorption of developed concrete mixes

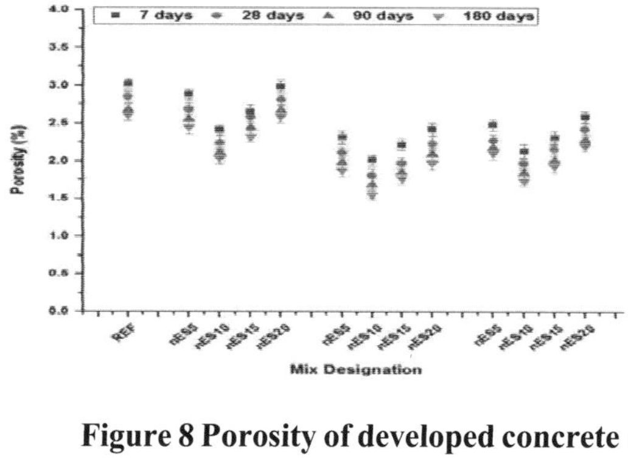

Figure 8 Porosity of developed concrete mixes

5.3 Electrical Resistivity

Figure 9 displays the electrical resistivity of silane-treated and untreated concrete at various ages. As the replacement level of the cement with nESP grew up to 10%, the electrical resistivity of the concrete mixtures increased as well. The improvement in electrical resistance at 7 days, 28 days, 90 days, 180 days for nESP10 mix is 27.8%, 32.3%, 37.1% and 45.1% respectively. Beyond this the resistivity levels considerably decreased but remained within the allowable range. The silane treatment produced concrete of typical quality, reducing the likelihood of corrosion even more. Due to the synergistic action of nESP and fly ash, the concrete mixes increased density and decreased porosity were the leading causes of this variation in electrical resistivity. The enhanced hydration products and additional CSH gel formation caused by the pozzolanic fly ash may be responsible for the increased electrical resistance with age. The decreasing porosity, tortuosity, and chemistry of the pore solution and pore network may have contributed to this increased electrical resistance with ageing.

5.4 Effect of optimum ESP substitution

As observed from the results, the optimum replacement of 10% nESP exhibits better properties than other substitution levels. The percentage improvement over the tests conducted is presented in figure 10. It can be clearly observed that the improved behavior at 28 days proves that the developed concrete mix with 10% nESP holds their implementation for concrete works. The early age improvement is also found to be satisfiable.

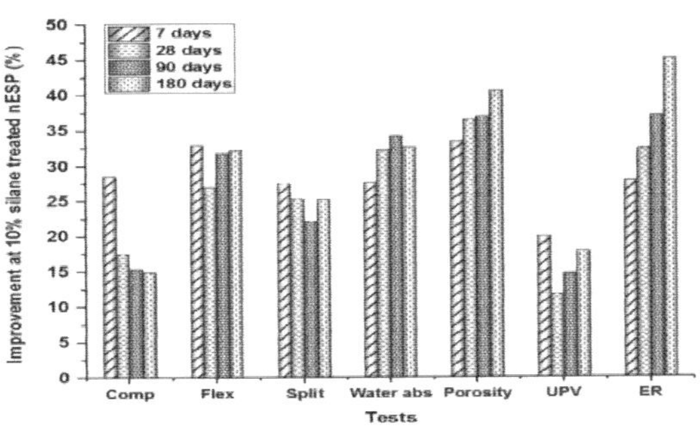

Figure 9 Electrical resistivity of developed concrete mixes

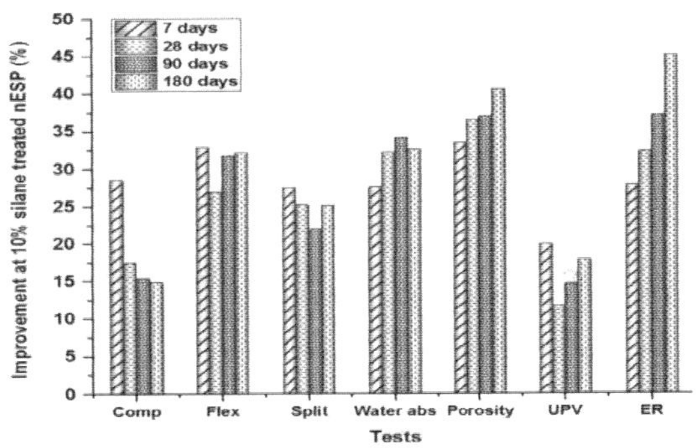

Figure 10 Effect of 10% nESP over the tests conducted on concrete

6. CONCLUSIONS

A thorough examination of the silane treatment and silane integration was carried out to enable the application of nESP in concrete. The experimental evaluation of the technical features is established from the development of concrete mixtures to the evaluation of strength and durability.

1. The strength aspects of the developed concrete mixes yielded better results when the developed mixes were silane treated than the silane added mixes. The silane treated mixes exhibited an optimum improvement of about 17.4%, 27%, and 25.2% for compressive, flexural and split tensile strength at 28 days respectively.
2. By filling up the existing spaces, the nESP effectively served as filler up to a 10% substitution, which made the internal structure of the concrete more packed and promoted the formation of higher strength. Ultra sonic velocities also decreased due to the eggshell powders more tremendous potential for absorption and the existence of micro pores. The 28 7.days UPV values were found to be maximum of about 11.7% for 10% nESP substitution as cement replacement in

concrete.

3. The filling capabilities of the nESP sealed pore sand permeable spaces, significantly lowering the water absorption values. The fly ash content created more secondary C-S-H gel in addition to being a excellent filler for the concrete mixture and adding additional calcium, which slows down the rate of water absorption. The optimum reduction in water absorption and porosity of about 32.2% and 36.5% is attained for 10% nESP at 28 days respectively. The enhanced hydration products and excess CSH gel formation brought on by the pozzolanic action may increase in electrical resistance with ageing. The improved resistance was provided by 10% nESP at 7 days, 28 days, 90 days, 180 days of about 27.8%, 32.3%, 37.1% and 45.1% respectively.

REFERENCE

1. Akid ASM *et al* 2021 Assessing the influence of fly ash and polypropylene fibre on fresh, mechanical and durability properties of concrete, *Journal of King Saud University-Engineering Sciences (InPress)* https://doi.org/10.1016/j.jksues.2021.06.005
2. Arabzadeh A *et al* 2017 Super hydrophobic coatings on Portland cement concrete surfaces, *Construction and Building Materials*, **141**,393401. https://doi.org/10.1016/j.conbuildmat.2017.03.012
3. Chen H *et al* 2018 The effect of super hydrophobic nano-silica particles on the transport and mechanical properties of hardened cement pastes, *Construction and Building Materials*, **182**,620-628. https://doi.org/10.1016/j.conbuildmat.2018.06.146
4. Christodoulou C *et al* 2013 Long-term performance of surface impregnation of reinforced concretestructures with silane, *Construction and Building Materials*, **48,** 708716. https://doi.org/10.1016/j.conbuildmat.2013.07.038
5. Courard L *et al* 2021 Influence of hydrophobic product nature and concentration on carbonation resistance of cultural heritage concrete buildings, *Cement and Concrete Composites*, **115**, 103860. https://doi.org/10.1016/j.cemconcomp.2020.103860
6. Dry C 2000 Three designs for the internal release of sealants, adhesives, and water proofing chemicals in to concrete to reduce permeability, *Cement and Concrete Research*, **30** (12), 19691977. https://doi.org/10.1016/S0008-8846(00)00415-4
7. Hamada H *et al* 2020 Effects of nano-palmoilfuelash and nano-eggshell powder on concrete",*Construction and Building Materials*, **261**, 119790. https://doi.org/10.1016/j.conbuildmat.2020.119790
8. Harihanandh M *et al* 2021 Comparative study on chemical and morphology properties of nano flyash in concrete, *MaterialsToday:Proceedings*, **45** Part 2, **3132-313,** https://doi.org/10.1016/j.matpr.2020.12.217
9. Her S *et al* 2022 Feasibility study on utilization of pulverized eggshell waste as an alternative to limestone in raw materials for Portland cement clinker production, *Construction and Building Materials*, **324**, 126589. https://doi.org/10.1016/j.conbuildmat.2022.126589
10. Heriyanto *et al* 2019 Effect of different waste filler and silane coupling agent on the mechanical properties of powder-resin composite, *Journal of Cleaner Production*, **224, 940–956.** https://doi.org/10.1016/j.jclepro.2019.03.269
11. Horgnies M and Chen J J 2014 Super hydrophobic concrete surfaces with integrated microtexture, *Cement and Concrete Composites*, **52, 81-90.** https://doi.org/10.1016/j.cemconcomp.2014.05.010
12. Khan S U 2014 Effects of different mineral admixtures on the properties of fresh concrete, *The Scientific World Journal*, **1–11**, https://doi.org/10.1155/2014/986567
13. Lei L *et al* 2020 Fabrication of super hydrophobic concrete used in marine environment with anti-corrosion and stable mechanical properties, *Construction and Building Materials*, **251**, 118946. https://doi.org/10.1016/j.conbuildma
14. Liu B *et al* 2020 Mechanical and permeability properties of polymer-modified concrete using hydrophobic agent, *Journal of Building Engineering*, **31**, 101337. https://doi.org/10.1016/j.jobe.2020.101337
15. Liu H *et al* 2019 Silica coating with well-defined micro-nano hierarchy for universal and stable surface super hydrophobicity, *Chemical Physics Letters,* **730**, **594-599.** https://doi.org/10.1016/j.cplett.2019.06.001
16. Liu P *et al* 2017 Super hydrophobic and self-cleaning behavior of Portland cement with lotus-leaf-like micro structure, *Journal of Cleaner Production*, **156**, **775-785.**

https://doi.org/10.1016/j.jclepro.2017.03.211

17. Mora E *et al* 2019 Control of water absorption in concrete materials by modification with hybrid hydrophobic silica particles, *Construction and Building Materials*, **221**, **210-218**. https://doi.org/10.1016/j.conbuildmat.2019.06.086
18. Muthuraman U *et al* 2022 Influence of silane treatment on the mechanical strength and durability of reactive powder concrete containing recycled fine aggregate, *Materials Today: Proceedings*, **62**(8), **5444-5451**. https://doi.org/10.1016/j.matpr.2022.04.093
19. Muzenski S *et al* 2015 Hydrophobic engineered cementitious composites for highway applications, *Cement and Concrete Composites*, **57**, **68-74**, https://doi.org/10.1016/j.cemconcomp.2014.12.009
20. Pan X *et al* 2017 A review on concrete surface treatment PartI:Types and mechanisms, *Construction and Building Materials*,**132,578–590.**https://doi.org/10.1016/j.conbuildmat.2016.12.025
21. Saha A K 2018 Effect of class F flyash on the durability properties of concrete, *Sustainable Environment Research* **28** (1), **25-31**.https://doi.org/10.1016/j.serj.2017.09.001
22. Sankar P L *et al* 2021 Studies on drying shrinkage and water permeability of fine fly ash high performance concrete, *Materials Today:Proceedings*, **46** Part1, **930-933**. https://doi.org/10.1016/j.matpr.2021.01.069
23. She W *et al* 2018 Biomimeti csuper hydrophobic surface of concrete: Topographic and chemical modification assembly by direct spray, Construction and Building Materials, **181,347–357**. https://doi.org/10.1016/j.conbuildmat.2018.06.063
24. She W *et al* 2020 Super hydrophobic concrete with enhanced mechanical robustness:Nano hybrid composites, strengthen Mechanism and durability evaluation, *Construction and Building Materials*, **247**, 118563. https://doi.org/10.1016/j.conbuildmat.2020.118563
25. Shen L *et al* 2019 Performance of silane-based surface treatments for protecting degraded historic concrete, *Progress in Organic Coatings*, **129**, **209-216**. https://doi.org/10.1016/j.porgcoat.2019.01.016
26. Sophia M and Sakthieswaran N 2019a Synergistic effect of mineral admixture and bio-carbonate fillers on the physico-mechanical properties of gypsum plaster, Construction and Building Materials, **204**, **419–439.** https://doi.org/10.1016/j.conbuildmat.2019.01.160
27. Sophia M and Sakthieswaran N 2019b Waste shell powders as valuable bio- filler in gypsum plaster – Efficient waste management technique by effective utilization, *Journal of Cleaner Production*, **220**, **74-86**.https://doi.org/10.1016/j.jclepro.2019.02.119
28. Song J *et al* 2017 Super-robust super hydrophobic concrete", *Journal of Materials Chemistry A*, **5**, **14542- 14550**.https://doi.org/10.1039/C7TA03526H
29. Subbiah K *et al* 2018 Development of water-repellent cement mortar using silane enriched withnano-materials, *Progress in Organic Coatings,* **125**, **48-60**, https://doi.org/10.1016/j.porgcoat.2018.08.021
30. Sun Y *et al* 2021 Prediction of compressive strength development for blended cement mortar considering fly ash fineness and replacement ratio", *Construction and BuildingMaterials*, **271**,121532 https://doi.org/10.1016/j.conbuildmat.2020.121532
31. Wang F *et al* 2020 Preparation and properties of foundrydust/Portland cement based composites and superhydrophobic coatings, *Construction and Building Materials*, **246**, 118466. https://doi.org/10.1016/j.conbuildmat.2020.118466
32. Xue X *et al* 2017 Asystematic investigation of the waterproofing performance and chloride resistance of a self- developed waterborne silane-based hydrophobic agent for mortar and concrete, *Construction and Building Materials*, **155**, **939-946**. https://doi.org/10.1016/j.conbuildmat.2017.08.042
33. Yang D *et al* 2022 Potential use of waste eggshells incement-based materials:A bibliographic analysis and review of the material properties, *Construction and Building Materials*, **344**,128143. https://doi.org/10.1016/j.conbuildmat.2022.128143
34. Zhang H F *et al* 2012 Application of wate rreducing agent in concrete of engineering materials, *Advanced Material Research*, **578**, **87–90**. https://doi.org/10.4028/www.scientific.net/AMR.578.87
35. Zhu Jand Liao K 2020 A facile and low-cost method for preparing robust super hydrophobic cement block", *Materials Chemtry and Pysics*, , 123064.ttps://doi.org/10.1016/j.matchemphys.2020.1230

Performance evaluation of HVFA content in cement concrete incorporating filler additives: *rheology and mechanical properties*

Saurav Kar[1], Pritha Das[2] and Anup Kumar Mondal[3]

[1] Ph.D. Research Scholar, Department of Civil Engineering, Techno India University, Salt Lake, West Bengal, India
[2] Postgraduate Research Scholar, Department of Civil Engineering, Techno India University, Salt Lake, West Bengal, India
[3] Professor, Department of Civil Engineering, Techno India University, Salt Lake, West Bengal, India

Corresponding author: Saurav Kar[1]

E-mail: sourav.kar@heritageit.edu

Abstract. For a better sustainable progressive world, construction industry turned wider gradually. It is well known fact that cement concrete is main component of construction for our modern civilization. Now days, fly ash is used as the most potential replacement material in cement concrete, as production of cement very badly harm our environment. Reports show that one tonne cement produces one tonne carbon dioxide in our environment and responsible for global warming. After successive research work, it is proven that replacement of high volume fly ash (above 50%) which is pozzolanic in material (consists of siliceous and aluminous compounds) in cement paste helps to achieve high strength, porosity, heat of hydration, micro structure as well as very economical and eco-friendly against normal cement paste. This paper mainly highlights on the detailed study of rheological and strength properties of high volume fly ash concrete (50%-70% replacement of cement in various grades of fly ash mixed as per IS standards) in addition with various additives and melamine based admixture. In addition to this, other properties such as consistency, hardening and stiffening, workability, mechanical properties are studied. This overall study include experimental program on various concrete specimen with varied cementitious content along with admixtures.

Keywords: high volume fly ash, OPC, silica fume, admixtures, flowability.

1. Introduction

At late 1980s High volume fly ash concrete invented by the Canadian Centre for Mineral and Energy Technology (CANMET) for structural uses [1]. In accordance with American Society for Testing & Materials (ASTM) C618, fly ash defined as Class F fly ash and Class C fly ash. These two types of fly ash are categorized by their amount of chemical (i.e. calcium, silica, alumina and iron) content in the ash. Fly ash mainly found in powder form having granulometry of 0.5 – 100 μm, mainly used as supplementary cementitious material (SCM) for Portland cement [2]. Thorough experimental work shows use of fly ash in concrete is highly advantageous such as it instigates improved mechanical, physical as well as rheological properties. Hence, the main focus for utilization of fly ash in concrete is to achieve satisfactory values of compressive strength at lower cement content. It is also observed that application of mineral admixtures and super plasticizer in fly ash concrete imparts beneficial effects to concrete, accelerate the high strength, workability, setting time, sulphate resistance [5].

This comprehensive literature study emphasizes on collected data on fly ash based concrete mixes including additives. The materials incorporated in various previous experimental works include nano-silica, condensed silica, improved HVFA properties up to 70% replacement. The overall review will assist to choose suitable combination of HVFA concrete with other mineral and chemical admixtures. This scrutiny will also help to develop HVFAC and their individual and combined effect of various additives, proper mix proportioning methodology and guidelines, helpful for construction industry in future.

2. Literature review

2.1. General

In this elaborate study, the main aim lies on detailed study on high volume fly ash. In concrete the percentage of replacement range from 20%- referred as low volume fly ash and above 40% as high volume fly ash [9]. This current analysis is based on a comprehensive study on changes of mixture after replacement of fly ash in concrete. Observations also analyse the behaviour of concrete after applying super plasticizer and different additives with fly ash and hydraulic binder at early age and later age.

2.2. History of Work

L.Lam et.al., (1998) showed high volume fly ash can be utilized in concrete with varied water-cementitious ratio, slump value, etc. Earlier experimental investigations emphasize improved strength parameters (42.5 MPa at 3-days curing stage and 85 MPa at 28-days curing stage) for replacement of 50%-70% (b/w) cement with fly ash in presence of various super plasticizers. Slump value changes with time [3]. It is also observed fly ash content below 30% (b/w) is unsuccessful in meeting expected levels of performance compared to 40%, 45% and 50% fly ash content with improved results in strength (after 28 days for R.C.C. constructions) and also shows great performances in high resistivity against abrasion and on chemical attack(i.e., chloride, sulphate and acid attacks) [4,17].

Lam L. et.al., (2000) presented with increasing amount of fly ash (high volume) in cement helps to increase porosity of paste level and also comes up with increased level of pore surface fractal dimension. In other hand it's mainly contribute during the process of heat of hydration [4]. But 45%-55% fly ash in paste unreacted in degree cement hydration reaction at higher curing age [10]. Research studies by **Cengiz D A et al (2003)** focussed on implementation of partial replacement of fine aggregates with fly ash. Utilization of fly ash as replacement of fine aggregate was studied [5,18,19].

Works of **M.J. Mc Carthy , R.K. Dhir (2005)** shows fly ash possess huge effect on environment and as well as takes a good lead in technical and economic benefits. High volume fly ash in high range can be used as combination with Portland cement for rapid hardening. Fly ash possesses a huge aspect on environment and as well as take a good lead in technical and economic benefits. High volume fly ash in high range can be used as combination with Portland cement for rapid hardening [6].

The study also indicates addition of silica fume (5%) in high volume fly ash concrete shows very impressive result in cylinder compressive strength and tensile splitting strength but not in cube. 15%-25% fly ash (b/w of cement) provides better effect on tensile strength but 45%55% fly ash replacement with cement display high larger fracture displacement and shows low load bearing capacity [9].

Jian – Hua Wu et.al., (2006) served high volume fly ash can be utilized in concrete with varied water cementitious ratio, slump value, etc. Earlier experimental investigations emphasized to improve strength parameters (42.5 MPa at 3-days curing stage and 85 MPa at 28-days curing stage) for replacement of 50%-70% (b/w) cement with fly ash in presence of various super plasticizers. Slump value changes with time [9]. It is also found out adding lime stone and gypsum with fly ash play role as a soil stabilizer which is better for environmentally friendly[10-12].

In addition to these studies, various authors such as **S. Mukherjee et.al., (2013)** conducted researches up to 60% replacement of fly ash in concrete create zero slump concrete that' s shows high compressive strength [11]

2.3. Objective:

Fly ash in green concrete has a immense aspect in sustainable development and step by step its research work growing that some work can be beneficial path in future i.e. from previous research studies, it is well known high volume fly ash content in cement concrete leads to enhanced mechanical strength properties on if extra additive such as silica fume (5%-10%) is added along with 1%-5% chemical admixture is added.

This present experimental investigation not only highlights the utilization of high volume fly ash in high performance concrete at various cement replacement levels (CRL) but also focuses on development of a new, rational, scientific and cost-effective mix design approach. This paper mainly deals with the strength and rheological effects due to various water/binder ratios of 0.25, 0.30, 0.33 and 0.40 with various dosage of chemical activators and mineral additives.

3. Experimental methodology

3.1. Objective

The class F fly ash required for the experimental purpose is obtained from Gujarat, India possesses fineness modulus (F.M.) 2.06. The fly ash obtained stored in controlled conditions. Chemical tests conducted to determine the chemical properties of FA and the oxide composition of fly ash and cement is as follows: SiO_2 -22%, Al_2O_3- 28.34%, Fe_2O_3- 12.5%, TiO_2-1.51%, CaO- 1.79%, MgO- 0.84%, SO_3- 2.25%, Na_2O- 0.42% and Loss of ingnition-2.42%. The test results for various chemical composition is given in figure 1 and figure 2.

CHEMICAL TEST RESULT OF PULVERIZED FUEL ASH

(*)TYPE OF PULVERIZED FUEL ASH:Siliceous
PERIOD OF TEST :08/04/2023-14/04/2023

Sr. No.	Type of Test	Test Result	Permissible Limits as per IS 3812 (Part 1)
1	Total Chloride, Percentage by Mass	0.02	Max. 0.05
2	Silicon Dioxide(SiO_2), Percentage by Mass	22	Min. 35
3	Magnesium Oxide (MgO), Percentage by Mass	0.32	Max. 5
4	Sulphur Trioxide(SO_3), Percentage by Mass	2.25	Max. 3

 # Test Parameter is calculated by Results under scope of NABL test parameters SiO_2, Al_2O_3 & Fe_2O_3.

Table: 3.1 Chemical Test Results

River sand of zone III used with fineness modulus 2.13(zone III) and specific gravity 2.55 and 20 mm nominal size coarse aggregate used, which fineness modulus 2.63 and specific gravity is 2.96. Commercial superplasticizer Sika® plastocrete® plus with composition of modified lignosulphonate is used as chemical admixture in concrete mix at 0.5 % to 5% b/w of total cementitious materials present.

Along with this in some tests, polycarboxylate ether (PCE) is also used. The specific gravity of the activator is 1.04. In some experimental series, silica fume is also used. The initial and final setting times of cement calculated in the laboratory and the values determined obtained as 150 minutes and 225 minutes respectively. Fine aggregate (F.A.) considered for this work- locally available(Zone-III) natural river bed sand. The coarse aggregates and fine aggregates utilized for concrete preparation kept oven dried maintaining S.S.D (Saturated Surface Dry) condition throughout. For designing mix proportions of various concrete mixes, following data is utilized as: nominal maximum size of aggregates adopted as 20 mm and degree of workability = 0.90 (C.F.). The target slump maintained at 100 to 120 mm. The degree of quality control maintained as good and type of exposure= mild. In this test program, normal cement binder (OPC 43 grade) used with specific gravity of cement= 3.15. The fine aggregates possess specific gravity of 2.55 and specific gravity of coarse aggregates (crushed angular) taken as 2.64.The water absorption adopted for coarse aggregates as 0.5 % and for fine aggregates as 1.0 %. Free (surface moisture) for coarse aggregates is considered nil and for fine aggregates as 2%.

3.2. Tests conducted

In this particular experimental study, compressive strength tests are conducted to determine the crushing strength values of various samples of 150 mm X 150mm X 150 mm sizes. All the experimental procedure conforms IS standards. To determine the workability properties, slump flow test, flow table test and compacting factor tests conducted in accordance with standard IS specifications.

4. Results and discussions

4.1. General

In all the experimental investigations conducted, primary materials used are ordinary Portland cement, class C and class F fly ash, pulverized fly ash available locally from thermal power plants, substituted with chemical and mineral admixtures.

4.2. Materials used in experimental work

Table 4.2.1: Detailed List of various materials involved in various experimental investigations

<table>
<tr><th colspan="2">Fine Aggregate</th><th>Coarse Aggregate</th><th>Chemical Admixture</th></tr>
<tr><td colspan="2">Sand (Fineness modulus 2.18 and Density 2.78 g/cm^3)</td><td>Crushed limestone (5-10 mm and 10-20 mm)</td><td>Sulpho Naphthalene based</td></tr>
<tr><td colspan="2">Sand (4.75 mm size)</td><td>Gravel (12.5 mm size)</td><td>Melamine based</td></tr>
<tr><td colspan="2">River sand + silica fume (5% b/w)</td><td>Crushed granites</td><td>None</td></tr>
<tr><td colspan="2">Sand</td><td>Gravel (10 mm max.)</td><td>Commercial Carboxylic</td></tr>
<tr><td colspan="2">Sand (5 mm max.)</td><td>Crushed limestone(16 mm)</td><td>Polycarboxylate based</td></tr>
<tr><td colspan="2">-</td><td>-</td><td>Lignosulphonate gluconate</td></tr>
<tr><td colspan="2">River sand</td><td>Crushed coarse (size 12 mm and 16mm)</td><td>Polycarboxylic ether (PEC)</td></tr>
<tr><td>Phase 1</td><td>River sand (0 4mm)</td><td>Crushed coarse (4-8mm and 8-16 mm)</td><td rowspan="2">Polycarboxylate ether (PEC)</td></tr>
<tr><td>Phase 2</td><td>River sand (0-4mm)</td><td>River coarse (4-8mm and 8-16 mm)</td></tr>
</table>

4.3. Results & Discussions

Table 4.2.2: Comparison of concrete rheological factors: Fluidity, slump and flow tables values of HVFA Concrete incorporating cement FA and additives

Notation	**Cement %**	**Fly ash %**	**(w/b) ratio**	**water Reducer %**	**Slump (mm.)**	**Spread (mm)**
C(100% cement)	100	0	0.3	1.5	60	525
F1(90% cement with replacement of 10%FA)	90	10	0.3	1.5	200	520
F2 (80% cement with replacement of 20% FA)	80	20	0.3	1.5	217	530
F3(70% cement with replacement of 30% FA)	70	30	0.3	1.5	240	570
F4 (60% cement with replacement of 40% FA)	60	40	0.3	1.5	260	690
F5(50% cement with replacement of 50% FA)	50	50	0.3	1.5	257	615
F6(40% cement with replacement of 60%FA)	40	60	0.3	1.5	242	610
F7 (30% cement with replacement of 70%FA)	30	70	0.3	1.5	240	600
GROUP 1						
F50.S2 (fly ash 50% with 2% sp)	194	194	0.35	2	100	650
F50.ES3 (50% fly ash with 3% sp)	197	197	0.31	3	75	500
F50.ES4 (50% FA with 4% sp)	200	200	0.29	4	65	440
GROUP 2						
F60.S2(60% FA with 2% sp)	153	230	0.38	2	95	630
F60.ES3(60% FA with 3% sp)	156	235	0.33	3	80	540
F60.ES4(60% FA with 4% sp)	158	236	0.31	4	75	470
GROUP 3						
F65.S2 (65% FA with 2% sp)	134	249	0.38	2	85	560
F65.ES3 (65% FA with 3% sp)	136	253	0.34	3	80	520
F65.ES4 (65% FA with 4% sp)	137	255	0.32	4	70	470

[FA = fly ash, sp= super plasticizer, w/b ratio= water/ binder ratio]

From above table 4.2.2 , it is clearly seen the slump of base concrete with no fly ash is as low as 60 mm and almost no spread. Addition of fly ash improves fluidity of concrete and possesses effective way on improving the plasticity of concrete[12]. Hence with increasing fly ash dosage the fluidity is improved accordingly.

According to table 4.2.2, specimens with 50% FA and 4% super plasticizer resulted in very less amount slump and spread value i.e. 65 mm & 440 mm. but other specimens gave satisfactory results with fly ash replacement and super plasticizer.

Table 4.2.3: Comparison of compressive strength values of HVFA Concrete incorporating cement +FA+ additives

(a.) At w/b ratio =0.25

Compressive strength in diff. days of curing (MPa)

Notation	Fly ash %	(w/b) ratio	sp content (%)	3-day	28-days	56-days
1C (100% cement)	0	0.25	2	65.5	85.3	94.7
1F5 (50% cement with 50% FA)	50	0.25	2	29.5	61.6	69.8
1F6 (40% cement with 60% FA)	40	0.25	2	28.3	62.7	68.4
1F7(30% cement with 70%FA)	30	0.25	2	28.4	58	65.2

[FA = fly ash, sp= super plasticizer, w/b ratio= water/ binder ratio]

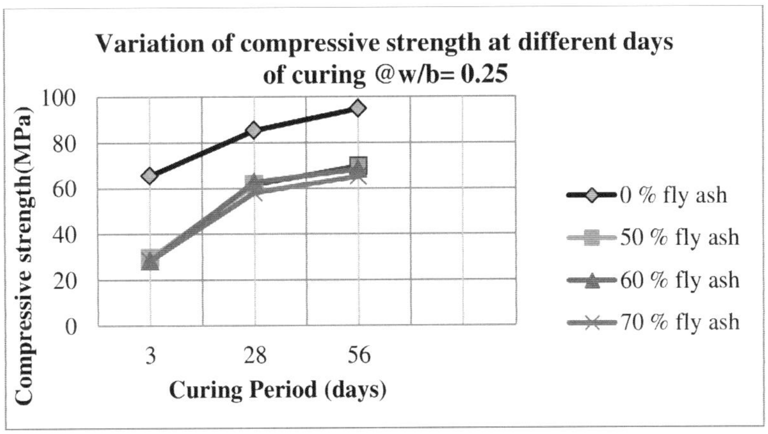

Fig. 1. Variation of compressive strength @w/b = 0.25

From the above table 4.2.3.(a) and the graph shown in figure 1, it is evident at very low w/b ratio at 0.25 the compressive strength values shows increasing trend. From 3-days to 56-days curing, for 0% replacement of fly ash and w/b ratio 0.25, the strength gradually increased from 65.5 MPa to 94.7 MPa. Respectively for 50%, 60% and 70% replacement of fly ash, the 3 days strength is very low compared to 3 days strength for 0% fly ash replacement [13]. Even the ultimate strength value at 56 days curing a found to be much lower for 50%, 60% and 70% replacement of fly ash.

At such low w/b ratio of 0.28, use of water reducing super plasticizer or high range water reducer (HRWR) is necessary and it is evident from the observation table [14]. The specimens with 50% FA +4% sp show slightly higher strength values compared to 56% replacement of FA. Tests results for 70% FA shows much lesser strength values compare to previous samples and is of nearly 40% lesser. Even though the initial strength values for 56% replacement of fly ash specimens showed lower strength compared to (50% FA and 4% sp), at the end of 1-year, the strength value are nearly same to each other.

(c.) At w/b ratio =0.3

Compressive strength in diff. days of curing (MPa)

Notation	Fly ash %	w/b ratio	sp content	3-days	7-days	28-days	56-days	90-days	180-days
C (100% cement)	0	0.3	2	58.8		79.8	83.3	-	-
F5 (50% cement with50% FA)	50	0.3	2	25.1		56.1	68.6	-	-
F6 (40% cement with 60% FA)	60	0.3	2	21.6		51.7	62.4	-	-
F7(30% cement with 70% FA)	70	0.3	2	19.1		48.4	58.2	-	-
S1-0-0 (0%FA with 0%silica fume)	0	0.3	0	64.9	75.5	86.8	87.2	95.7	97.7
S1-15-0 (15%FA with 0% silica fume)	15	0.3	0	52.1	66.4	86	94.8	99.6	106.3
S1-25-0(25%FA with 0% silica fume)	25	0.3	0	48	65.7	85.4	90.4	95.4	107.8
S1-45-0 (45%FA with 0% silica fume)	45	0.3	0	34.1	49.2	71.8	85.4	87.7	97.7
S1-55-0 (55%FA with 0% silica fume)	55	0.3	0	22.3	36.4	57.4	66.6	72.8	79.9
S1-0-5 (0%FA with 5% silica fume)	0	0.3	5	58.3	75.5	87.8	93.1	93.6	99.3
S1-20-5(20%FA with 5% silica fume)	20	0.3	5	46.3	65.6	78.5	85.8	90.3	95.9
S1-40-5 (40%FA with 5% silica fume)	40	0.3	5	30.5	48.6	71.1	80	83.4	88.3

[FA = fly ash, sp= super plasticizer, w/b ratio= water/ binder ratio]

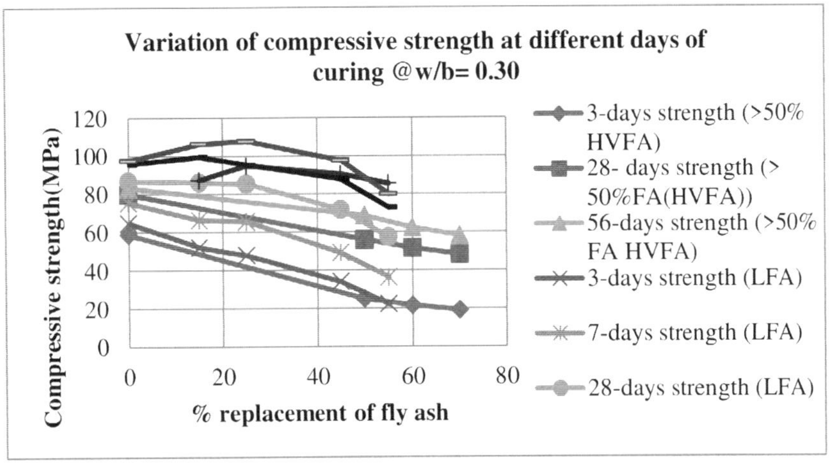

Fig. 2. Variation of compressive strength @w/b = 0.3

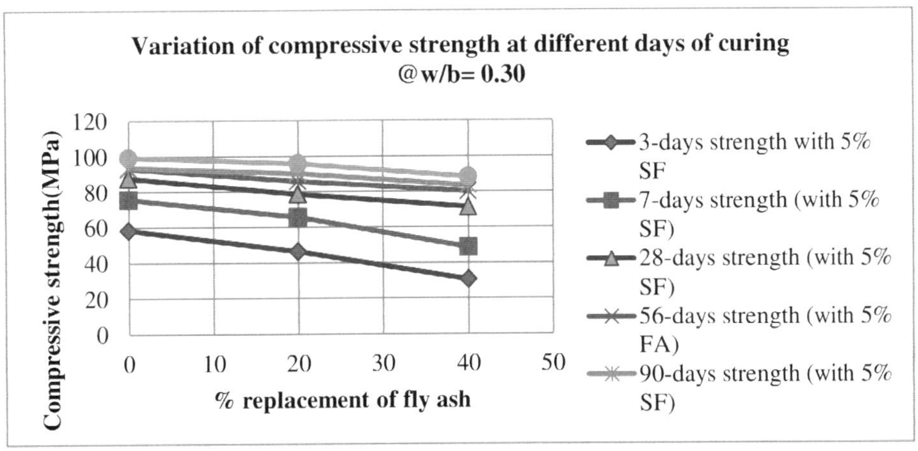

Fig. 3. Variation of compressive strength @w/b=0.3

The above table highlights utilization of w/b ratio of 0.30 found to be much popular in majority of all experimental investigations. From figure 2 and figure 3, it is observed that low FA content (LAF 0%, 15%, 25%, 45% and 55% FA) shows much higher strength values at a later stage of curing i.e. 180 days compare to all the specimens. Among all the various sample specimens 3-days strength, 7-days strength for HVFA and LFA shows lower values. Considering the 28-days strength for LFA, 90-days strength, 56-days strength and 180-days strength shows higher value[15]. In fig. 3, LFA specimens with 5% silica fume is represented. Similar to the previous figure, the general trend from early days curing to later days curing, the strength values correspondingly increased. In corporation of 5 % silica fume in all concrete specimens the strength values ranged from nearly 60MPa to nearly 100Mpa. While, the same behaviour s represented for concrete specimens with LFA and HVFA without silica fume content [16].

For w/b ratio for 0.33, very limited literature is available. The 3 days strength for 50% FA is slightly more compare to 7 days strength for 60% FA + 3% sp. It is prominent enough to view 365 days strength for 50% FA and 60% FA+ 3%sp achieved compressive strength around 60 and 70MPa respectively.

For w/b ratio 0.34, very limited literature available. Compared to w/b of 0.33, the 7 days strength for 50% and 60% FA shows lesser strength values for w/b ratio 0.34. Comparing w/b ratio of 0.3,0.33 and 0.34, 0.3 w/b ratio is recommended for flowable and pump able concrete with h strength values.

(g.) At w/b ratio =0.4

Compressive strength in diff. days(MPa)

Notation	% FA	w/b ratio	sp %	3 days	7 days	28 days	56 days	90 days	180 days	365 days
S2-0-0(0%FA with 0% silica fume)	0	0.4	0	35	48.4	60.7	67.1	70.5	70.6	-
S2-15-0(15%FA with 0% silica fume)	15	0.4	0	29.3	39.9	56	63.4	68.5	72.1	-
S2-25-0 (25%FA with 0% silica fume)	25	0.4	0	24.7	33.7	49.3	60.8	66.2	70.2	-
S2-45-0 (45%FA with 0% silica fume)	45	0.4	0	14.5	20.3	43.9	54.1	61.2	63.7	-
S2-55-0 (55%FA with 0% silica fume)	55	0.4	0	13.6	19.8	37.3	47.1	52.9	63.2	-
S2-0-5(0%FA with 5% silica fume)	0	0.4	5	37.3	53	69.4	72.1	73.7	74.5	-
S2-20-5(20%FA with 5% silica fume)	20	0.4	5	28.9	42.1	62.3	69.9	72.4	76	-
S2-40-5(40%FA with 5%silica fume)	40	0.4	5	14.5	20.5	44.6	55.3	59.1	68.4	-
M-2(40% fly ash)	40	0.4	-	-	17	26.7	-	-	-	38.6
M-3 (45% fly ash)	45	0.4	-	-	15.3	24.7	-	-	-	34.4
M-4(50% fly ash)	50	0.4	-	-	14.7	23.1	-	-	-	32.1
X-50(50% FA)	50	0.4	-	-	-	36.3	-	-	-	54.2
X-50 (50% FA)	50	0.4	-	-	19	28	-	41	-	55
X-60(60% FA)	60	0.4	-	-	16	27	-	36	-	50

[FA = fly ash, sp= super plasticizer, w/b ratio= water/ binder ratio]

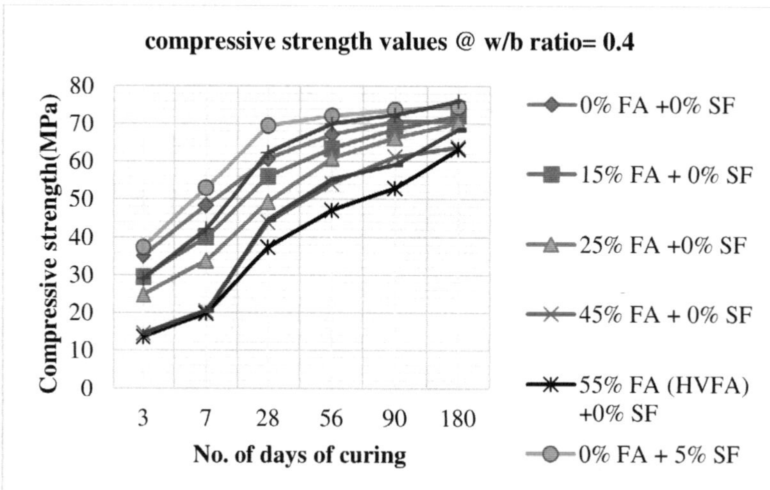

Fig. 4. Variation of compressive strength @ w/b=0.4

Previous experimental studies prominently highlight use of w/b ratio 0.4. Extensive experiments are conducted taking 0.4 w/b ratio. From the fig. 4 it is observed the compressive strength values gradually increased from 3 days ,7 days , 28 days , 56 days , 90 days and 180 days for 15% FA, 25% FA , 45% FA and 55% FA respectively. Addition of 5% silica fume to 0%, 20% and 40% FA shows immense increasing strength values over the period of 3 days to 180 days curing. From the observation table it can be ascertain that with no content of silica fume or condense silica, fly ash replaced concrete specimens (from 0% to 60%) attains very lower strength even at 365 days curing.

5. Conclusions

Performance evaluation of physical, mechanical and rheological properties of HVFA concrete with replacement of fly ash with/without additives studied in details. Detailed experimental study reveal, HVFA concrete usually tends to attain minimum early strengths and performs slow at pozzolanic reaction compare to cement concrete. Another striking feature highlights replacement of fly ash beyond 80% by weight fails to develop satisfactory strength and applicable for structures but uses of many additives in HVFA concrete could be possible to achieve improved performance of concrete.

In overall scenario, HVFA concrete shows increasing values in the mechanical properties of including compressive strength, flexural strength, splitting tensile strength and elastic modulus of concrete. Also we can conclude that replacement of FA in concrete great effect on lower drying shrinkage, and higher resistance against chloride, sulphate and acid attacks. For more uses of HVFA concrete need large-scaled further studies needed to determine the actual behavior of HVFA concrete in structural aspect. The review has identified that the nature of the fly ash and additives in concrete, the fineness of fly ash and effect of additives, that determine the mechanical properties of HVFA concrete effectively manner.

Acknowledgement

The authors are thankful to the Department of Civil Engineering, techno India University for full support of this experimental research work. The whole experimental program is fully funded by Techno India University and Heritage Institute of Technology, Kolkata, West Bengal, India.

References

[1] Malhatra VM 1986 *Super plasticized fly ash concrete for structural application Concr. Int.***8**, pp 28-31
[2] Feldmen RF 1990 *Studies on mechanism of development of physical and mechanical properties of HVFA cement pastes. Cem. & Conc. Comp.* **12** pp 245-251
[3] Lam L *et al* 1998 *Cem. & Conc. Comp* **28** pp 271-283
[4] Lam L *et al* 2000 *Cement and Concrete Research* ***30*** pp 747 – 756
[5] Cengiz D A *et al* 2003 *High volume fly ash concrete with high strength and low drying shrinkage. J. Mat. in civil eng. ASCE* **153**
[6] Dhir R K *et al* 2005 *Development of HVFA cements for use in concrete construction Fuel* **84** pp 1423–1432
[7] Jian Hua W *et al* 2006 High performance Concrete with HVFA *Key Eng. Mat.* pp 470-478
[8] Qiang Z *et al* 2012 *Cem. and Conc. Res.* **42** (2) pp 194–204
[9] J.H. Wu *et al* 2012 *Key Eng. Mat. Trans Tech Publishers*
[10] Ondova M *et al* 2012 *Procedia Engineering* **42** pp 2044 – 2054
[11] Mukherjee S *et al* 2013 *Comparative study on physical and mechanical properties of high slump and zero slump HVFA Global NEST Journal* 15 pp 578-584
[12] Yoshitake I *et al* 2013 *Tensile properties of high volume fly-ash (HVFA) concrete with limestone aggregate Constr. Build. Mater.* **49** pp 101–109.
[13] Yoon S *et al 2014 Constr. Build. Mater.* **54** pp 432–442.
[14] Dragas J *et al* 2016 *Mag. Conc. Res.* **68(12)**, pp 632–645
[15] Yu J L and Leung CKY 2018 *Constr. Build rate* **161** pp 509-518,
[16] Feng J *et al* 2018 *J.Adv. Civ Eng* **4023178**
[17] Charith H *et al* 2020 *J.Cons. and Build. Mat.* **258** 120606
[18] Siddique R 2020 *Cem. and Conc. Res.* **34** (2020) pp 487 – 493
[19] Hafiz A *et al* 2020 *J.Build. Eng*. S2352-7102(**16**)30202-9
[20] Wichian C *et al* 2021 *J.Conc. Struc. and Mat.* 15**44**

Fuzzy Logic Model For Prediction Of Properties Of Exfoliated Vermiculite based Light weight concrete

N Kiran Kumar[1]*, B Damodhara Reddy[2], and C.Sasidhar[3]

[1]Research Scholar (PhD), JNTUA, Anantapur &Assistant Professor,SVCET(A), Chittoor, AP, India
E-mail: kirsir.kumar@gmail.com
[2]Professor, SVCET (Autonomous), Chittoor, Andhra Pradesh, India
[3]Professor, Jawaharlal Technological University, Anantapur, AP, India

Abstract: Concrete is highly sought material in construction industry and is made with cement, Natural Sand and gravel. There is lot of advancements done in the field of concrete. A model based on fuzzy logic is used to arrive at fresh as well as indurate properties of Exfoliated vermiculite related Light weight concrete (EVLWC) containing Exfoliated Vermiculite, Granite Fines and Silica Fume has been developed. Materials investigated experimentally possess 0 %, 5 %, 10 %, 15 % and 20% weight of Exfoliated Vermiculite (EV) and fixed 10% Granite Fines (GF) both replacing Natural Sand and fixed 5% Silica Fume (SF) replacing cement in each concrete. Water/cement ratio and content of super plasticizer were chosen as 0.40 % and 1.0 % of content of used respectively. The slump property of fresh concrete and hardened properties of concrete like Stress-strain relationship, compressive strength and acid resistance are arrived for said concrete.

Keywords: Concrete, Natural Sand, Exfoliated Vermiculite, Silica Fume, Granite Fines and Fuzzy Logic

1. Introduction:

Concrete is a heterogeneous material formed with combination of cement, aggregates, water, and some admixtures. In order to arrive at the strength and durability achieved by concrete members it is a need to ascertain certain properties of concrete matrix over different age periods. The concrete strength is observed from the first day of curing to the 28 th day and fuzzy logic is applied pertaining to the 28-day compressive strength and durability pertaining to acid resistance of cement concrete under water curing.

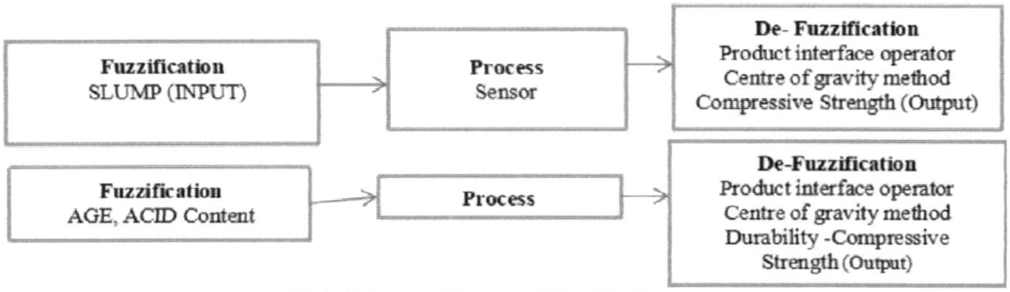

Fig 1: Schematic Diagram of Fuzzification process

Light Weight concretes (LWC) were developed with seven proportions, each of which had fixed Silica Fume (5%) and Granite Fines (10%) and several Exfoliated Vermiculite (EV) ratios (0–20%). The compressive strength were checked at the end of 3rd, 7th, and 28 days. A fuzzy logic (FL) was developed keeping in mind input variables such as contents of Sand, Exfoliated Vermiculite, Granite Fines,, Silica Fume, cement, and age) and the output variable such as slump, compressive strength and durability as shown in figure 1.

After concrete cubes are casted, experimental set up was established, process for measurement of data commenced using signals and their responses were recorded. The response of signals depend on hydration process the concrete material undergone and stiffness of the concrete material.

2. Literature Review

Fuzzy logic was first coined by Loffi Zadeh in 1965, which varies between completely true and completely false condition between integer value 0 and 1 and these fuzzy models represents a mathematical presentation of confusion existing in system and locate the truth in the system to exact degree.[1]

As per the studies of **Gokmen Tayfur et.al** five different binders with varying silica fume between 0 to 15% for age of 3, 7 and 28 days were adopted and total 60 number of concrete cubes were prepared. With input like contents of silica fume, cement and age and with output as compressive strength which were fuzzified and thus fuzzy logic was successfully applied.[2]

Uygunoglu et.al In studied about effect of fly ash content, water/cement ratio and concrete age on the compressive strength was evaluated using fuzzy logic and a model was developed for varying amount of fly ash of 0%,10%,20% and 30% and for age period of 3,7,28,90, 180 and 365 days and optimum content of Fly Ash content, water/cement ratio and properties like compressive strength was arrived for concrete.[3] Similar study on High Strength concrete was performed by Paratibha Aggarwal et.al, (2013) who applyied fuzzy logic for arriving at a model for predicting the compressive strength having different cementitious material. They further attempted to optimize mateials by reducing their cost of targeted mixes.[4]

The properties like temperature during hydration, design mix of materials and Concrete compressive strength, are evaluated using fuzzy logic and were compared with physical testing of concrete cubes and cylinders by developing a fuzzy logic-based model. This model took into consideration data taken from maturity method as well as data obtained from universal testing machine.[5]

3. Materials and Methods

3.1 *Popular Fuzzy System Methods:*

3.1.1 Mamdani rule System: Fuzzification

Out of the various Fuzzy systems available, Mamdani rule System involving fuzzifying input values are converting into fuzzy functions and applying rules and arriving at calculate fuzzy output functions which are later later defuzzify the fuzzy output functions resulting with output values for given input values.

While fuzzification two values considered are '0' and '1', The value '0' is not to be considered in the defined set with value '1' is said to be highest value in the set. For example consider Slump, generally it may be Low or high. 'Low' is considered to be zero and 'high' condition is said to be having value of '1'. The condition in between low and high condition is termed to be 'medium' and value varies between 0 to 1. The Blue portion represents to low Condition, red triangle corresponds to high and yellow color triangle represents 'medium' condition which is shown in figure 2. For example the results show that it is "slightly medium" then value can vary between 0.1 to 0.3 and moderately medium means value ranges between 0.4 to 0.6 and "higher medium" has values between 0.7 to 0.9.

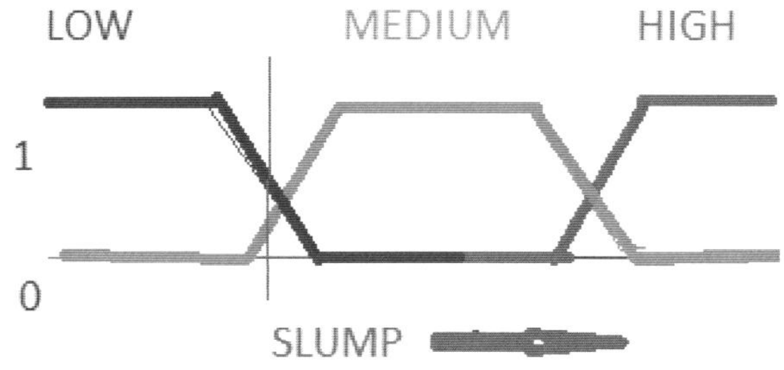

Fig 2: Fuzzy logic Slump

3.1.1.1 Fuzzy operators

Fuzzy set represented triangle shaped curve has an increasing slope from 0 to 1 and decreasing slope between 1 to 0. IF-THEN concept is applied to determine compressive strength based on slump (Slump high, compressive strength low and vice versa).

3.1.1.2 Defuzzification

After fuzzification there is a need for obtaining truth value which are exactly matching with the "intention" shown in the truth value. From obtained truth values of slump, an actual slump is to be caluculated that exactly mentions 'least', 'medium' and 'highest'.[From the obtained values defuzzification is applied.

3.1.2 Takagi-Sugeno-Kang (TSK)

The Takagi-Sugeno-Kang varies from Mamdani technique only that execution of fuzzy rules consists of defuzzification expressed in polynomial function.
IF Slump Is very low=2
IF Slump Is very low and Water cement ratio is high=2 x Slump + 1 x (water cement ratio).
The prime benefit of this method is ease in calculations and easy control of usage of algorithms.[6]

3.1.3 Sensors

The fuzzy logic can be applied using sensors. The sensor used is of brand xcluma whose specifications are revealed in table 1.

Table 1: Specifications of Sensor used

Characteristics	Values/Details
Model	VL53L0X
Type of Sensor	Laser Ranging Sensor
Voltage	2.8 Volts
Distance covered by sensor	2 meters
Current/Power	150mA
Resolution	1 mm

EFFICACY Concrete Spacers of 20/25/40MM RCC Cover Block was used in 50 numbers along with Bosch GMS120 Professional Detector for measuring signals are used which are depicted in figure 3.

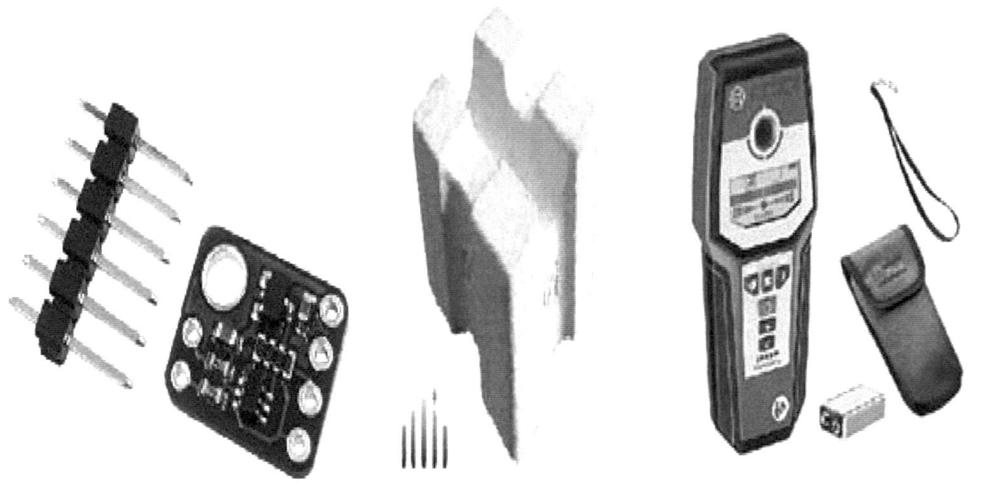

(a) Sensor (b) Concrete spacer (c) Professional Detector

Fig 3: Sensor, concrete spacer and professional detector

4. Results and Analysis:

While inducing compressive loading, the strain is measured using the strain gauge by obtaining the cross head displacement. The minute change in electrical resistance was arrived utilising the four probe method, while utilizing DC current ranging between 0.5 to 3.5 Amperes. The details of compressive stress using sensor and CTM instrument and strain are shown in Table 2, Figure 4-6.

Table 2: Mix proportions and Electrical Resistivity of Various concrete Mixes

Sample	Electrical Resistivity (R) (Ω cm)	Compressive Stress-Sensor (MPa)	$(R-R_0)/R_0$	Compressive Stress (MPa)	Strain
NC	1.7×10^5.	34	10.5	35	1.75
SFC	3.1×10^5.	35	11	37	1.72
GF-SFC	1×10^{12}.	36	11.25	38	1.69
GF-SFC-EV5	1.2×10^{12}.	37	11.5	38.5	1.65
GF-SFC-EV10	1.9×10^{12}.	37.5	11.9	38.75	1.55
GF-SFC-EV15	1.6×10^{12}.	36.75	12.2	37.5	1.6
GF-SFC-EV20	1.4×10^{12}.	35.9	12.75	36.5	1.65

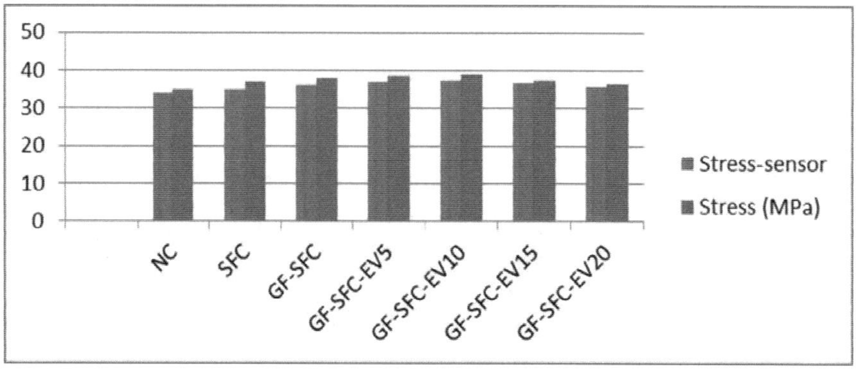

Figure 4: Stress (vs) Strain of various concrete mixes

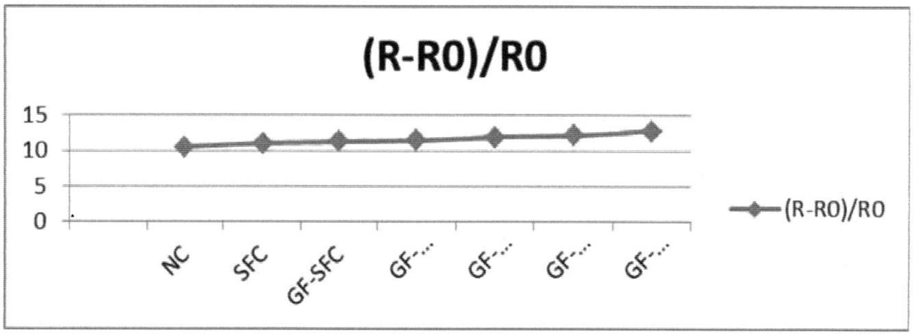

Figure 5: $(R-R_0)/R_0$ of various concrete mixes

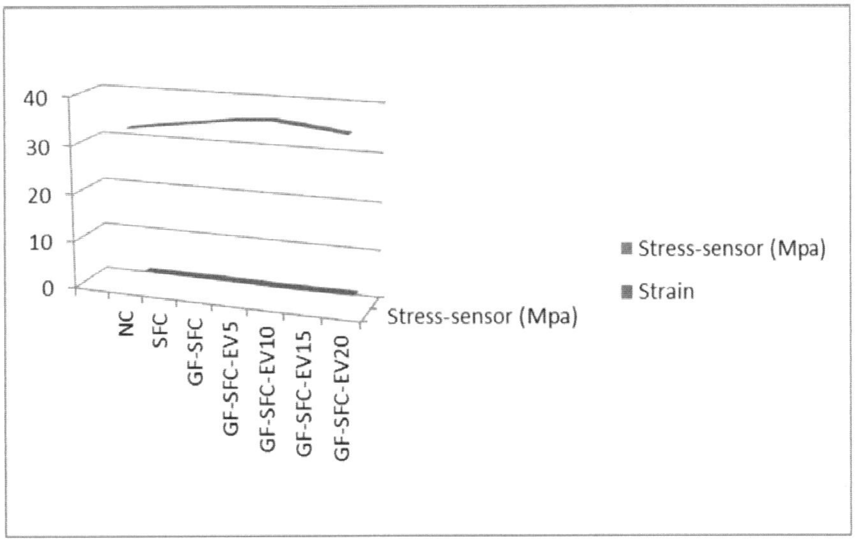

Note: ρ=R(A/d) where ρ =Density, R=Resistance, A=Area of Cross section and d=4 cm for compressive testing.

Figure 6: Stress-sensor (vs) strain of various concrete mixes

Table 3: Mix proportions and Electrical Resistivity of Various concrete Mixes relating to various tests

Test	Sample	Electrical resistivity (Ωm)
Slump	NC	**0.0003**
	SFC	**0.0004**
	GF-SFC	**0.0005**
	GF-SFC-EV5	**0.0002**
	GF-SFC-EV10	**0.0004**
	GF-SFC-EV15	**0.0007**
	GF-SFC-EV20	**0.0004**
Compressive Strength	NC	**170000**
	SFC	**31000**
	GF-SFC	**1E+12**
	GF-SFC-EV5	**1.2E+12**
	GF-SFC-EV10	**1.9E+12**
	GF-SFC-EV15	**1.6E+12**
	GF-SFC-EV20	**1.4E+12**
Acid Resistance	NC	**150000**
	SFC	**29000**
	GF-SFC	**9.00E+11**
	GF-SFC-EV5	**1.1E+12**
	GF-SFC-EV10	**1.7E+12**

	GF-SFC-EV15	**1.5E+12**
	GF-SFC-EV20	**1.1E+12**

The Electrical resistivity (ER) values and various ER ratios for slump test, compressive strength test and Acid resistance test are shown in table 3-4, Figure 7-11

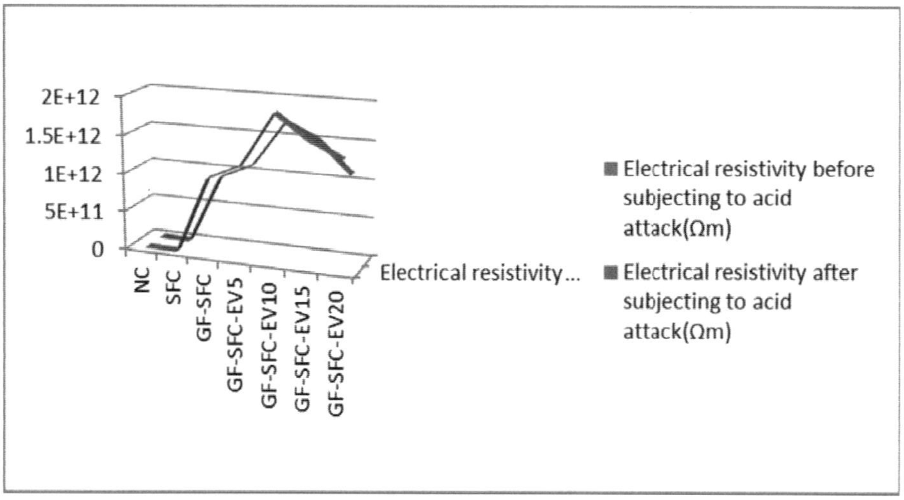

Fig 7: Electrical Resistivity of Concrete samples before and after soaking in Acid

Table 4: Compressive strength (CTM & Sensor) and $\Delta R/R_0^+$ of Various concrete Mixes relating to various tests

Test	**Sample**	**Compressive Strength-CTM(MPa)**	**$\Delta R/R_0^+$.**	**Compressive Strength-Sensor (MPa)**
Compressive Strength	NC	**35.6**	**69.5**	**35.5**
	SFC	**38.8**	**71.3**	**38.7**
	GF-SFC	**39.8**	**72.7**	**39.9**
	GF-SFC-EV5	**41.5**	**74.1**	**41.7**
	GF-SFC-EV10	**42.7**	**75.6**	**42.9**
	GF-SFC-EV15	**41.9**	**73.4**	**41.2**
	GF-SFC-EV20	**41.3**	**72.9**	**39.9**
Acid Resistance	NC	**31.7**	**63.5**	**31.6**
	SFC	**36.3**	**69.8**	**36.2**
	GF-SFC	**38.5**	**69.1**	**38.3**
	GF-SFC-EV5	**39.4**	**71.5**	**39.9**
	GF-SFC-EV10	**40.8**	**73.7**	**41.4**
	GF-SFC-EV15	**39.9**	**72.6**	**39.8**
	GF-SFC-EV20	**38.7**	**71.3**	**38.6**

IOP Conf. Series: Materials Science and Engineering 1282 (2023) 012005 doi:10.1088/1757-899X/1282/1/012005

Note: R-Resistance, R_0-Original Resistance

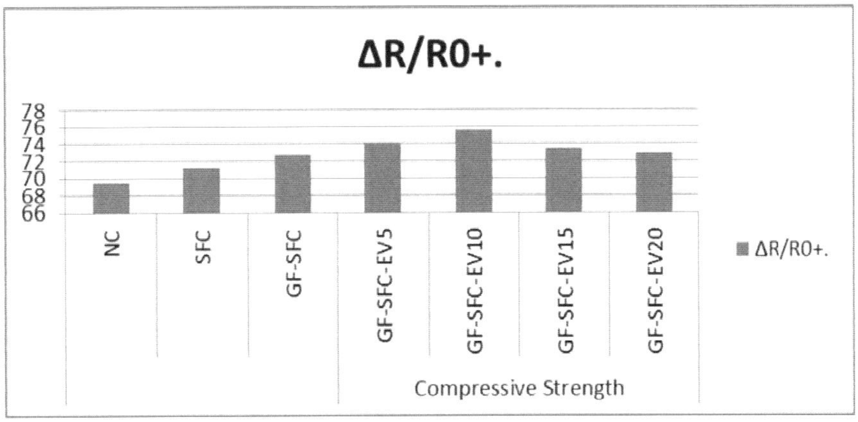

Fig 8: ΔR/R0⁺ (Vs) Compressive strength of concrete samples before exposing to Acid

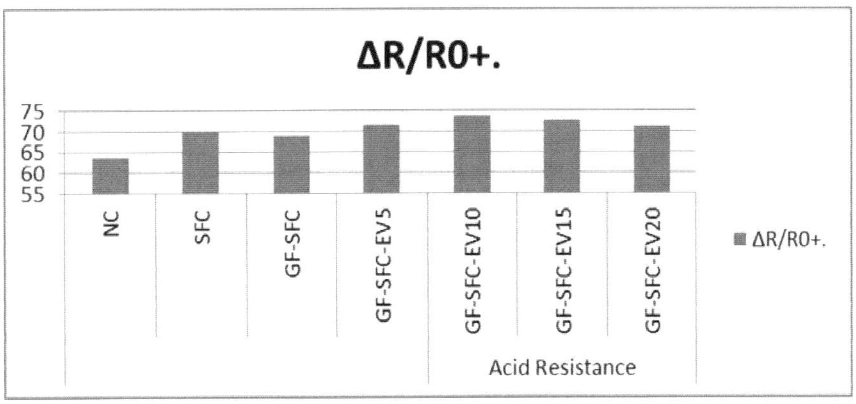

Fig 9: ΔR/R0⁺ (Vs) Compressive strength of concrete samples before exposing to Acid

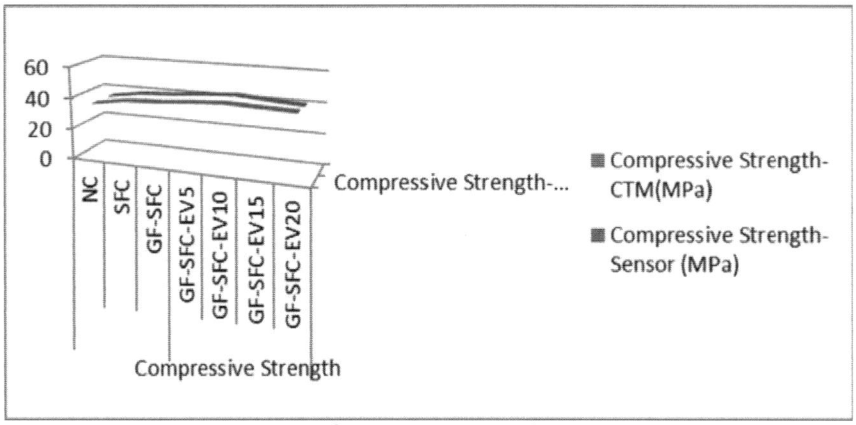

Fig 10: Compressive strength for concrete samples before exposing to Acid

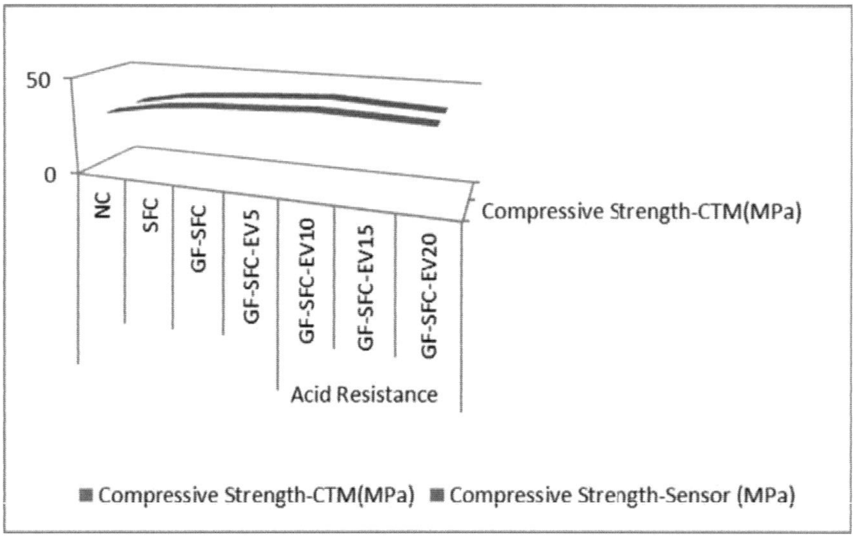

Fig 11: Compressive strength for concrete samples after exposing to Acid

All the tests are conducted in accordance to ASTM standard procedure. For compressive testing, according to ASTM C-39-83b, concrete cylinder specimens of diameter 100 mm diameter and 200 mm length are used as shown in Table 5. The results pertaining to compressive strength are in accordance to studies performed by sedat akkurt, Gokmun tayfur, Uygunoglu et.al, Paratibha Aggarwal et.al, and Tareen N. The results achieved are in conformance with studies predicted by Vardhan Nagarkar et.al[7]. The results pertaining to slump, stress, strain, Acid resistance tests and Modulus of elasticity are in accordance with the codes referred [8-9].

Table 5: Stress, Strain and Modulus of Elasticity of various cylinder specimens

Mix specifications for cylinders	Stress (N/ mm^2)	Strain	Modulus of Elasticity(N/ mm^2)
NC	14.7	0.0003	49006
5%SF	15.3	0.0004	38215
5% SF&10%GF	15.7	0.0005	31400
5% SF&10%GF& 05% V	16.4	0.0002	49249
5% SF&10%GF& 10% V	20.4	0.0004	51000
5% SF&10%GF& 15% V	15.3	0.0007	21858
5% SF&10%GF& 20% V	6.79	0.0004	16984

5. Conclusion:

The sensors Fuzzy Logic can opportunely be used to arrive at the mechanical properties like slump, stress, strain and modulus of elasticity, compressive strength and durability of Light Weight Concrete. The input variables like Sand content, Exfoliated Vermiculite content, Granite Fines content, Silica Fume content, cement binder content, and age were adequate to achieve precise results. The operators utilized in the Fuzzy Logic were identified as suitable for determination of slump. Stress, strain and other mechanical properties like modulus of elasticity, compressive strength and durability prediction. In future other PZT material, and other CNTs, can be used as sensors for developing concrete to meet strength and durability requirements towards alkali, sulphates and chlorides.

6. References:

1. Sedat Akkurt, Gokmen Tayfur, and Sever Can 2004, Fuzzy logic model for the prediction of cement compressive strength, *Cement and concrete research*, vol 34 pp 1429-1433.
2. Gökmen Tayfur, Tahir Kemal Erdem and Önder Kırca 2014, Strength Prediction of High strength concrete by Fuzzy Logic and Artificial Neutral Networks, *Journal of materials in civil engineering*, Vol 26
3. Uygunoglu, and Tayfun unal 2006. A new approach to determination of compressive strength of fly ash concrete using fuzzy logic., *Journal of Scientific& industrial research*, vol 65 pp 894-899.
4. Paratibha Aggarwal, Yogesh Aggarwal, Rafat Siddique, Sakshi Gupta & Harshit Garg 2013, Fuzzy logic modeling of compressive strength of high-strength concrete (HSC) with supplementary cementitious material, *Journal of Sustainable Cement-Based Materials*, 2:2, 128-143,5. .
5. Tareen N, Kim J, Kim WK, Park S, 2021, Fuzzy Logic-Based and Nondestructive Concrete Strength Evaluation Using Modified Carbon Nanotubes as a Hybrid PZT-CNT Sensor. *Materials (Basel)*. Vol 14(11) p2953.
6. https://en.wikipedia.org/wiki/Fuzzy_logic
7. Vardhan Nagarkar, Preeti kulkarni and Shreenivas Londhe 2020, Prediction of Concrete Compressive Strength Using Fuzzy Logic and Model Tree, *Advances in Structural Engineering*, vol 74, pp 231-240.
8. IS 456:2000
9. IS 800:2007

An experimental study on structural performance of reinforced concrete beams strengthened for combined flexure and shear by using NSM-GFRP strips

Aishwarya K M [1], Nikitha V Anand [2] and Balaji N C [3]

[1] Post Graduate, Dept. of Civil Engineering, The National Institute of Engineering (NIE), Mysuru.

[2] Assistant Prof., Dept. of Civil Engineering, The National Institute of Engineering (NIE), Mysuru.

[3] Associate Prof., Dept. of Civil Engineering, The National Institute of Engineering (NIE), Mysuru.

E-mail: *balajinallaval@gmail.com*

Abstract: Near surface mounted fiber reinforced polymer (NSM-FRP) technique is the latest & most effective techniques used in strengthening the existing structural components, which requires strengthening against increased service loads, faulty design, environmental effects, etc. In the past, many researches on NSM technique have been carried out by using different cross-sectional shapes & dimensions of FRP materials to strengthen for shear & flexure. In the present experimental study, the behaviour of R.C. beams strengthened for combined flexure & shear studied. The beams of 150x150x700 mm size with M30 grade of concrete & Fe500 steel were considered. The control beams (unstrengthened) & strengthened beams using GFRP (Glass fibre reinforced polymer) strips were subjected to three point loading to study the behaviour of beams in combined shear & flexure. The main objectives were to assess the behaviour of strengthened beams with control beams & to study the influence of the orientation of shear strips. The beams were studied for the failure characteristics, yield & ultimate load carrying capacity, stress-strain behaviour, deflection, & crack patterns. From the test results, it's been observed that, the load carrying capacity of the strengthened beams were increased, the improved in deflection, when compared to the control beams. The ultimate load carrying capacity of beams with shear strip orientation of 45°, 60° & 90° were increased by 50.3%, 39.3% & 25% respectively, compared to the control beams. This NSM strengthening technique can be advantageous & suitably adopted to enhance the load carrying capacity of the beams.

Keywords: Near Surface Mounting (NSM), Glass Fiber Reinforced Polymer (GFRP), Shear & flexural strengthening, Concrete.

1 Introduction

Numerous real-world projects all around the world are employing the Externally Bonded Reinforcing (EBR) approach for strengthening the existing structures [1][2]. Unlike EBR technique, Near Surface Mounting (NSM) technique is an unique & effective strengthening technique, which was earlier called the "grouted reinforcement" or the "embedded reinforcement" technique [3][4][5]. FRP composites are widely used materials for strengthening & their characteristics rely on their orientation, & have superior mechanical qualities in the direction they are orientated. FRP composites are available in different size, shape, dimensions & of different materials[6][7][8]. The type of groove filler & groove size affect the effectiveness of the NSM technique. Based on studies, epoxy as a groove filler provided better mechanical performance when compared to cement based fillers[9][10]. The bond & ultimate load-bearing capability of the strengthened members enhance as the groove size increases [11][12][13]. The RC members may be strengthened more effectively for shear [14][15][16] [17][18] & for flexure[19][20][21][22] using the NSM approach.

In the experimental study by Ummar Shariff et. al., [23], an enhancement of 30-50 % in the ultimate load carrying capacity was observed when GFRP strips were used for flexural strengthening. Also, there was a delay in the appearance of 1st crack in NSM strengthened beam compared to unstrengthened beam. To increase the flexural capacity of R.C. beams using GFRP strips as a strengthening FRP material, Kusuma et al. [24] experimentally compared the B-NSM (Bottom NSM) & S-NSM techniques (Side NSM). Beam Load carrying capacity was increased by 87.5 % in beams with B-NSM, 62.5 % in beams with S-NSM where the strips are oriented horizontally & 81.25 % in beams with S-NSM where the strips are placed vertically. As per the results the B-NSM is effective when compared to S-NSM. The load carrying capability of R.C. beams strengthened for only shear using NSM-GFRP strips was experimentally planned and studied by M Rashmi et al. [25] by orienting the shear strips at 45° & 90° with varied strip spacing's. The load carrying capability of the beams strengthened with GFRP strips increased significantly from 48 to 93.5 % as compared to the control beam. T Rashmi et, al [26] compared the strength between the control beam & the beam strengthened for shear using NSM-GFRP strips by orienting the strips at an angle of 60° & 75° with different spacing. According to the investigations done by M Rashmi et al. [25] & T Rashmi et, al [26] , the load carrying capacity of the strengthened beams has significantly increased as the angle of strips orientation with respect to the horizontal has decreased & as the strips spacing reduced. The R.C. beams when strengthened for shear or flexure, there is an enhancement in the ultimate load carrying capacity. In the current work, an attempt is made to strengthen the R.C. beams against both flexure & shear, which is a feasible scenario where three of the beams' faces are accessible for strengthening. To determine the influence of the NSM strips orientation angle for the shear strengthening, the orientation of the shear strips is varied.

2 Experimental programme

Present study involves the use of GFRP strips for improving the structural performance of reinforced concrete beams by adopting NSM technique. The NSM approach entails cutting a groove in the concrete cover of the beams, followed by filling the groove using an adhesive & GFRP. Twelve beams of size 150x150x700 mm, with M30 grade concrete & Fe500 steel were grouped into two groups shown in Table 1. Group A comprises of three beams that were not reinforced using the NSM approach & were known as the control beams (CB) or the reference beams (Figure 1). Group B composed of nine beams that were reinforced against combined flexure & shear by utilising the NSM-GFRP technology. For all the group B beams, same amount of shear strips & flexure strips were employed as reinforcement. The size of GFRP strips used for flexural & shear strengthening is 10 mm wide & 2 mm thick. At the soffit of the beams, one flexural strip of 700x10x2mm (long, x wide, x thick) is provided. Twelve shear strips, six each on one side face of the beams were provided to account for shear strengthening as shown Figure 2. The influences of three different angles of orientations of shear strips were studied.

2.1 Material Properties

The R.C. beams were casted using M30 grade concrete, for which OPC 53 cement, coarse aggregates of 20 mm & 12 mm down size, M-sand (crushed stone aggregates) as fine aggregates & potable water were used. The beams were designed as an under reinforced beam to take the load of 30 kN. Two rebar's of 10mm diameter as the longitudinal tension reinforcements at the bottom and two rebar's of 8mm diameters at top were provided as compression reinforcements. Two legged vertical stirrups of 8 mm diameter with 100mm c/c spacing were provided as shear reinforcement, with an all-round clear cover of 25mm. The GFRP strips of 10mm wide & 2mm thick were used for both shear & flexural strengthening. According to the manufacturer, tensile strength, modulus of elasticity & Poisson's ratio of the GFRP strips are 750MPa, 165MPa, & 0.27 respectively. As the groove filler qualities play a significant role on bond efficiency [27], epoxy was used as the groove filler. The epoxy used consist of two contents i.e., resin & hardener, which were mixed in a ratio of 9:1. The tensile & shear strengths of epoxy are 14.8 & 15 MPa, respectively.

2.2 Groove dimensions & groove spacing

Based on the literatures [28] the minimal groove size should be 3x times the strip thickness & 1.5x times the strip width. The thickness & width of the strips used in the present are 2 mm & 10 mm respectively. The minimum groove width & depth needed to meet the requirements of the literature studies are 6 mm & 15 mm, respectively. The groove width & depth considered were 6 mm &17 mm respectively. As per the literature review, the GFRP spacing incorporated in the experimental investigation was set to ensure the spacing between grooves was at least twice the width of the strips & that the spacing between the GFRP shear reinforcement is not exceeding L_{net} [28]. The spacing followed in the present study is shown in Table 1.

$$L_{net} = [L_b - \frac{2C}{sin\alpha}]$$

Where, L_{net} is net length of GFRP strips, L_b is actual length of GFRP strips, C is the clear cover of beam & α is orientation of NSM-GFRP.

2.3 Spacing between groove & the beam edge

A minimum of 20mm distance should be maintained between the edge of the support & the NSM-GFRP in order to prevent splitting failure of the concrete cover. Cracks were not observed at the edges of the members strengthened using NSM technique when a minimum edge distance of 30 mm was maintained [25]. Taking this into account, an edge distance of 30 mm was followed in the present study.

Table 1: Beam designations considered in the experimental study

Group	Nomenclature	Beam Case	Grove orientation (degree)	Length of strips (mm) used for		GFRP strip Spacing
				Flexure	Shear	
A	CB	Control Beam	-	-	-	-
B	B90	Beams strengthened with NSM	90	700	150	90 mm
	B60		60		173.2	100 mm
	B45		45		212.1	100 mm

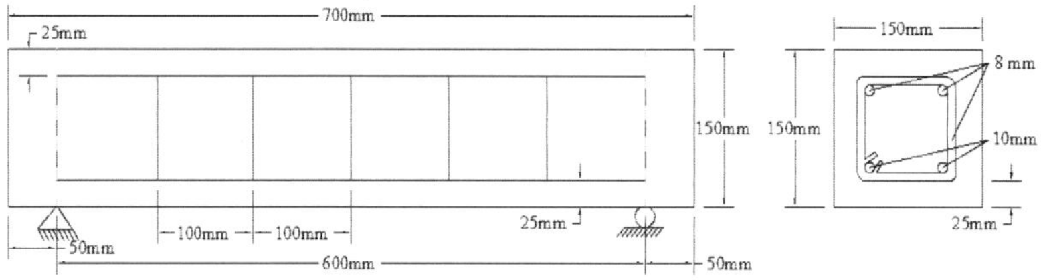

Figure 1: Typical details of RC beam (CB)

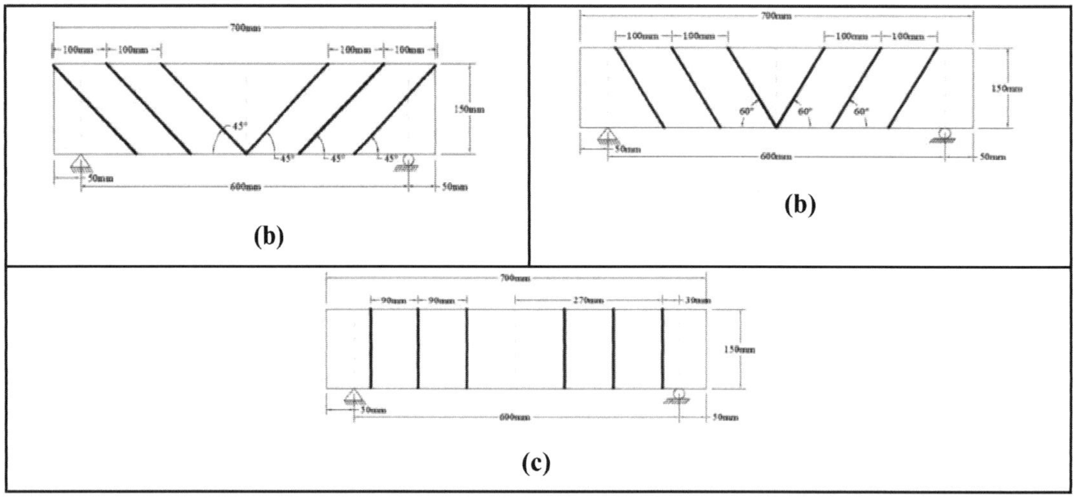

Figure 2: Longitudinal sectional details of strengthened beams (a) B45, (b) B60 & (c) B90

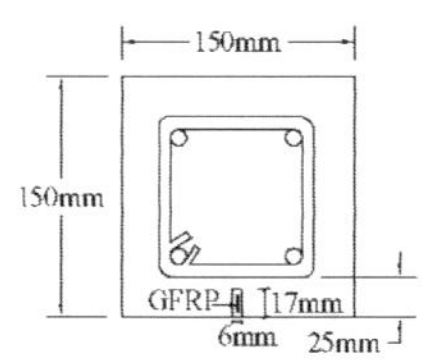

Figure 3: Cross sectional details of strengthened beams B45, B60 & B90

2.4 Strengthening

The NSM-GFRP strengthened beams being cured for 28 days. The following steps were followed-(a) cutting the grooves into the concrete cover region, (b) groove cleaning, (c) Trimming &cleaning the GFRP (d) preparing epoxy material as per the specifications prescribed by the manufacturer (e) filling three fourth of the grooves with epoxy (f) inserting GFRP into the groove made, & slightly applying the force to make the adhesive to flow between the GFRP (g) levelling the surface after the removal of excess adhesive (h) air curing the strengthened beams for 7 days.

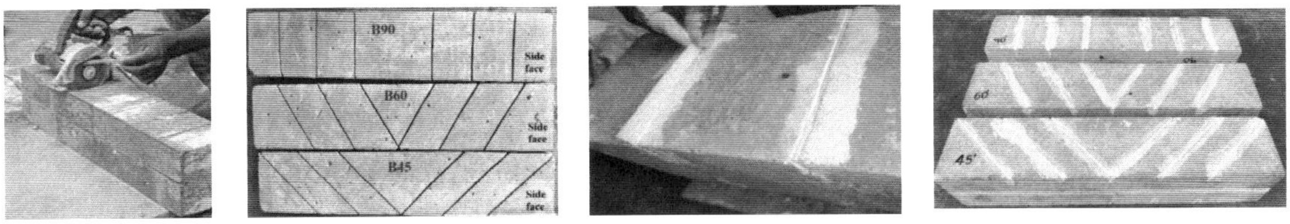

(a) (b) (c) (d)

Figure 4: (a) groove cutting, (b) beams with groove orientation of 90°, 60° & 45°, (c) Placing of GFRP inside the groove filled with epoxy, (d) strengthened beams

2.5 Testing

The beams were tested using three-point loading test procedure on the Universal Testing Machine which has the loading capacity of 100 tonne. The span of the specimen considered or the distance between the supports is 600 mm. The experimental setup of the beam testing is shown in Figure 5. The beam deflections at the mid & quarter spans were measured using the dial gauges having a least count of 0.01mm. Three strain reference points were considered as the point of interest at the tension zone, compression zone & at the zone close to neutral axis. The strain across the reference points were measured using the “Demountable Mechanical Strain Gauge” having a 0.002mm least count. The dial gauge & DEMEC gauge readings were noted at an interval of 5 kN increment of the load.

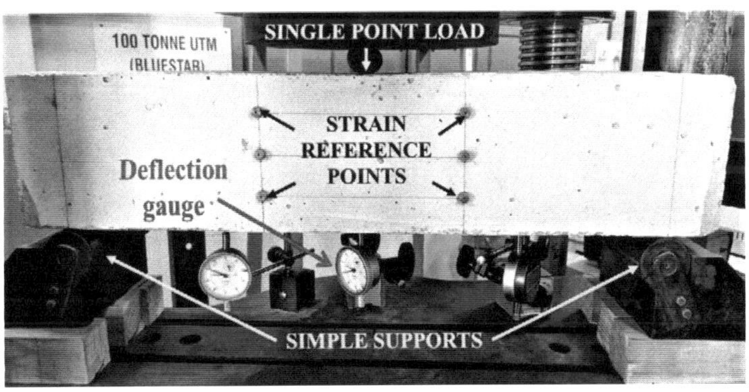

Figure 5: Three-point loading test set up

3 Results & discussions

All beams were tested for three-point loading. Deflection at midspan & quarter span, as well as linear deformation, was measured at every 5 kN load increment using dial gauge & DEMEC gauge. The failure patterns, deflection, crack patterns& the stress strain variation of the beams are discussed below. The unstrengthened beams were related with the strengthened beams to assess the structural behaviour of the beams.

3.1 Crack patterns & load deflection behaviour

3.1.1 Control beam (CB)

The control beams are the one which are not strengthened using NSM method. At 40 kN load, the first crack started to propagate from the tension zone in all the control beam specimens & extended towards the neutral axis of the beams as load increases. The dial gauges used to measure deflection were removed at 75 kN load as the crack widened. At a 75 kN load, the avg. deflection values at the mid & quarter spans of the beam were 3.13mm & 1.61mm, respectively. At an ultimate load of 80kN, the combined effects of bending & shear caused the beams to fail at the mid-span. The load corresponding to 2.4mm deflection was considered as the service load ("According to IS 456:2000, clause 23.2, & the maximum allowable deflection is [span / 250] limited to beam"). Hence, the load at 2.4 mm deflection is 57.5 kN. The failure of the beams was caused by the yielding of steel & crushing of concrete at the point of loading. The crack pattern of the control beam CB is shown in Figure 7.

3.1.2 Strengthened beam(B45)

The beams were strengthened using NSM method with 45° shear & flexural strips at the side & soffit of beam respectively. At 50 kN load, the first crack was formed in all of the specimens in the tension zone. The cracks started to propagate towards the neutral axis (towards centre) of the beam as load increases. The dial gauges were removed at 90 kN load as the crack widened. Prior to the removal of the dial gauge, at 90kN load, the average deflection values at the mid & quarter spans of the beam are 3.4 mm & 2.89 mm, respectively. The service load carried by the beam is 83.2 kN. At an ultimate load of 120.2kN, the combined effects of bending & shear caused the beams to fail at the mid-span. The cracks propagated after 65 kN load were nearly perpendicular to the GFRP strips orientation, which were placed for the purpose of shear strengthening. The crack propagation was arrested by the GFRP strips to certain extent. The beams failed due to steel yielding, failure of GFRP strips both on side faces & at the beam soffit & also due to crushing of concrete at the point of loading. The shear cracks & failures of GFRP are shown in Figure 6.

3.1.3 Strengthened beam(B60)

The 1st crack appeared & was initiated at a load of 50kN in all the strengthened beam samples in the tension zone. The cracks started to propagate towards the neutral (towards centre) axis of the beam as the load increases. The dial gauges were removed at 85kN load as the crack widened & the average deflection measured at the mid & quarter spans of the beam were 3.2mm & 1.17mm respectively. The service load carried by the beam was 81.9 kN. At an ultimate load of 111.4kN, the combined effects of bending & shear caused the beams to fail at the mid-span. The beams failed due to steel yielding, failure of GFRP strips & concrete crushing at the loading point. The crack pattern observed in the beam B60 is shown in Figure 6.

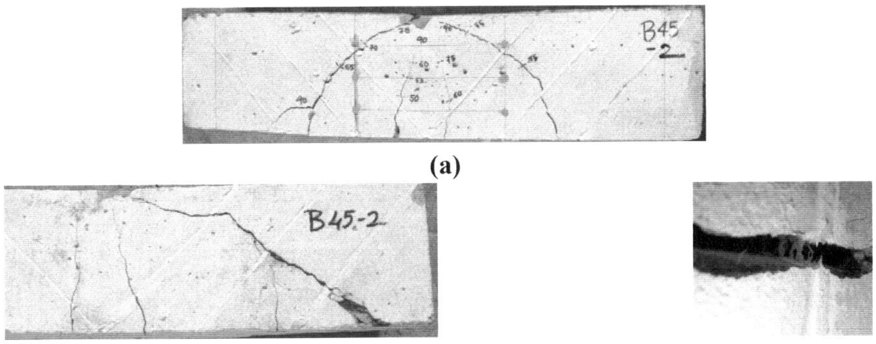

(a)

(b) **(c)**

Figure 6: (a) & (b) Crack patterns observed in B45 beams, (c) Failure of GFRP at the soffit of the beam

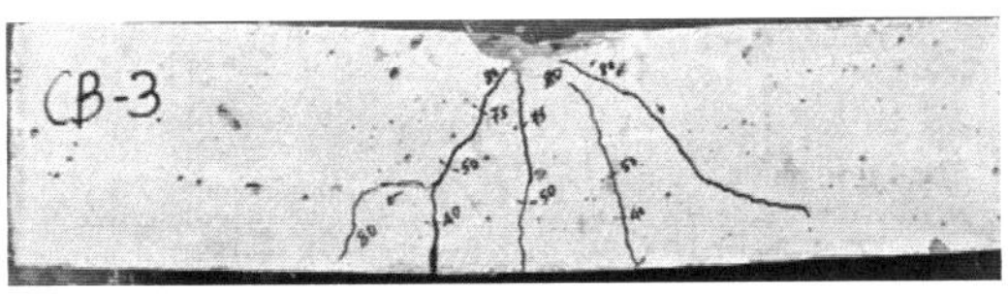

Figure 7: Crack patterns observed in control beams

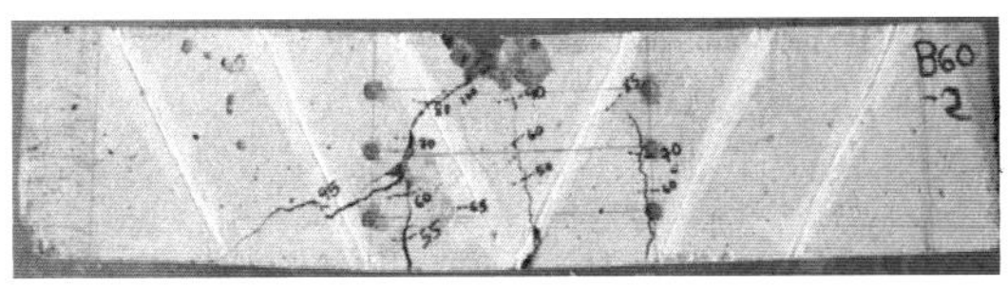

Figure 8: Crack patterns observed in B60 beams

3.1.4 Strengthened beam (B90)

The 1st crack appeared & was initiated at a load of 45kN in all the B90 beams in the tension zone. The cracks started to propagate towards the neutral (towards centre) axis of the beam as the load increased & at 65 kN load vertical crack at the junction of epoxy & concrete was observed. The crack propagation is due to bond failure at the junction of epoxy & concrete (Figure 7 (b)). As the crack widened, the dial gauges were removed at a load of 85kN, & the average deflection values at the mid & quarter spans of the beam were 3.2mm & 1.17mm, respectively. The service load carried by the beam was 80.68 kN. At an ultimate load of 100kN, the combined effects of bending & shear caused the beams to fail at the mid-span. The beams failed due to steel yielding, failure of GFRP strips at the soffit & crushing of concrete at the load point. The main reason for debonding at the interface of epoxy & concrete is due to the propagation of cracks at the mid span, which were parallel to the orientation of the GFRP strips.

Table 2: Experimental results

Beam designation	First crack load (kN)	Service load(kN)	Ultimate load (kN)	Mid span deflection at 75 kN load (mm)	Percentage increase over the Control beam	
					Service load	Ultimate load
CB	40	57.5	80	3.13	-	-
B45	50	83.2	120.2	1.68	44.7	50.3
B60	50	81.9	111.4	1.9	42.4	39.3
B90	45	80.7	100	2.04	40.3	25

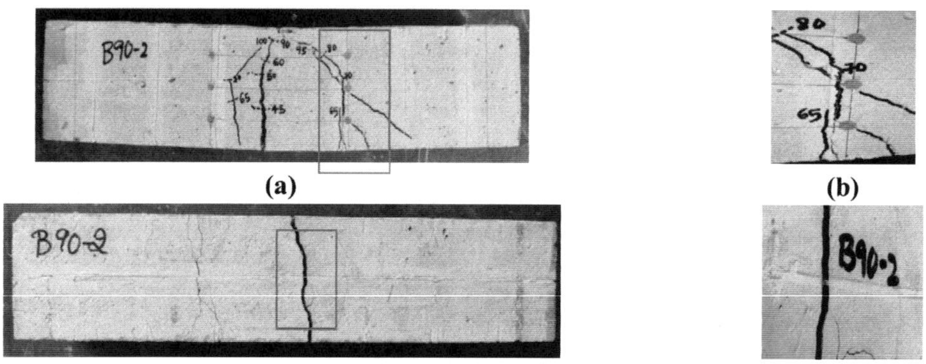

(a) (b)
(c) (d)

Figure 9: (a) crack pattern on the side face of B90, (b) Debonding failure at the concrete and epoxy interface, (c) Crack pattern of B90 at the soffit, (d) Failure of GFRP strips at the soffit

3.2 Comparison of the control beam & strengthened beams

3.2.1 Control beam & B45

The 1st crack in CB was observed at 40 kN load, whereas in the beams strengthened for combined flexure & shear with 45° orientation of shear strip, the first crack started to propagate at a load 1.3 times more than that of CB, i.e., at 50 kN. Also, the service & ultimate loads carrying of the strengthened beams were increased by 44.7 % & 50.3 % respectively. Since the dial gauges were removed at a load of 75kN in the CB, the deflection at 75 kN is taken as the reference. The deflection of 3.13mm & 1.68mm were observed at the mid span of the CB & B45 respectively. Similarly, the deflection of 1.61mm & 0.97mm were observed at the quarter span of the CB & B45 respectively. Compared to CB, the deflection values of the B45 beams were lower at the mid & quarter spans because of the introduction of GFRP strips as the strengthening material increased the stiffness of the beam.

3.2.2 Control beam & B60

The first crack in CB was observed at 40 kN load, whereas in the beams strengthened for combined flexure & shear with shear strips oriented at an angle of 60° the first crack started to propagate at a load 1.3 times higher than that of CB, i.e., at 50 kN. Also, the service & ultimate loads of the strengthened beams were increased by 42.4 % & 39.3 % respectively. The deflection values at 75 kN load observed at the mid span of the CB & B60 were 3.13mm & 1.9mm respectively & the deflection of 1.61mm & 0.98mm were observed at the quarter span of the CB & B60 respectively. The deflection values of the B60 beams were less at mid & quarter spans due to the increase in stiffness of the beam due to the introduction of GFRP strips as the strengthening material.

3.2.3 Control beam & B90

The first crack in CB was observed at 40 kN load, whereas in the beams strengthened for combined flexure & shear with shear strips oriented at an angle of 90° showed the 1st crack at a load 1.1 times higher than that of CB, i.e., at 45 kN. Also, the service & the ultimate loads of the strengthened beams were increased by 40.3 % & 25 % respectively. The deflection of 3.13mm & 2.04mm were observed at mid span of the CB & B90 respectively & the deflection of 1.61mm & 1.18 mm were observed at the quarter span of the CB & B90 respectively for 75 kN load. The deflection values of the B90 beams were less at mid & quarter spans due to the increase in stiffness of the beam.

3.3 Effect of different angle of orientations of shear strips

The appearance of the 1st crack was delayed & the load carrying capacities of the beams were enhanced when the angle of orientation of the shear strips with respect to vertical was decreased. The first crack in the beams with 60° & 45° shear strip orientation appeared at a load 1.1 times more than the load on beams with 90° shear strip orientation. The service load was increased from 80.68 kN to

81.9 kN on changing the NSM orientation from 90° to 60° & from 80.68 kN to 83.2 kN when the NSM orientation was changed from 90° to 45° with respect to horizontal. Comparing the beams with 45° & 60° shear strips to the beams with 90° shear strips, there was an average increase in ultimate load carrying capacity of 20.2 % & 11.4 %.

Table 3: Strain values at the strain reference points at 40 kN load

Beam designation	Tensile strain ($x10^{-4}$)	Strain at Neutral axis (10^{-4})	Compression strain ($x10^{-4}$)
CB	1.130	1.047	-0.200
B90	0.300	0.265	-0.140
B60	0.173	0.140	-0.800
B45	0.093	0.086	-0.800

Figure 8 shows the variation of load-deflection for CB, B45, B60 & B9. The deflection values of B45 were less than B60 & the deflection values of B60 were less than B90, which infers that the beam B45 is stiffer than B60, & B60 is stiffer than B90. In comparison to beams strengthened with 90° NSM orientations, those with 45° inclined GFRP strips showed significantly better improvement in load carrying capability. It was also shown that 45° inclined strips were more efficient than 60° & 90° GFRP strips. This is primarily because shear failure cracks propagated were nearly perpendicular to the direction of 45° inclined strips, which prevents tension-induced shear crack development & delays the propagation of shear cracks.

3.4 Stress strain relationship

The first crack in the CB was observed at 40 kN load which is taken as the reference. The shear stress & flexural stress values corresponding to 40 kN loads were 1.33 N/mm^2 & 10.67 N/mm^2. The strain values are tabulated in Table 3. It can be observed that as the angle of orientation of strip is varied from 90 to 45 degree, the strain in the member is reduced, along with reduction in the lateral deformation. Compared to the group A with group B beams has exhibited a better result in terms of lateral deformations.

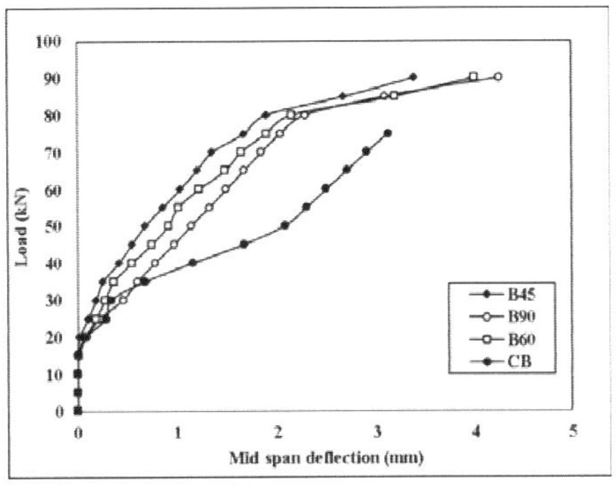

Figure 10: Load deflection curve of CB, B45, B60 & B90

4 Conclusion

The study was conducted to assess the structural performance of reinforced concrete beam strengthened for combined flexure & shear using GFRP strips by NSM technique by orienting the shear strips at different orientation angles. The results are drawn from the present study are as below:

- When the strengthened beams were compared to the control beams, it was found that the 1st crack load, service load, ultimate load, & deflection characteristics were enhanced.
- Comparing all of the strengthened beams to the control beams, a delay in the onset of the 1st crack was observed.
- Compared to the control beams, all of the strengthened beams exhibited lower deflection values. Particularly, the observed deflection values of B45 beams were lower than those of B60 beams & the observed deflection values of B60 beams were lower than those of B90 beams.
- When compared to 90° NSM strip orientation of shear strips, the beams with 60° NSM orientation has shown the better results, similarly the beams with 45° NSM orientation has shown better results compared to 60°. It can be inferred that as the orientation angle is reduced to 45° from 90° the load carrying capacities of the beams is enhanced.
- The service load carrying of the beams was increased by 44.7% in B45, 42.4% in B60 & 40.3% in B90 beams when compared to reference beams (CB). Similarly, the ultimate load carrying of the beams was increased by 50.3% in B45, 39.3% in B60 & 25% in B90 beams compared to control beam.
- The ultimate load carrying capacity of the beams was increased by 20.2 & 11.4 % when the NSM orientation angle of shear strips was changed from 90° to 45° & 60° with respect to the horizontal.
- The strain values of the strengthened beams at the tension zone, zone near the neutral axis & compression zone were less compared to the control beams.

5 References

[1] W. Sun, X. Peng, and Y. Yu, "Development of a simplified bond model used for simulating FRP strips bonded to concrete," *Composi. Struct.*, vol. 171, pp. 462–472, 2017.

[2] W. Sun, X. Peng, H. Liu, and H. Qi, "Numerical studies on the entire debonding propagation process of FRP strips externally bonded to the concrete substrate," *Constr. Build. Materi.*, vol. 149, pp. 218–235, 2017.

[3] L. De Lorenzis and J.G. Teng, "Near-surface mounted FRP reinforcement: An emerging technique for strengthening structures," *Composi. Part B Eng.*, vol. 38, no. 2, pp. 119–143, 2007.

[4] S. S. Zhang, T. Yu, and G. M. Chen, "Reinforced concrete beams strengthened in flexure with near-surface mounted (NSM) CFRP strips: Current status and research needs," *Composi. Part B Eng.*, vol. 131, pp. 30–42, 2017.

[5] J. Sena Cruz, "Strengthening of Concrete Structures with Near-Surface Mounted CFRP Laminate Strips," 2005.

[6] I. Shakir Abbood, S. aldeen Odaa, K. F. Hasan, and M. A. Jasim, "Properties evaluation of fiber reinforced polymers and their constituent materials used in structures – A review," *Mater. Today Proc.*, vol. 43, pp. 1003 – 1008, 2021.

[7] A. Carolin, "*Carbon fibre reinforced polymers for strengthening of structural elements"*, vol. 1992, no. 10. 2003.

[8] I. Abbood, S. A. Odaa, K. Hasan, and M. A. Al-Kubaisi, "Properties evaluation of fiber reinforced polymers and their constituent materials used in structures – A review," *Mater. Today Proc.*, vol. 43, pp. 1003–1008, 2021.

[9] S. M. Soliman, E. El-Salakawy, and B. Benmokrane, "Bond Performance of Near-Surface-Mounted FRP Bars," *J. Composi. Constr.*, vol. 15, no. 1, pp. 103–111, 2011.

[10] M. R. A. E. Emara, "Bond Behavior of NSM FRP Strips in Concrete under Sustained Loading Mohamed Reda Ali Elewa Emara," 2018.

[11] D. Galati and L. De Lorenzis, "Experimental Study on the Local Bond Behavior of NSM-FRP Bars to Concrete," in *Proceedings of 3rd International conference on FRP composites in civil engineering (CICE 2006)*, 2006, pp. 95–98.

[12] S. M. Soliman, E. El-Salakawy, and B. Benmokrane, "Bond performance of near-surface-mounted FRP bars," *J. Composi. Constr.*, vol. 15, no. 1, pp. 103–111, 2011.

[13] R. Parretti and A. Nanni, "Strengthening of RC members using near-surface mounted FRP composites: Design overview," *Adv. Struct. Eng.*, vol. 7, no. 6, pp. 469–483, 2004.

[14] J. A. O. Barros and S. J. E. Dias, "Near surface mounted CFRP laminates for shear strengthening of concrete beams," *Cem. Concr. Composi.*, vol. 28, no. 3, pp. 276–292, 2006.

[15] A. Rizzo and L. De Lorenzis, "Behavior and capacity of RC beams strengthened in shear with NSM FRP reinforcement," *Constr. Build. Materi.*, vol. 23, no. 4, pp. 1555–1567, 2009.

[16] M. Jalali, M. K. Sharbatdar, J.-F. Chen, and F. J. Alaee, "Shear strengthening of RC beams using innovative manually made NSM FRP bars," *Constr. Build. Materi.*, vol. 36, pp. 990–1000, 2012.

[17] A. Khalifa and A. Nanni, "Rehabilitation of rectangular simply supported RC beams with shear deficiencies using CFRP composites," *Constr. Build. Materi.*, vol. 16, no. 3, pp. 135–146, 2002.

[18] J. Sim, G. Kim, C. Park, and M. Ju, "Shear strengthening effects with varying types of FRP materials

and strengthening methods," in *7th International symposium on fiber reinforced polymer (FRP) reinforcement for concrete structures*, 2005, pp. 1665–1680.

[19] N. T. K. Al-Saadi, A. Mohammed, R. Al-Mahaidi, and J. Sanjayan, "A state-of-the-art review: Near-surface mounted FRP composites for reinforced concrete structures," *Constr. Build. Materi.*, vol. 209, pp. 748–769, 2019.

[20] F. Al-Mahmoud, A. Castel, R. François, and C. Tourneur, "Strengthening of RC members with near-surface mounted CFRP rods," *Composi. Struct.*, vol. 91, no. 2, pp. 138–147, 2009.

[21] L. De Lorenzis and J. G. Teng, "Near-surface mounted FRP reinforcement: An emerging technique for strengthening structures," *Composi. Part B Eng.*, vol. 38, no. 2, pp. 119–143, 2007.

[22] S. E. El-Gamal, A. Al-Nuaimi, A. Al-Saidy, and A. Al-Lawati, "Efficiency of near surface mounted technique using fiber reinforced polymers for the flexural strengthening of RC beams," *Constr. Build. Materi.*, vol. 118, pp. 52–62, 2016.

[23] F. U. Shariff, Nikitha V. Anand, and N. C. Balaji, "Flexural strengthening of RC beams using near surface mounted-Glass fiber reinforced polymer strips," *A.I.P Conf. Proc.*, vol. 2327, no. February, 2021.

[24] Y. Kusuma, T. Rashmi, Nikitha V. Anand, and N. C. Balaji, "An experimental study on flexural performance of RC beams strengthened by NSM technique using GFRP strips for a resilient infrastructure system," *Mater. Today Proc.*, no. 52, 2022.

[25] M. Rashmi, Nikitha V. Anand, and N. C. Balaji, "Shear strengthening of RC beams using near surface mounted technique with glass fiber reinforced polymer," in *A.I.P conference proceedings*, 2021, vol. 2327, no. 1, p. 20012.

[26] T. Rashmi, Y. Kusuma, Nikitha V. Anand, and N. C. Balaji, "Shear Strengthening of RC Beams by NSM Technique Using GFRP Strips for Sustainable and Resilient Infrastructure," in *Recent Advances in Civil Engineering*, Springer, 2022, pp. 15–26.

[27] L. Torres, I. A. Sharaky, C. Barris, and M. Baena, "Experimental study of the influence of adhesive properties and bond length on the bond behaviour of NSM FRP bars in concrete," *Jour. Civ. Eng. Manag.*, vol. 22, no. 6, pp. 808–817, 2016.

[28] K. Soudki and T. Alkhrdaji, "*Guide for the Design and Construction of Externally Bonded FRP Systems for Strengthening Concrete Structures" (ACI 440.2R-02)*. 2005.

Effect of Depth of Water and Container Diameter in Seismic Analysis of R. C. Elevated Circular Water Tank

Chitte Chetan Jaiprakash[1], Shrikant Charhate[2], S. Sangita Mishra[3]
[1,2,3]Amity School of Engineering and Technology, Amity University Maharashtra, Mumbai, Maharastra 410206, India
[1]Research Scholar, [2]Professor, [3]Associate Professor
Corresponding Author:chetanjchitte@gmail.com

Abstract: Safety of water tanks is very important as they are very important life living structures. Tanks shall always be functional for the access to drinking purpose as well as for the requirement of fire fighting in the area of occurrence of earthquake. The study focus on analysis of RC water tank of circular shape in accordance with IS 1893 (Part-2):2014. The analysis of tank is carried out for zones II, III, IV and V and Rocky or Hard Soil, Medium Stiff Soil and Soft Soil conditions in accordance to Indian standard. Further three different depth of water with constant diameter ratios with tank full, half-filled and tank empty condition are considered for analysis. From the analysis, it is found that tank design is governed in full tank condition. Values of seismic horizontal design coefficient in impulsive mode (Ahi) are less than those for convective mode (Ahc), for full tank condition. This is due to lower values of time period in impulsive mode. For the same soil condition, values of Ahi and Ahc increases for higher seismic zone. The values of highest shear and moment at bottom of container are found to be governing in full tank condition as compared with half-filled and empty tank condition.

Keywords: elevated water tank, seismic analysis, maximum shear at base, maximum moment at base, impulsive and convective modes, seismic coefficient

1. Introduction

Safety of water tanks is very important as they are very important life living structures. Tanks shall always be functional for the access to drinking purpose as well as for the requirement of fire fighting in the area of occurrence of earthquake. Earthquake loads are infrequent loads [9, 10] acting on the structure. So it is not advisable to allow structure to remain elastic or unharmed which in result will be improvident. In common design practice it is consent some structural and non-structural damages at the time of occurrence of earthquake. Ultimately this shall not lead to complete fall of structure so that human life is saved in the worst case [4, 6, and 7], however tolerable destruction of structural components is acceptable. In tolerable destruction of structural components in case of earthquake, it is be designed for earthquake forces significantly lower than predicted in occurrence of earthquake if structure is expected to be elastic linear [8, 10]. Actual lateral base shear is lowered to determine the design lateral force by using Response reduction factor, R [11].

Design lateral force for the elastic structure as per Indian seismic code [1-3] is determined by dividing by 2R. Reduction factor is the amount of ductility & over-strength of a structure [12, 13]. For construction of a structure for considerably less force than expected to be at the time of strong shaking, over strength, redundancy and ductility factors shall be considered for prevention of complete collapse of structure.

In the occurrence of earthquake water mass inside container vibrates as two dissimilar modes. Impulsive mass is the amount of water vibrating with the container and convective or sloshing mass is the part of water vibrating relative to tank container. Impulsive as well as convective mass of liquid contributes to hydrodynamic force. Water in elevated tanks are modelled by two approaches [5, 6, and 7] as Single DOF 2-DOF system. In SDOF system, total water weight is taken as impulsive in addition to container and staging mass. Housner in 1963 [5] recommended an equivalent mechanical model for 2-DOF system. It consist of first degree of freedom for impulsive mass along with container & staging mass and second degree of freedom for convective mass. As per model recommended by Housner [5], total hydrodynamic force embraces of convective and impulsive components [14].

2. Provisions of Indian Code

2.1 Base Shear

Impulsive and convective modes are considered in IS1893 (Part-2) [3]. Combined lateral shear for water tanks shall be calcualted with consideration of impulsive and convective mode lateral base shear using square root of sum of squares rule and is give as:

$$V= \sqrt{(Vi^2 + Vc^2)} \quad \text{......Eq.1}$$

Just above the staging base, shear in impulsive mode, is given as:

$$V_i = (Ah)_i \, . \, (m_i + m_s) \, . \, g \quad \text{......Eq.2}$$

and shear in in convective mode is given as:

$$V_c = (Ah)_c \, . \, m_c \, . \, g \quad \text{......Eq.3}$$

Where,

$(Ah)_i$ = impulsive mode seismic horizontal coefficient;

$(Ah)_c$ = convective mode seismic horizontal coefficient ;

m_i= impulsive water mass;

m_s = container + one third staging mass ;

m_c, = convective mode water mass;

g = acceleration due to gravity.

Design seismic coefficient in horizontal direction, Ah is individually calculated in impulsive $(Ah)_i$ &S convective $(Ah)_c$ modes.

$$V = Ah \times W \quad \text{......Eq.4}$$

and,

Coefficient of base shear Ah, is given by,

$$A_h= ([Z/2][I/R])/[S_a/g] \quad \text{......Eq.5}$$

In above equation,

S_a/g = response acceleration coefficient; Z = zone of seismic factor; W = Weight of container plus $(1/3)^{rd}$ staging weight; I = Factor of Importance (1.5); R = Response factor. 'R' values ranges from 2 to 4 for water tank.

2.2 Base Moment

As per Indian code [3], the base moment 'M' for tank, shall be considered as combining the impulsive mode moment (M_i) and convective mode base moment (M_c) using SRSS and is given as follows

$$M^*=\sqrt{M_i^{*2} + M_c^{*2}} \quad \text{......Eq.6}$$

Impulsive moment at base, is calculated as:

$$M_i^* = (Ah)_i \times [m_i \times (h_i^* + h_s) + m_s \times h_{cg}] \times g \quad \text{......Eq.7}$$

and convective mode base moment in is calculated as:

$$M_c^* = (Ah)_c \times m_c \times (h_c^* + h_s) \times g \quad \text{......Eq.8}$$

Where

h_i^* = impulsive mass height above tank wall bottom

h_c^* = convective mass height above tank wall bottom

h_s = staging elevation;

h_{cg} = C.G. of empty container of above footing

3. Methodology

In the present study of RC circular water tank in accordance with Indian standard is analyzed for seismic loads.The analysis of tank is carried out for zones II to V with Rocky or Hard Soil, Medium Stiff Soil and Soft Soil conditions as per Indian standard. Further three different depth of water with constant diameter ratios with tank full condition, tank half-filled condition and tank empty condition are considered.

Seismic analysis of water tank is performed using following parameters.

Circular Water container capacity = 200Cu.m, Height of container, H =5m,
Free board =0.3m, Depth of water, h = 4.7m (Tank full), 2.5m (Tank Half full), 0 (Tank empty)
Diameter of container, D = 7.5m (Constant in all cases),
Seismic zone = Zone II to V, Factor of Importance, I =1.5, Response factor, R =4,
Soil Site condition = Rocky or Hard, Medium stiff & Soft soil
For detailed geometry of circular water tank refer Figure 1.

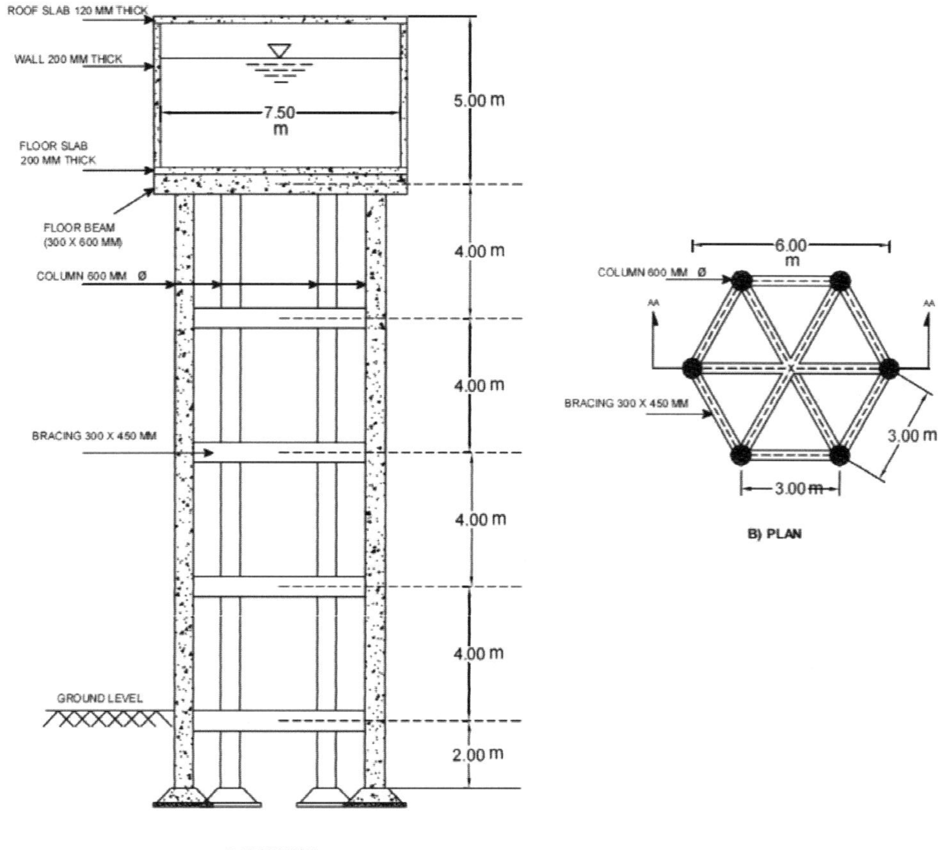

Figure 1: Circular Tank

4. Results

The values of maximum shear as well as moment are found out in full tank,half-filled and empty condition for different zone factors, soil conditions, tank full, half-filled and empty condition. Table 1, 2 and 3 presented in summary of various calculated parameters.

IOP Conf. Series: Materials Science and Engineering 1282 (2023) 012007 doi:10.1088/1757-899X/1282/1/012007

Table 1: Analysis Results of 200Cu.m Capacity Tank_Full Tank Condition

Zone Factor	*Soil Condition*	*Full Tank Condition*			
		Seismic Coefficient in impulsive mode (Ahi)	*Seismic Coefficient in convective mode (Ahc)*	*Maximum Base Shear (kN)*	*Maximum Base Moment (kN.m)*
II	Rocky or Hard Soil	0.0107	0.0112	38	767
III		0.0172	0.0180	61	1228
IV		0.0258	0.0270	92	1841
V		0.0387	0.0404	137	2762
II	Medium Stiff Soil	0.0146	0.0153	52	1044
III		0.0234	0.0244	83	1670
IV		0.0351	0.0367	124	2504
V		0.0526	0.0550	187	3757
II	Soft Soil	0.0179	0.0188	64	1281
III		0.0287	0.0300	102	2050
IV		0.0431	0.0450	153	3075
V		0.0646	0.0675	229	4613

Table 2: Analysis Results of 200Cu.m Capacity Tank_Half Filled Condition

Zone Factor	Soil Condition	Tank Half Filled Condition			
		Seismic Coefficient in impulsive mode (Ahi)	*Seismic Coefficient in convective mode (Ahc)*	*Maximum Base Shear (kN)*	*Maximum Base Moment (kN.m)*
II	Rocky or Hard Soil	0.0129	0.0107	31	622
III		0.0207	0.0172	50	996
IV		0.0311	0.0257	72	1493
V		0.0466	0.0386	112	2240
II	Medium Stiff Soil	0.0176	0.0146	42	846
III		0.0282	0.0233	68	1354
IV		0.0422	0.0350	102	2031
V		0.0634	0.0525	153	3046
II	Soft Soil	0.0216	0.0179	52	1039
III		0.0346	0.0286	83	1663
IV		0.0519	0.0430	125	2494

V		0.0778	0.0645	187	3741

Table 3: Analysis Results of 200Cu.m Capacity Tank_Empty Condition

Zone Factor	Soil Condition	*Seismic Coefficient in impulsive mode (Ahi)*	*Maximum Base Shear (kN)*	*Maximum Base Moment (kN.m)*
II	Rocky or Hard Soil	0.0146	22	434
III		0.0234	35	694
IV		0.0351	52	1041
V		0.0527	78	1562
II	Medium Stiff Soil	0.0199	29	590
III		0.0318	47	944
IV		0.0478	71	1416
V		0.0716	106	2124
II	Soft Soil	0.0244	36	725
III		0.0391	58	1159
IV		0.0586	87	1739
V		0.0880	130	2608

Graphical reprentation of results from Table 1, 2 and 3 is given as below.

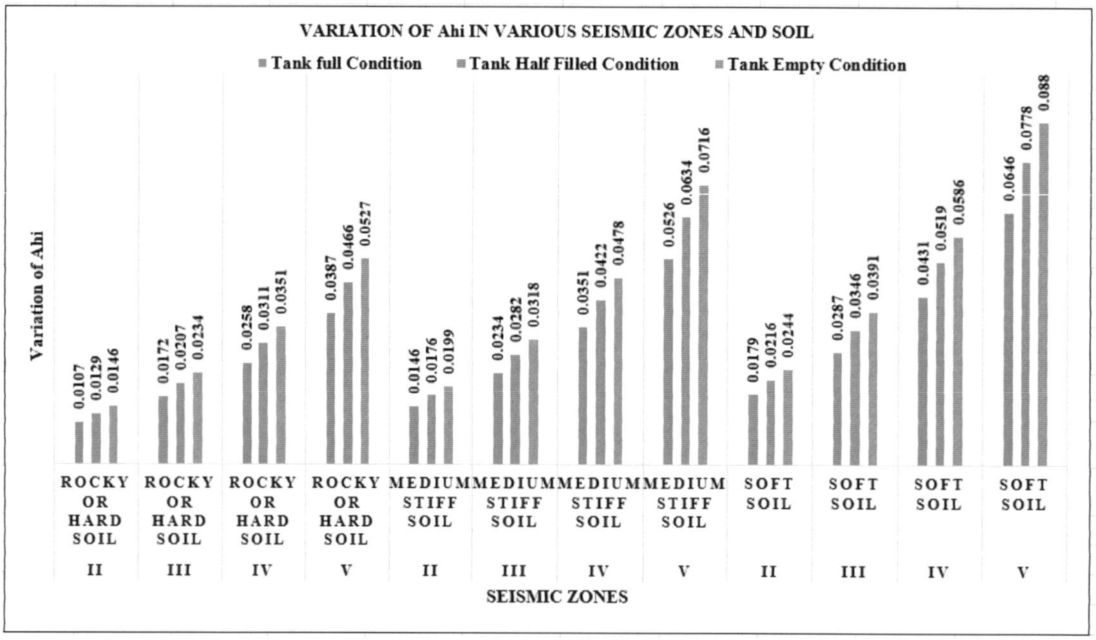

Figure 2: Variation of Ahi in various seismic zones and soil

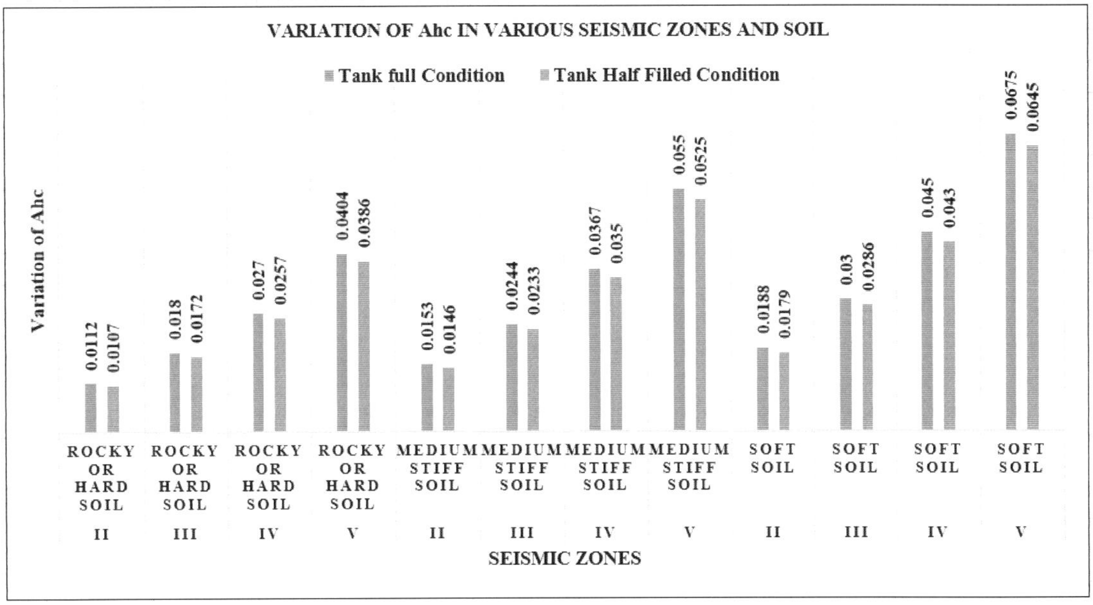

Figure 3: Variation of Ahc in various seismic zones and soil

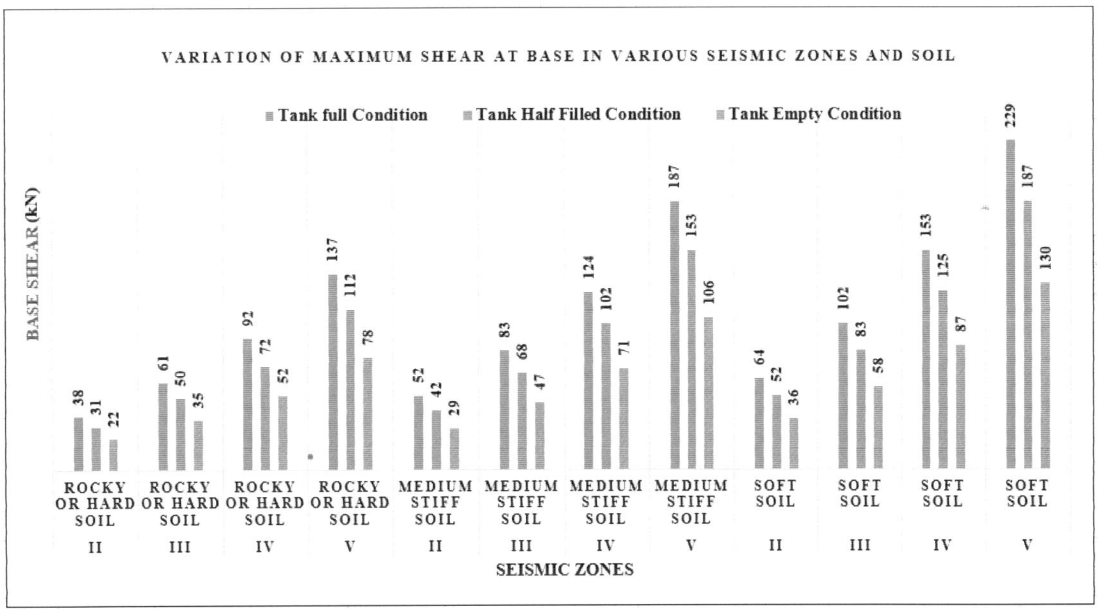

Figure 4: Variation of Maximum Shear at base in various seismic zones and soil

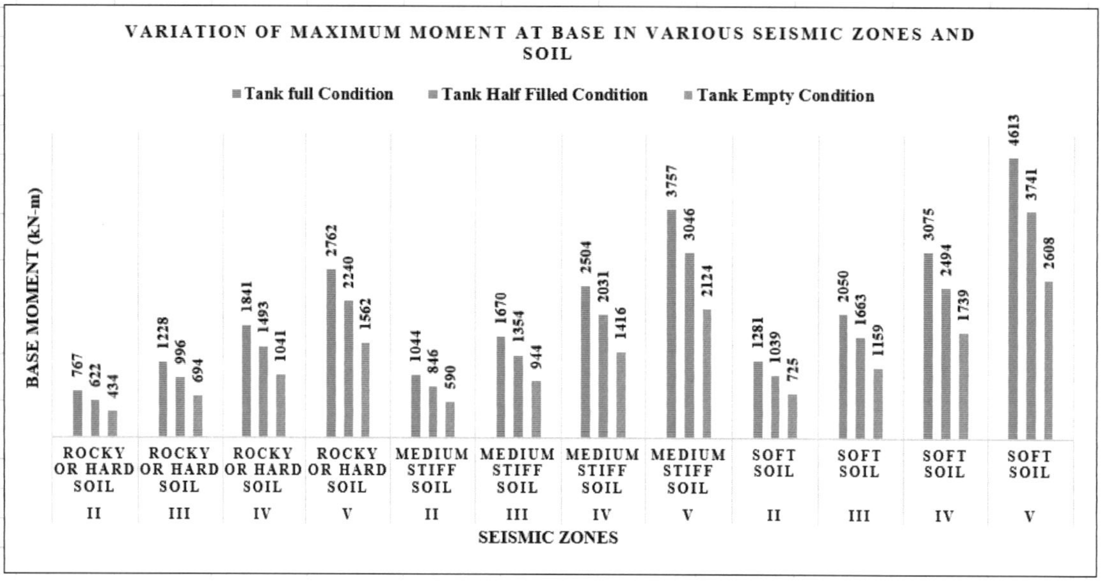

Figure 5: Variation of Maximum Moment at base in various seismic zones and soil

5. Conclusions & Discussions

From seismic analysis of circular tank with 200Cu.m capacity , the results are represeneted in Table No.1, 2 ,3 and Figure 2 to 5. From these results, following conclusions can be drawn.

- Values of horizontal seismic coefficient in impulsive mode (Ahi) are lesser than those for convective mode (Ahc), for tank full condition. This is due to lower values of time period in impulsive mode.
- Values of horizontal seismic coefficient in impulsive mode (Ahi) increases in tank half filled condition and tank empty condition as compared with tank full condition.
- Values of horizontal seismic coefficient in impulsive mode (Ahi) are more than the values for convective mode (Ahc), for tank half-filled condition. This is due to lower values of impulsive mass of liquid. [For tank half-filled condition, mass of water is less than that of tank full condition].
- For the same soil condition, values of horizontal seismic coefficient in impulsive mode (Ahi) and convective mode (Ahc) increases as zone factor increases [higher seismic zone]
- As soil conditions, changes from Rocky hard to medium stiff and soft soil, horizontal seismic coefficient value in impulsive mode (Ahi) and convective mode (Ahc) increases for higher zones.
- Maximum shear & moment is higher at full tank with that for tank half-filled and tank empty condition.
- Maximum shear & moment increases with higher seismic zone.
- Maximum shear & moment values are governing in rocky hard soil condition as compared with medium stiff and soft soil.

6. References

[1] BIS IS 1893, 1984—Criteria for Earthquake Resistant Design Structures, Part 2, New Delhi (India): Bureau of Indian Standards, 1984
[2] BIS IS 1893, (Part 1) 2016—Criteria for Earthquake Resistant Design Structures, Part 1, New Delhi (India): Bureau of Indian Standards,
[3] BIS IS 1893, (Part 2) 2014—Criteria for Earthquake Resistant Design Structures, Part 2, New Delhi (India): Bureau of Indian Standards,

[4] IITK-GSDMA, (2007)—Guidelines for Seismic Design of Liquid Storage Tanks – Provisions with Commentary and Explanatory Examples,‖ Indian Institute of Technology Kanpur, Kanpur, India,.
[5] Housner, G.W. (1963) "The Dynamic Behavior of Water Tanks." Bulletin of the seismological society of America 53.2: 381-387
[6] Newmark N, Hall W. (1982). "Earthquake spectra and design." Berkeley, CA: Earth- quake Engineering Research Inst.
[7] S. K. Jain, and S. U. Sameer,(1993) —A Review of Requirements in Indian Codes for Aseismic Design of Elevated Water Tanks, Bridge & Structural Engineer, vol. 23 no.1,
[8] S. K. Jain, C. V. R. Murty, N. Chandak , L. Seeber , and N. K. Jain, (1993)—The September 29, 1993, M6.4 Killari, Maharashtra Earthquake in Central India,‖ EERI Special Earthquake Report, EERI Newsletter, vol. 28, No.1, 1994.
[9] Rai, D. C., (2002). Elevated tanks, in Bhuj, India, Earthquake of January 26, 2001: Reconnaissance Report, Earthquake Spectra, Supplement A to Volume 18, S. K. Jain et al. _Eds._, pp. 279–295, Earthquake Engineering Research Institute, Oakland, CA
[10] D. C. Rai, (2003)—Performance of Elevated Tanks in Mw 7.7 Bhuj Earthquake of January 26th, 2001,‖ Journal of Earth System Science, vol. 112, no. 3, pp. 421-429
[11] O. R. Jaiswal, D. C. Rai, and S. K. Jain,(2007) —Review of seismic codes on liquid-containing tanks,‖ Earthquake Spectra, vol. 23, no. 1, pp. 239-260,
[12] Bhavin Patel and Dhara shah, (2010), Formulation of Response Reduction Factor for RCC Framed Staging of Elevated Water Tank, World Congress on Engineering, London ,U.K
[13] S. Bozorgmehrnia, M.M. Ranjbar and R. Madandoust, 2013, "Seismic Behavior Assessment of Concrete Elevated Water Tanks"Journal of Rehabilitation in Civil Engineering 1-2 (2013) 69-79
[14] Suraj O. Lakhade, Dr. Ratnesh Kumar, Dr. O. R. Jaiswal, (2017),"Effect of Modified Provisions of IS 1893 (Part 2):2014,on Design Base Shear of Elevated Water Tanks", International Journal of Engineering Research in Mechanical and Civil Engineering (IJERMCE) Vol 2, Issue 3, March 2017, ISSN (Online) 2456-1290, 429-433

Influence Of Helical Confinement On Crack Development In RCC Beams

Bahaa Tarek Amin Ahmed Abdelbaki[1], Sachin Kuckian[2], Dr. Omar Al Obaidi[3], Dr. Hussin Yahia[4]
[1]Student, Middle East College, Muscat, Oman
[2]Lecturer, Middle East College, Muscat, Oman
[3]Assistant Professor, Baghdad University, Baghdad, Iraq
[4]Assistant Professor, Middle East College, Muscat, Oman
E-mail: 14F12877@mec.edu.om
skuckian@mec.edu.om
oalobaidi@gmail.com
hyahia@mec.edu.om

Abstract. Structures made of reinforced concrete are susceptible to developing cracks when they are in the path of earthquakes, strong winds, or increased structural loading. It is therefore possible that it will be necessary to manage structural cracking in the Service Limit State to maximise the lifetime and strength of the structural components under any stress. Helical containment is superior to rectangular containment in terms of its ability to strengthen and ductile the Reinforced Concrete (RC) structural member. In the current work, six distinct beams were each subjected to a series of tests to see how they behaved in terms of cracking when subjected to bending loads. Each of these tests used a unique helical pitch space distance. The beams all measured 150 mm by 150 mm and 750 mm in length. They were designed in accordance with the criteria of Euro code 2, with the identical dimensions. Both 50 and 100 mm were used to take the measurement for the helical pitch. The findings indicate that the utilisation of helical elements does influence the cracking behaviour of the beams. The most important finding was that the measurement of the crack's thickness was slightly reduced when the beams were restricted to a helical zone with a closer spacing between them. The fact that the theoretical crack is longer than the actual crack that was measured experimentally demonstrates that the Euro code 2 standard provides a more accurate forecast as well as a higher factor of safety. This is demonstrated by the fact that the theoretical crack is larger than the actual crack that was measured.

Keywords: Strength, ductility, fracture breadth, failure profile, tomography.

1. Introduction

The rapid rise of the world's population has led to an increase in the number of new building constructions that satisfy the people's demands. There are several basic features that must be met for a building to be considered for better serviceability conditions. The construction industry is facing day to day challenges that includes the design of sustainable materials [2]. In building, this material is frequently utilized because of its cheapness and flexibility to be molded into various sizes and shapes. The material's strength can withstand significant compression stresses as well [1]. Due to its brittleness, this material cannot be used as a stand-alone component in the building. To be able to bear tensile loads, this material needed to be strengthened. The use of "reinforced cement concrete" was introduced because of this perspective. Reinforced cement concrete, often known as reinforced concrete, is a combination of concrete and steel bars that is commonly used in the construction of buildings. Steel bars are used to counteract the tensile forces that are generated by the ductility of the material [3]. Reinforced concrete structures consist of slabs, beams, columns, frameworks, footings, and walls. As a focal point, the beam is featured in this article. Beams made of reinforced concrete resist vertical loads from the structure's walls, slabs, and their own weight. In earthquake-prone areas, it is also utilized to provide additional structural support to existing structures. Deflection in this structural component mostly occurs via bending. It deflects in a manner that compresses the top and tensions the bottom of the beam when it is subjected to large loads. However, the compression area where the concrete is effective, and the steel bars are tension resistant. Beams can be either a single or double reinforced depending on the strength of the weights being applied.

There are steel bars at the tension zone of the single-reinforced beam and Steel bars are placed in both tension and compression zones of the doubly reinforced beam [4]. On the other hand, Reinforced structures are frequently plagued by cracks. According to Tammo, Kristian [5], structural elements acquire flexural cracks when their tensile strength deteriorates. It takes a lot of tension for secondary cracks to form since the primary ones aren't strong enough to resist the increasing tension. The corrosion of the reinforcement was caused by deep fissures in the structural part, which in turn led to failure [5]. To lessen the width of cracks in reinforced concrete, helical confinement was tested in this paper. The main research highlights are:

- To investigate the cracking behavior of Helical containment beams under bending loads.
- To demonstrate that helical containment influences beam cracking behavior, with the crack thickness measurement being significantly smaller when beams are constrained to a helical zone with a tighter spacing [19].

2. Methodology

Rebar size, shear connection, and nominal cover were all considered when designing six reinforced concrete beams in accordance with Eurocode 2 [6] regulations. The beams were 150 mm x 150 mm x 750 mm long, with the same dimensions. The pitch distances were 50mm and 100mm. The details of each beam's steel reinforcement are shown in the table below.

Table 1. Details of the Steel Reinforcement.

	Longitudinal Reinforcing steel		Shear Reinforcing			
			Shear Span		Constant Moment Zone	
Beam Designation	Top	Bottom	Bar diameter	Bar Spacing	Helical diameter	Helical spacing
Beam 1	2H 6	4H6	8	50		
Beam 2					6	50
Beam 3					6	100
Beam 4						
Beam 5					6	50
Beam 6					6	100

Concrete mixes consist of fine and coarse aggregate with cement, water, and chemical admixture. the chemical admixture known as GLENIUM 388RM was employed to improve workability of the C30 concrete, The British Standard design was used to prepare the concrete mixture. The concrete's components are listed in the following table

Table 2. Concrete mix designs

Materials	Quantities per beam(kg)	Quantities for 6 beams(kg)
Cement	42	247
Water	106	631
Coarse aggregate Fine aggregate	118 221	703 1321

2.1. *Specifications for the Beam case configurations, characteristics, and detailed description*

Table 3. Specifications for the Beam case

Sr. No.	Beam	Width (in mm)	Height (in mm)	Length (in mm)	Cover (in mm)	Diameter (in mm)	Zone	Helical spacing (in mm)
1	B1-C-U					6 and 8		Nil
2	B2-S50-U					6 and 8		50
3	B3-S100-U	150	150	750	10	6 and 8	tensile	100
4	B4-C-O					8		Nil
5	B5-S50-O					8		50
6	B6-S100-O					6		100

Table 3 gives the dimensions and details of the six beams that have been used in the experiment. A particular code was assigned to each of the six beams depending on the design type, i.e., balanced type as U, over-reinforced type as O, and spacing of stirrup in mm. There are multiple shuttering areas depending on the distance between helical confinements. A cross-sectional view of the beams is seen in Figure 1-6.

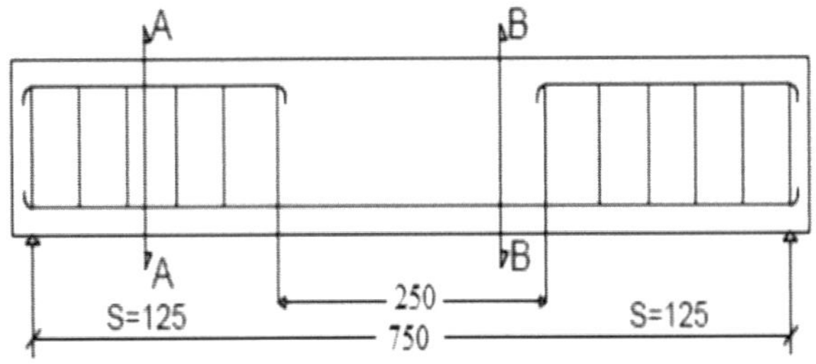

Figure 1. Elevation and cross-sectional perspectives of BM1-CU

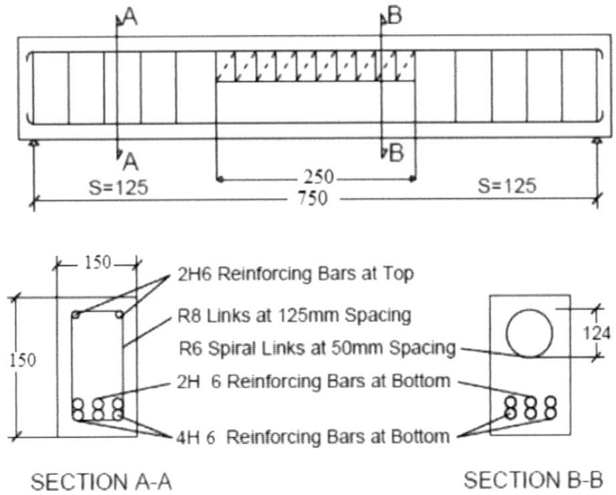

Figure 2. Elevation and cross-sectional perspectives of BM2-S50U.

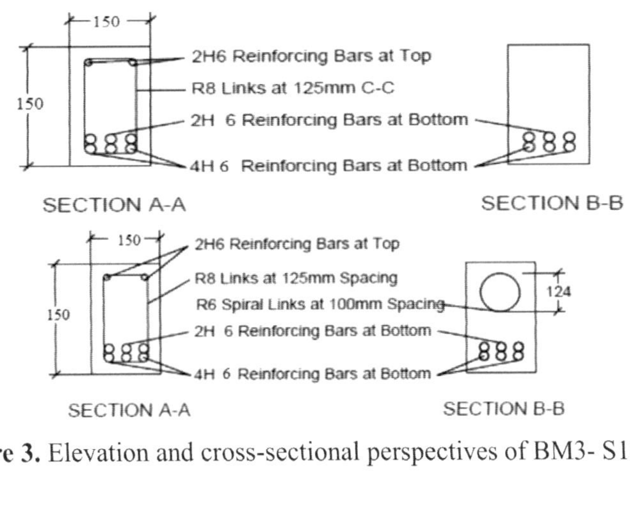

Figure 3. Elevation and cross-sectional perspectives of BM3- S100U

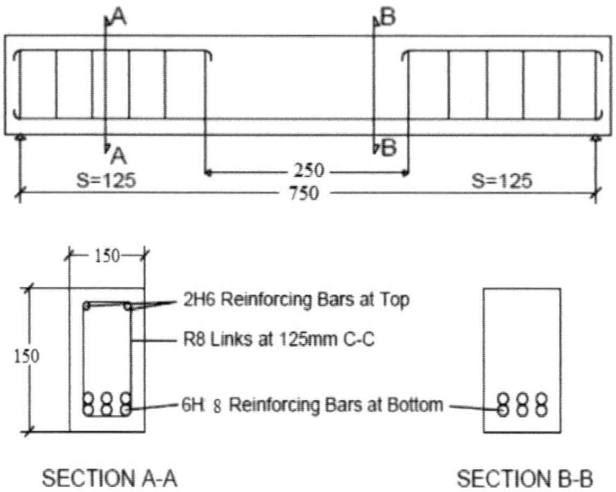

Figure 4. Elevation and cross-sectional perspectives of BM4-CO.

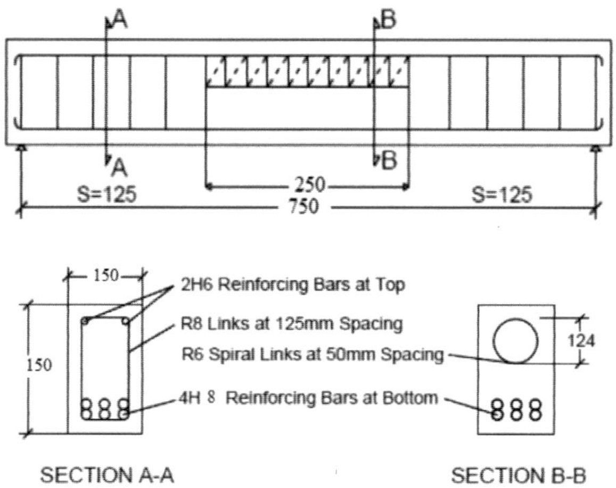

Figure 5. Elevation and cross-sectional perspectives of BM5-S50O.

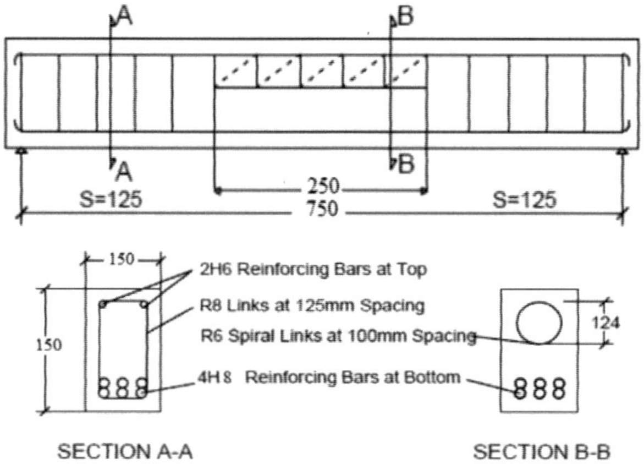

Figure 6. Elevation and cross-sectional perspectives of BM6-S100O.

2.2. *The Slump test*

The workability or consistency of a concrete mix can be evaluated using the slump cone test, which can be performed in a laboratory or directly on the construction site as the work advances. To guarantee that the quality of the concrete remains stable during the entirety of the construction process, a slump test is carried out on each new batch of concrete. [15]. Figure 7 shows the slump test [7] that used to determine the workability of fresh concrete.

Figure 7. Slump experiment

2.3. *Test of compressive force*

The strength of concrete determined by a compression test, which involves smashing cubical concrete in a compression testing machine is known as compressive strength. It assesses the ability of concrete to bear a load before failing [16] and twelve cubes [8] were cast with 150 mm dimensions, and the concrete's compressive strength was measured. As shown in figure 8, compression test of cubes after water curing for 7, 14 and 28 days.

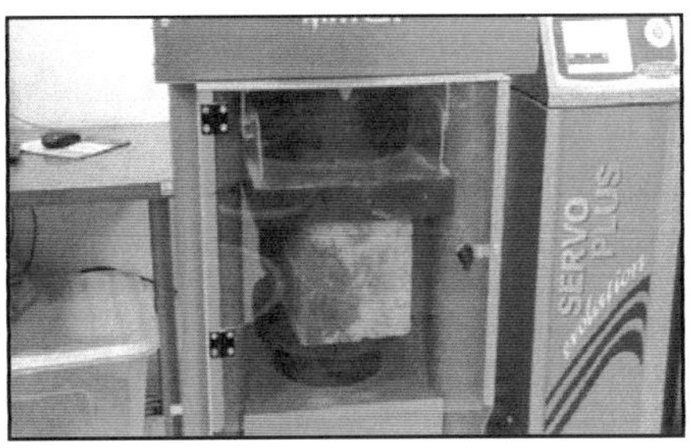

Figure 8. Compression strength testing.

2.4. *Flexural Test*

Flexural strength is the amount of stress and strain that an unreinforced concrete slab, beam, or other structure can withstand without bending [17]. Flexural strength is sometimes referred to as flexural bearing capacity. After 28 days of curing, flexural 4-pin loading was applied to each of the six reinforced concrete beams to perform an evaluation, as shown in figure 9. The purpose of this experiment was to determine the maximum amount of force necessary to bend the beam and to determine its fracture length.

Figure 9. Measuring the flexure strength.

2.5. *Measurement of breadth of crack*

Crack width measurements were made to examine the cracking pattern because of the beam being loaded. In the tension or shear zones, a crack is formed, and a failure pattern was forecasted. The microscope is used to measure the crack's breadth [22].

2.6. *Tests for tomography*

Using a CT scan, any concrete flaws were found in the beams. When concrete pieces are subjected to extreme loads or are frequently used, microcracks could occur. As an active measurement device, tomography [9], demonstrates that a bob is placed on the surface of the members and sensors are used to collect data. The Tomography gadget is depicted in Figure 10.

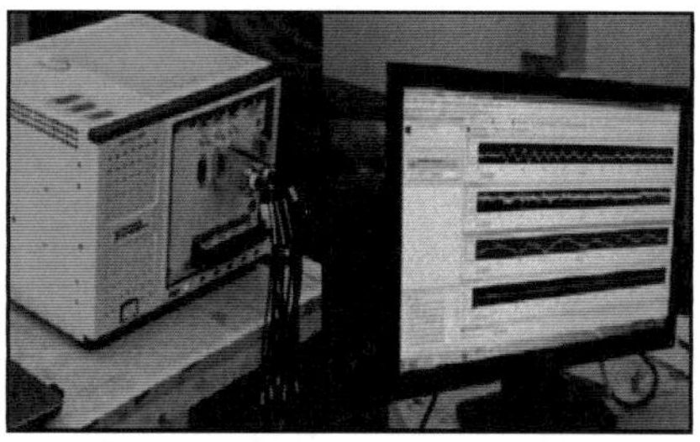

Figure 10. Tomography Equipment

3. Discussion and findings

3.1. *Test of compressive force*

Table 4. Results of the compressive investigation

Age	Compressive Strength (N/mm^2)			
7 days	49.72			
14 days	59.59			
28 days	60.11			

Compressive strength tests were performed on 12 concrete cubes at cure times of seven, fourteen and twenty-eight days. Table 4 displays the average concrete cube strength. The average strength of cubes grew as the curing period progressed, according to a study. At 28 days of cure, the average concrete strength was greater.

The results in the table 4 showed that the strength of cubes has been increased with the age of curing. The strength of cubes at 7days has been increased by 20% at 14 days and 21% at 18 days.

3.2. *Flexural Test*

Table 5. The average width of the cracks in the beam

Load (KN)	Under-reinforced					Over-reinforced
	Beam 1	Beam 2	Beam 3	Beam 4	Beam 5	Beam 6
20						
40	0.02					
60	0.03	0.01	0.01	0.01		0.01
80	0.03	0.02	0.02	0.01	0.01	0.01
100	0.04	0.02	0.03	0.02	0.02	0.02
120	0.05	0.04	0.04	0.04	0.02	0.03
140	0.06	0.04	0.04	0.05	0.03	0.04
160	0.08	0.04	0.05	0.06	0.03	0.04
180	0.09	0.04	0.07	0.07	0.04	0.05
200	0.10	0.05	0.07	0.06	0.05	0.06

Average crack widths for the six beams are shown on Table 5. it can be seen from these results that the beams 4, 5, and 6 are more capable of resisting the weight than beams 1, 2, and 3. Beams 4 and 6 were where the cracks first appeared at a force of 60 kN. Beams 4, 5, and 6 each had a crack width of 7 x 10^{-2}mm, 5 x 10^{-2} mm, and of 6 x 10^{-2} mm at 200 KN, respectively. In contrast, the crack widths of beams 1, 2, and 3 at 200 KN were 0.10 mm, 0.05 mm, and 0.07 mm in diameter. As a further detail, Beam 1 cracked at 40 kN, whereas beams 2 and 3 cracked at 60 kN. It appears that the helical 50 mm spacing is effective in lowering the crack breadth in Beam 2, which is having the narrowest average crack breadth among Beams 1 and 3. Due to lack of zero-shear blocking, beam 1 shows a larger average crack width than beams 2 and 3.

3.3. *Analysis of Tomography*

Non-destructive test software [10] and piezoelectric sensors [23] were used to determine the recorded wavelengths to produce technical images of the contained concrete. During the experiment [11], sensors measured the amplitude (A), which was recorded for each picture. The cracks in the following beams can be seen before and after loading, as shown in the figure 12.1 and 12.2:

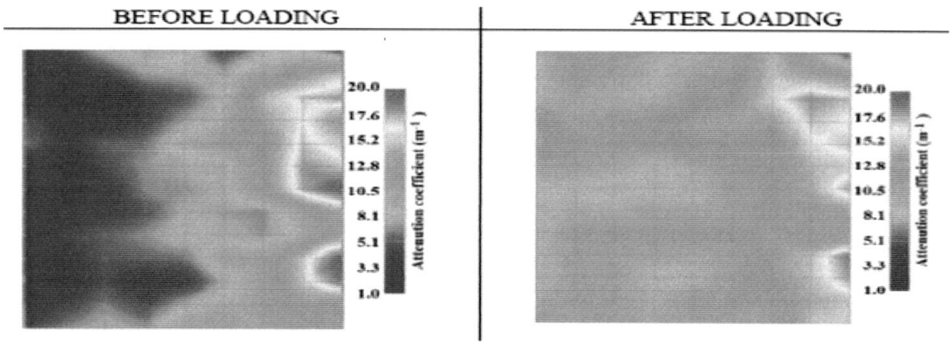

Figure 11. Beam No.1 pre- and post-loading.

Using a CT examination [24], beam 1's concrete appears to be in good condition; however, a large side portion of the beam's concrete was damaged prior to loading. Poor compaction may lead to this damage. Despite this, the quality of the concrete is excellent, the flaws are confined to cover-point of the correct side and the design of beam inside is adequate. Following loading, the fracture spreads outward from the beam side, and the concrete's exterior surface fails first.

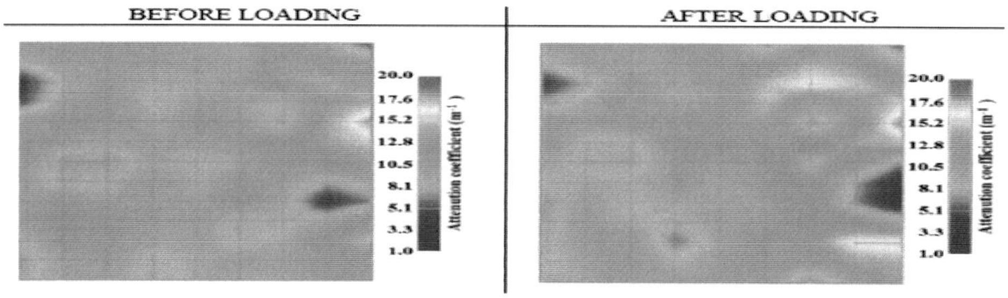

Figure 12. Beam No.2 pre- and post-loading.

Images in figure 13 are showing medium attenuation of critical concrete areas in beam 2 tomographic analysis because of sensor noise induced by the sensitivity [20]. on the other hand, Interior confinements adequate, such that before loading very minor cracks observed in the concrete cover. When loading applied, the image reveals deterioration of the internal concrete and a crack pattern on the beam right side that extends into the beam. It's possible to simply see a minor change before and after loading. It's clear from testing that this beam's failure section only extends to the right side of the beam's centerline, owing to helical confinement explaining why the spreader's load is right side distributed. An experimental study conducted by Whitehead and Ibell [12] also proved that in beams that helical confinement is more effective than a rectangular tie.

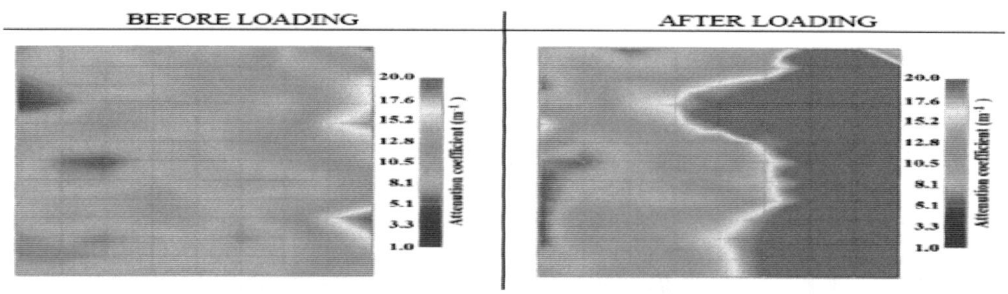

Figure 13. Beam No.3 pre- and post-loading.

The topographical scan reveals that the inner cement of beam No. 3 seems to be in excellent condition, with just a few areas exhibiting symptoms of degradation prior to loading. Attenuation in concrete could be attributing to poor compaction and shrinkage microcracks. After loading, the failure of concrete beam on the tension side can be seen because of yielding the steel bar. Due to the high attenuation ratio, the concrete on the outside of Beam No. 3 is still in good condition, but the concrete on the inside of the beam is crumbling. Since the beam collapse is situated on the proper location, spreader bears a greater strain on the correct hand [18].

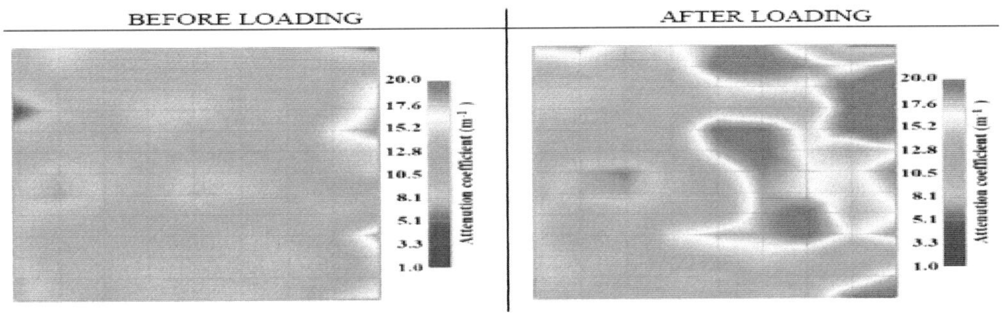

Figure 14. Beam No.4 pre- and post-loading.

Using the pre-loading picture of Beam 4, the inner bulk of cement is in excellent condition and that there are just a few areas of possible concrete shrinkage on the side of the beam. The constant attenuation implies quality concrete, which is passable in strength. Failure on the beam's correct side indicates the spread of load cracks after loading [21].

Owing to absence of helical confinement, Beam No. 4 has a substantial diminution rate, which directs a crack. If the beam fails on the proper side, this might indicate that the spreader is distributing its weight more evenly on the correct side than the left. Elbasha and Hadi [13] also tested experimentally tested helically confined high strength concrete beams and found similar results which clearly establishes present analysis is in coherence with literature with updated technology used in present experimental work.

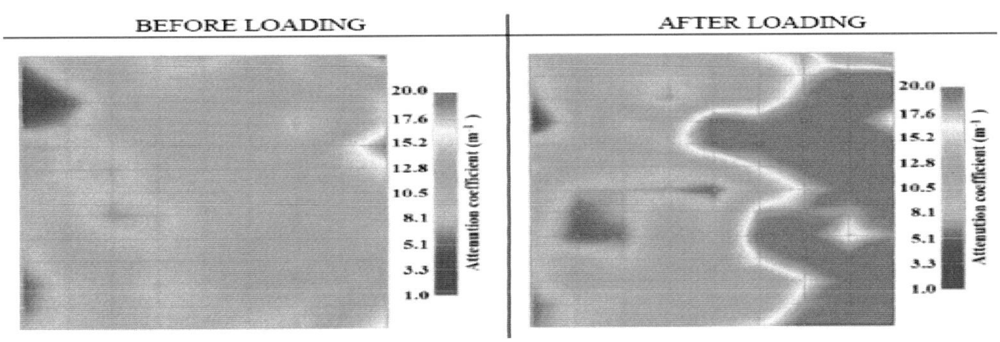

Figure 15. Beam No.5 pre- and post-loading.

Images of beam number 5 before loading reveal that the concrete has a medium coefficient of attenuation, indicating that the concrete is properly contained to an inner area without injuring any areas. If a shrinkage crack appears on the beam's correct side, it can reveal the curing method. As seen, concrete structures have microcracks due to load, creep, and shrinkage [14]. As a result of this, the beam's inner concrete remains unaffected by this point in time. Failure of the right-hand wall upon loading reveals exposed reinforcement bars. Owing to the concrete element of Beam No. 5 is in excellent shape, it is a brittle collapse. As a result of the high attenuation factor in the inner concrete failure, this is the case. Spreader load is concentrated on the right-hand side of beam, indicating that it was not evenly distributed on both sides.

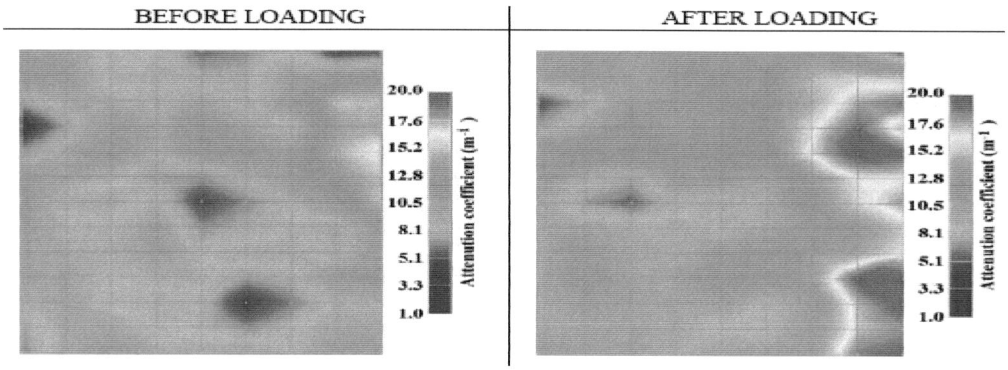

Figure 16. Beam No.6 pre- and post-loading.

The tomographic examination of beam 6 that was performed prior to loading showed that the medium attenuation coefficient was applied to the concrete, which implies that the inner region is correctly restricted without damage. Additionally, the research confirms that there are no critical areas on the beam side, which indicates that there is no initial creep. After loading, the shot illustrates how the reinforcing bars have spread out across the cover, which finally results in the concrete splitting apart. Due to the unusually low attenuation coefficient of Beam No. 6, some of the concrete in the beam is still in great condition, and the inner confining concrete has not been weakened. This makes Beam No. 6's failure a remarkable achievement. Failure occurred on the proper side of the beam, and the weight from the spreader is distributed more uniformly on the correct side than it is on the left side of the beam.

4. Conclusion

Contained Helical Beam Cracking Behavior in Context is the primary objective of this article. Beams bound to a helical region that was closer in distance lowered the assessment of crack thickness marginally, according to presented work. When the tensile area of the beam is contained by helical containment, the ductility increases, and the shear disruption occurs.

Because of its high split width, the beam with no helical containment is over-reinforced and balanced section. It is because the helix delivers a consistent radial stress along the concrete part, while the rectangle confines the concrete, especially at the corners, that it is more effective than the tie. Due to their lower stiffness than the four corners, the tie's sides will tend to flex outwards, reducing the effective concrete surface area in the cross section.

5. Acknowledgement

We thank the management and lab staffs of the Middle East College for providing support in smooth conduct of the experiments. We thank other faculty members in contributed to the project directly and indirectly.

References

1. Kar S and Biswal K C, 2020 Shear strengthening of RC beams with basalt fiber reinforced polymer (BFRP) composites. *Advances in concrete construction*, 10(2), pp.93-104.
2. Kuckian S and Dalvi A 2020 Strength Possibilities on Fly ash Based Interlock Pavers. *In Applied Mechanics and Materials*." Trans Tech Publications Ltd, Vol. 897, pp. 12-19.
3. SWAROOP G V and REDDY, T. P, 2016 A Study on Higher Order Load Analysis and Design of Combined Footing Using STADD Foundation.
4. Krishna N K., Sandeep S, and Mini K M (2018, February). Study on reinforced concrete beams with helical transverse reinforcement. In IOP Conference Series: Materials Science and Engineering (Vol. 310, No. 1, p. 012046). IOP Publishing.
5. Naotunna CN, Samarakoon S S M. and Fosså, K T, "Experimental investigation of crack width variation along the concrete cover depth in reinforced concrete specimens with ribbed bars and smooth bars." Case Studies in Construction Materials, 2021, 15, p.e00593.
6. UGORJI H, EPHRAIM M., and ADETILOYE A (2020). EFFECTS OF VARYING WATER-CEMENT RATIOS ON DIFFERENT GRADES OF CONCRETE USING LOCALLY MATERIALS. Journal of Natural Sciences Engineering and Technology, 19(1), 92-100.
7. Aguayo F, Torres A., Kim, Y. J and Thombare O (2020). Accelerated carbonation assessment of high-volume fly ash concrete. Journal of Materials Science and Chemical Engineering, 8(3), 23-38.
8. Valagiannopoulos C, Marengo E A, Dimakis A G and Alù A (2018). Aharonov–Bohm-inspired tomographic imaging via compressive sensing. IET Microwaves, Antennas and Propagation, 12(12), 1890-1894.
9. Ljubomirović M D (2015). Software for calculation statically determinate beams. Tehnika, 70(5), 809-812.
10. Duan D, Ouyang L, Gao W, Xu, Q, Liu, W, and Yang, J. (2022). Fire Performance of FRP-RC Flexural Members: A Numerical Study. Polymers, 14(2), 346.
11. Deng, M, Zhang, M, Ma, F, Li, F, and Sun, H. (2021). Flexural strengthening of over-reinforced concrete beams with highly ductile fiber-reinforced concrete layer. Engineering Structures, 231, 111725.
12. Krishna N K., Sandeep S, and Mini K M. (2018, February). Study on reinforced concrete beams with helical transverse reinforcement. In IOP Conference Series: Materials Science and Engineering (Vol. 310, No. 1, p. 012046). IOP Publishing.
13. Salim O J and Kuckian S (2020). Solidification of Recycled Concrete using Plastic waste. Solid State Technology, 63(6), 19506-19529.
14. Li, Y, Mu, J, Wang, Z, Liu, Y, and Du, H (2021). Numerical simulation on slump test of fresh concrete based on lattice Boltzmann method. Cement and Concrete Composites, 122, 104136.
15. Mohammed A., Rafiq S, Sihag P, Kurda, R, and Mahmood W (2021). Soft computing techniques: systematic multiscale models to predict the compressive strength of HVFA concrete based on mix proportions and curing times. Journal of Building Engineering, 33, 101851.
16. Wright P J F., and Garwood F (1952). The effect of the method of test on the flexural strength of concrete. Magazine of Concrete Research, 4(11), 67-76.
17. Iwaszenko S, Munk J, Baron S, and Smoliński A (2021). New Method for Analysis of the Temporomandibular Joint Using Cone Beam Computed Tomography. Sensors, 21(9), 3070.
18. Krishna N K., Sandeep S, and Mini K M (2018, February). Study on reinforced concrete beams with helical transverse reinforcement. In IOP Conference Series: Materials Science and Engineering (Vol. 310, No. 1, p. 012046). IOP Publishing.
19. Verburg, S A, and Fernandez-Grande E (2021). Acousto-Optical Volumetric Sensing of Acoustic Fields. Physical Review Applied, 16(4), 044033.

20. Moura M A, Carvalho M A, and Ferreira G (2020). Influence of measuring points in ultrasonic tests on the tomographic image quality of concrete.
21. Huang L, Zheng F, Deng Q, Thi Q. H, Wong L. W, Cai, Y, and Zhao J (2020). In situ scanning transmission electron microscopy observations of fracture at the atomic scale. Physical Review Letters, 125(24), 246102.
22. Roy G, Panigrahi B, and Pohit G (2021). Crack identification in beam-type structural elements using a piezoelectric sensor. Nondestructive Testing and Evaluation, 36(6), 597-615.
23. Khormani M, Jaari V R K., Aghayan I, Ghaderi S H and Ahmadyfard A. (2020). Compressive strength determination of concrete specimens using X-ray computed tomography and finite element method. Construction and Building Materials, 256, 119427.

COMPARISON OF SEISMIC RESPONSE OF MULTI-STOREY BUILDING HAVING SHEAR WALLS WITHOUT OPENING ON STAAD-PRO AND ETAB

Pinaki[1] and A.K. Dwivedi[2]
[1]M. Tech. Student, Rajasthan Technical University (RTU), Kota, Rajasthan, India
2Professor, RTU, Rajasthan Technical University (RTU), Kota, Rajasthan, India
E-mail: [1]pinaki11singh@gmail.com [2]akdrtu@gmail.com

ABSTRACT. In India's main cities, high-rise reinforced concrete residential buildings with more than 15, 16, or 17 stories are widespread these days, which are aligned parallel to the Indian government's "Pradhan Mantri Awas Yojna". The development of an economical and eco-friendly R-C building by providing a limited area of shear walls at the diagonal corners of the structure is evaluated in STAAD-PRO and ETAB in this paper. This work also focuses on the location and minimal area of shear walls along with maximum strength, and its 3-D analysis gives conclusions. The paper concludes that the design of 20-story buildings in STAAD and ETAB with minimum shear walls gives safe results, and Horizontal forces increase the structure's strength and stiffness requirements.

Keywords: ETAB, STAAD-PRO, Multi-storey building, Shear wall, Stiffness, Strength

1. INTRODUCTION

To make the buildings robust enough to support the stresses exerted on them, shear walls are used in the construction of many high-rise buildings in India [1], [3]. The seismic loads acting on the building's shear wall are influenced by its size and form. Given the design of the building, a critical component is a seismic analysis, for which shear walls are used [2], [5]. For an elevated 20-story building, shear walls are included in the building's diagonal corners only, so that the cost of the building could be reduced and we could afford to make an affordable residential building.

2. METHODOLOGY AND ANALYSIS

The most vital thing to keep in mind is to emphasise accurately calculating the seismic load and select the best techniques for giving the building the requisite pliancy [7], [4]. The seismological burden on the structure is described as the arbitrary changes in bearing, extent, lengths, and organisation as well as increasing velocities and speed when the structure is significantly impacted by earthquakes [8]. Based on external activity and the behaviour of the structural materials, the examination can be conducted. [9]. In such a scenario, a comparison of the shear walls on the structure's diagonal corners in the STAAD and ETAB is required to maintain the structure's strength and make it more cost-effective.

2.1 Equivalent Lateral Force Analysis

The most of lateral loads are active loads, with the horizontal force acting on the structure as their main component. Examples of typical horizontal loads include an earthquake, a wind load against a facade, etc. The static or similar static technique is one way to find design lateral pressures in buildings. This approach is the simplest; it requires fewer computations and focuses on the analysis statistics provided in the instructional code. Dynamic inspection is not necessary, though. The shear force is initially required to register for the entire structure, and this is relevant to the entire building [10].

2.2 Loading, Analysis, & Design

IS 875 Part 3 and IS 1893, respectively, evaluate wind load and earthquake load in both the X and Y directions. Post conclusion of modelling, ETABS, and STAAD produce and disseminate code-compliant stacking. The government can designate a varied number of load cases and combinations based on schemes. Elements and frameworks will be automatically sized by the design highlights for configuration layout strengthening schemes, and the government can modify the structure as desired for enrolment measures [11]. We may contribute to affordable housing by aligning with the government's ongoing schemes. The important question, though, is which software generates the best outcomes and why. As a consequence, as shown in Fig. 1, both software and hardware were utilized for analysis and design.

2.3 Structural Configuration

As demonstrated in Tables 1, these models include geometric information as well as the types of loads used for both ETAB and STAAD software, respectively. Case (1.1) was created in ETAB, whereas Case (1.2) was created in STAAD. As seen in Fig. 1, shear walls are installed at diagonal corners. As a result, no opening has been created because we are highly interested in viewing the comparison outcomes supplied by both the software under the specified circumstances.

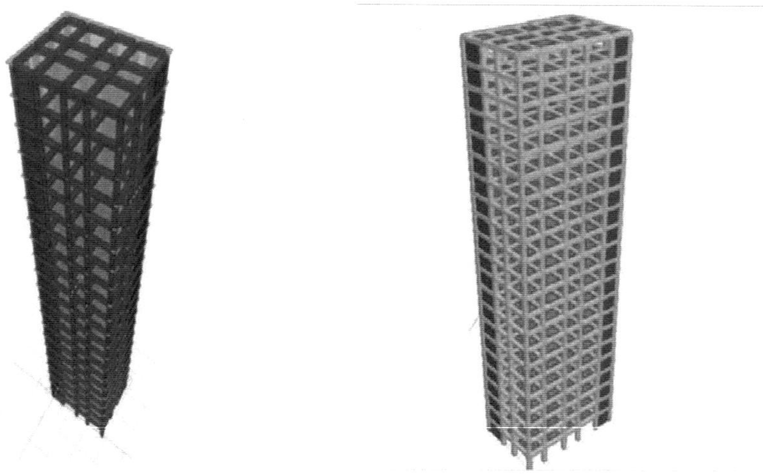

Fig.1. The three-dimensional design for case-1: (ETAB) and case 2 (STAAD)

2.3.1 ETAB Loading Diagram (Earthquake Loading and Wind Loading)

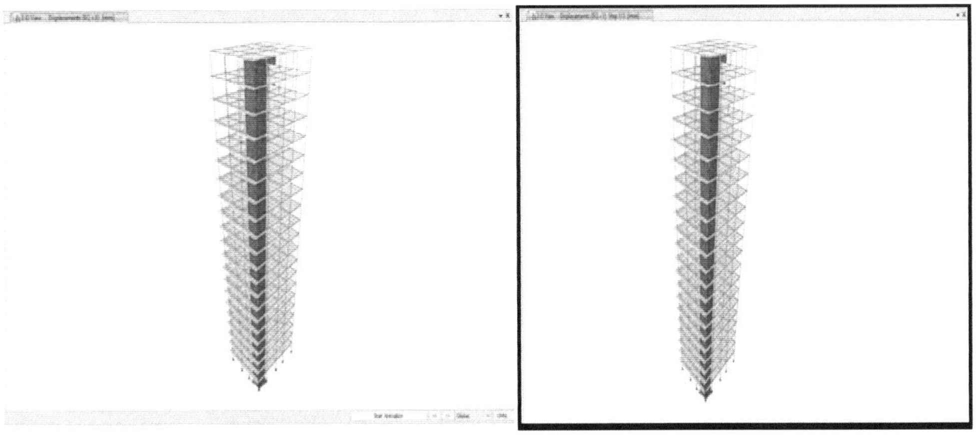

Fig.2. Direction of seismic loading in X and Z

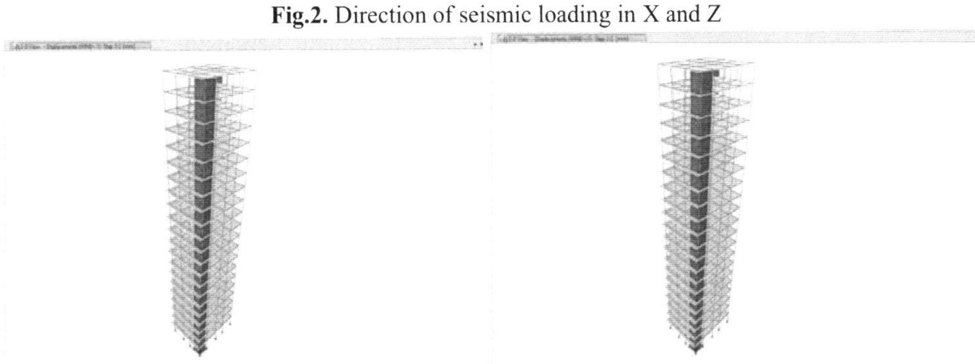

Fig. 3. Directions X and Z for lateral loads

IOP Conf. Series: Materials Science and Engineering 1282 (2023) 012009 doi:10.1088/1757-899X/1282/1/012009

2.3.2 STAAD Loading Diagram (Earthquake Loading, Wind Loading)

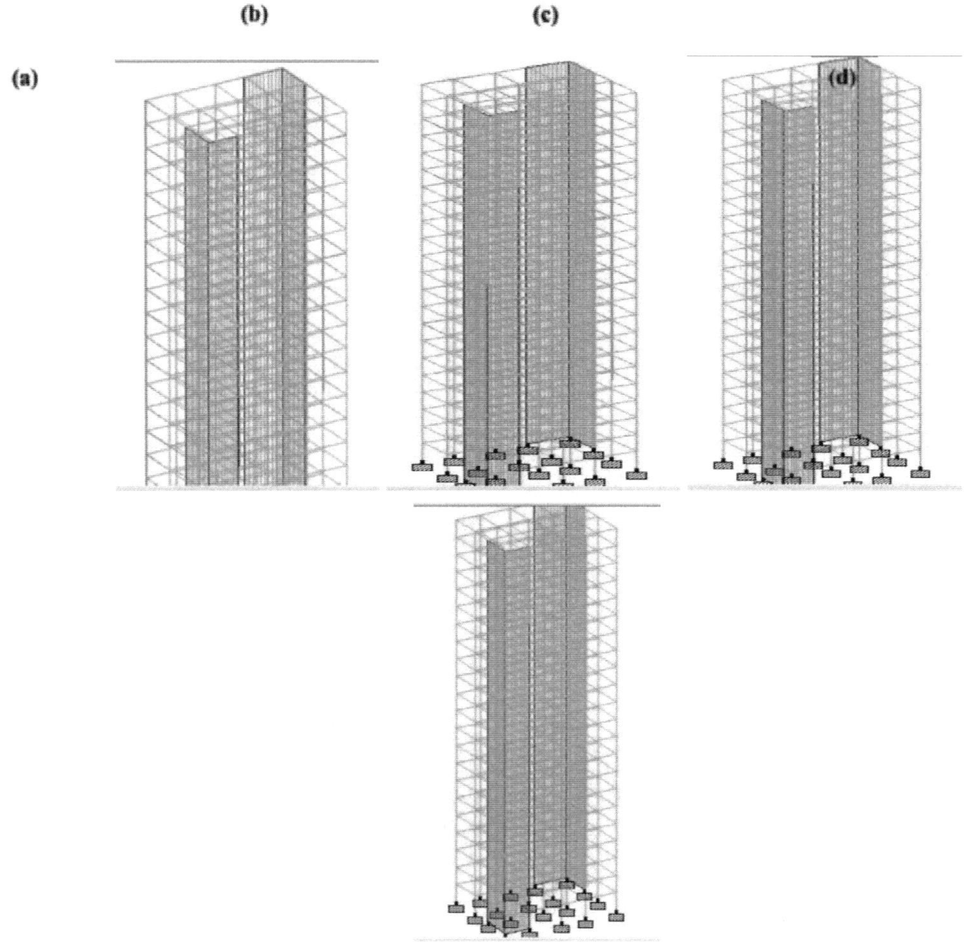

Fig. 4. Direction of seismic loading in (a) X, (b) Z-direction; and lateral loading (c) X, (d) Z-direction

Table.1 Detailed specifications of building

Details		**G+20 storey Residential building**
Beam member size	:	230*600 mm
Column member size	:	300*800 mm
Beam Nominal cover	:	60 mm
Column Nominal cover	:	40 mm
Roof thickness	:	140 mm
Classic storey height	:	3m
Imposed load	:	2KN/m2
Floor covering	:	0.701KN/m2
Roof covering	:	0.701KN/m2
Size of the shear wall	:	300 mm (M25)
Wind direction	:	X and Y
Wind speed	:	47 m/sec
Terrain category	:	4
Risk coefficient factor	:	1
Topography	:	1
Earthquake Direction	:	X and Y

Seismic Zone	:	zone IV
Soil type	:	Type II (Medium soil)
Materials	:	M25
Importance factor, I	:	1

3. RESULTS AND DISCUSSION

IS 1893:2016 Clause 7.8 deals with the calculation of drifting narratives and shifts in seismic structures, and IS 875, part 3. Clause 5.3 deals with the design wind speed (Vz) calculations. The frameworks of both the NDC and the LDC are subjected to a dynamic inspection under this clause [14]. The Response Spectrum approach is used to dissect the structure, and Clause is used to calculate the plan shear stresses. No. 7.11.2 of IS 1893 (Part 1): 2016 [12]. When we developed the model in ETAB and STAAD, we used parameters such as Zone factor 0.24, importance factor 1, response reduction factor 3, seismic zone "ZONE IV,"and basic wind speed "47 m/sec." [6]. For the purposes of dynamic analysis of reinforced concrete buildings, the value of damping for dwellings may very well be taken as a percentage of the critical, respectively, from Clause 7.8.2 of IS 1893 (Part 1). The data has been kept constant throughout both programs. Even though the section of shear walls delivered is smaller than the maximum previously specified, the software passed the loading condition fairly. After thoroughly analyzing it, we discovered that story drift grows progressively from the first to the twentieth story, resulting in a 0.33 mm X-direction deflection and a 0.50 mm Y-direction deflection. The maximum story drift that is attributable to the design lateral force must not exceed 0.004 times the height of the story, per to 1893:2002, section 7.11.1. [13]. The maximum permissible story drift is 0.004 x 63 m, or 252 mm. As a result, these needed parameters may be aligned for the urban housing project " Pradhan Mantri Awas Yojana (Urban), which aims to give the urban poor a better quality of life at a reasonable price.

4. CONCLUSIONS

1. Providing shear walls at the frame's diagonal corners yields satisfactory results in terms of construction parameters as well as economical parameters.
2. When shear walls are used in the construction, the story displacements are reduced.
3. Shear wall frameworks have higher story stiffness than non-shear wall structures.
4. Lateral resisting constructions generally have lower narrative drift than non-shear wall structures.
5. The strength and stiffness of the structure can be improved by the presence of lateral load resisting constructions.
6. In seismic activity areas, shear walls are advised due to their higher resistance to seismic stresses.
7. Shear walls are recommended in high wind situations because to their greater resistance to wind stresses.
8. We are capable of designing a specific structure with the above-mentioned specific detailing for the Pradhan Mantri Awas Yojna (urban)

5. Competing Interests Declaration

According to the author, there are no known financial or personal conflicts of interest that might have appeared to have an impact on the research presented in this study.

References

[1] Suruchi Mishra and Rizwanullah. Comparative Analysis of Regular and irregular Buildings with and without Shear wall, Int. J. Res. Sci. Innov. (IJRSI), 2017,4, pp 56-60.
[2] Kiran Tidke, Rahul Patil and Dr. G.R.Gandhe. Seismic Analysis of building with and without Shear wall. Int. J. Innov. Res. Sci. Eng. Technol., 2016,10,pp 2319- 2323
[3] V.Abhinav, Dr.Sreenatha Reddy, M. Vasudeva Naidu and S. Madan Mohan. Seismic Analysis of Multi Story RC Building with Shear Wall Using STAAD PRO, Int. J. Innov. Technol. Res., 2016,4, pp 3776-3779.
[4] A.P. Nagendra babu and SK Jain Shahab. Comparative Study of Shear Wall in Multi-Stored R.C Building, Int. J. Mag. Eng. Technol. Manag. Res. 2017, 6, pp 2348-2351

[5] Sonali Pandey, Dr. Krishna Murari, Ashish Pathak and Chandan Kumar.A Review on Shear wall in High Rise Building, Int. J. Eng. Inv., 2017, 19, pp 2278-2281.
[6] P. Kalpana, R.D. Prasad, B. Kranthi Kumar, Analysis of Building with andwithout Shear Wall at various Heights and Variation of Zone III and Zone V, Int.J. Eng. Res. Appl. 8 (2016) 631–635.
[7] Khushboo K. Soni and Dr. Prakash S. Pajgade. Design of Multi storied Regular R.C Buildings With and Without Shear Wall. Int. J. Adv. Eng. Res. Dev., 2015, 2, pp 777-781. [8] P.P. Chandurkar, P.S. Pajgade, Seismic Analysis of R.C.C Building with and without Shear Wall, Int. J. Modern Eng. Res. 3 (2013) 231–235.
[9] Masato sakurai, Hiroshi kuramoto, Tomoya matsui and Tomofusa akita. The 14th world conference on earthquake engineering, 2008,7, pp 538-542.
[10] Swamy nadh Vandanapu, K.Muthumani. Properties of structural lightweight concrete under treated oil palm shells as coarse aggregate in concrete, Asian J.
Civ. Eng., 2018,19, pp 673-678
[11] Bhruguli H. Gandhi. Effect of opening on behaviour of shear wall, Constr. Eng. Manag., 2015, 3, pp 655-659.
[12] Ashok Kankuntla, Prakarsh Sangave, Reshma Chavan. Effects of opening in shear wall. IOSR J. Mech. Civil Eng. (IOSR-JMCE), 2016,13, 2256-2262.
[13] Alhamd Farqaleet1 1 Student of M.tech (Earthquake Engineering), Jamia Millia Islamia (A Central University), New Delhi
[14] V. Naresh Kumar Varma and U. Praveen Kumar, Seismic response on multi-storied building having shear walls with and without openings, Materials Today: Proceedings, https://doi.org/10.1016/j.matpr.2020.05.827

Investigation on Mechanical Properties of Concrete by replacing palm Ash and fine seashell powder as PartialCement replacement

Janat Almaamari[1] , Rahma Al Fahdi[2] , Fatma alzahraa Al Jabri[3] and Sachin Kuckian[4]
[1]Student, Middle east college, Muscat, Oman
[2] Student, Middle east college, Muscat, Oman
[3]Student, Middle east college, Muscat, Oman[4]Lecturer, Middle east college, Muscat, OmanE-mail:19f19174@mec.edu.om 19F19086@mec.edu.om 19F19180@mec.edu.om skuckian@mec.edu.om

Abstract. Manufacturingcement causes harm to the environment due to Co2 emissions. CO2 emission by the industry could be reduced by recycling natural waste materials such as palm ash and seashell powder; since seashell has 95% of calcium carbonate and palm ash can improve the mechanical properties. 15 cube, 3 beams were subjected to several tests to find the mechanical characteristics in a certain period. The tests were compression strength test, flexural test , workability, SEM tests, and sieving analysis. Thebeam dimensions measurements are 150mm*150mm*650mm, and 150mm*150mm*150mm for cubes.OPC was the main finding of compression strength which gained the highest strength, but in 28 days 25% of cement replacement gained 31.7Mpa when compared with other specimens.On the other hand, OPC and 5% of partial cement replacement were classified as high workability. In addition, flexural strength test showed that 15% of cement replacement achieved the highest strength compared with other specimens. SEM test showed the surface texture of PA and SP affect the hydration,therefore it will affect the strength of concrete.

Keywords: Palm ash, Seashell powder, SEM Test, Compressive test, Partial Replacement, waste material, flexural test, sieving analysis.

1. Introduction

Concrete is commonly used in the construction industry because of it is cost effective. Cement has a high value on construction, therefore; the production of cement has increased which causes environmental effects.This experiment is necessary because worldwide countries are facing environmental issues that impact human life such as Co2 rise in the atmosphere. Previous research has proven that optimum percentage of replacing cement with palm ash is 10% and 5% of seashell powder. On the other hand, seashell powder reduces the permeability of concrete and it's a cost-effective concrete[1], and seashell also has a similar percentage of calcium carbonate as limestone powder and so it can be replaced with cement to minimize Co2 emission [2]. There are about 9million date palm trees in Oman that produce yearly around 83.2% of palm leaves waste. As a solution, the study suggests replacing cement with alternative waste materials that have similar chemical and mechanical compositions such as date palm ash (PA) and fine seashell powder (SA) with percentages of (5,15 & 25%). Thus, the chemical composition of PA and SA have calcite and silica that provides concrete with workability, flexure, and compressive strength. As alternative to that,the study recommends replacing concrete with palm ash and seashell which will benefit the construction industry by improving workability and strength without the massive CO2 emission. Hence, the use of palm ash and seashell powder has an advantage on both the economy and the environment.

This research aims to improve the mechanical properties of concrete through waste materials and tominimize cement costs by

- Examining the chemical composition of the palm ash and seashell fine powder.
- Evaluating the mechanical properties with partial cement replacement of palm ash and seashellfine powder.
- Finding the optimum percentage for seashell and palm ash replacement.

2. Literature Review

Previous researchers have proven that cement production causes Co2 emissions to the environment[3]. Thus, the use of alternative cement by recycling natural waste such as seashell that has huge amount of calcium carbonate willstrengthen concretemix as pointed out byBhagyasree&Santhakumar[3]. However, the test that has been used was compressive strength test and workability test with a grading of M35. The results of seashell powder showed that 7 to 28 days of curing has increased in the 20% within the specimen and decreased in 30%, 25%[4].

According to Al-Kutti[5], there are other waste materials that can be manufactured with cement , such as palm ash, by studying properties of cement mortar and specimens which have compressive strength and chemical composition analysis. The palm ash was sieved with 425 µm in the percentage of 20%,30%, and 10% as a partial replacement for 3,7,14,28days within 55mm cube and 0.4W/C ratio. The obtained results were palm ash with 35.93% of $SiO2$, and OPC had 63%of CaO. On the other hand, PA obtained 56.8MPa of strength compared with another specimen such as OPC 27.5 MPa.

Ofuyatan et al.[6] mentions that objective of the paper was to obtain mechanical properties of the partial replacement of cement with palm ash and seashell powder, as they conduct workability test and compressive strength for 7 and 28 days. Although, the results showed that in workability test, as the PA and SP increases, the workability decreases. In compressive strength, the relation showed that the replacement percentage of 10 and 50% cause increase in strength.

Soltanzadeh et al., [7],in this research paper, seashell is used as areplacement of cement according to these percentages(5, 6, 7, 8, 9, 10, 15, 20, and 30%) to study thechemical properties, SEM and flexural strength within 3,7,and 28 days. As results showed, 8% of replacement was the optimum percentage of seashell in flexural strength. SEM test showed that it has similar chemical composition of sandstone which can be replaced with limestone.

3. Methodology& materials

3.1 Material Elaboration

3.1.1 Seashell Powder

Seashell waste of about 4Kg was collected from Wadi-Shab, Masirah Island, and Manoma beach. After that, seashells were cleaned and dried, then heated to 1640 Celsius for 7 hours as well as crushedLos-Angeles machines and sieved with 90 µm.

Figure 1. Sieving seashell powder.

3.1.2 Date Palm ash

The date palm was gathered from farms of Barka and Suwaq in Oman, after that, it was burned into ash and sieved with 90 μm, moreover; around 6Kg of the ash was collected.

Figure 2. Date palm leave.

Figure 3. Sieving palm ash

3.2 Mixture Design

Volume=L*W*D, Concrete mass=(density of concrete*volume)+[(density of concrete*volume)×15%], Cement and fine aggregate=1/4×Concrete mass, Coarse Aggregate=1/2×Concrete mass. Though the dimensions of cube molds are 150mm*150mm*150mm, and beams are 150mm*150mm*650mm. The mixture has been prepared manually by tray and trowel within 12 cube mold and 3 beams including control specimen. Two water-cement ratios were chosenas 0.5 for cubes and 0.6 for beams according IS 10262:2009, in grade of M25 (refer. Table 1.). The following symbol SP represent Seashell powder and PA as Date palm ash that will be mentioned in paragraphs and figures. This experimental mixture proportions were cured for 7, 14, and 28 days. In addition, concrete density was taken as a standard around 2400Kg/m^3.

Table 1. Mixed proportion.

Specmen	Cement Type	W/C	Concert Density-Kg/m³	Density of water-L	Mass of concrete -Kg	SP %	PA %	Mix proportion	Weight of Seashell powder (Kg)	Weight of Palm ash (Kg)
Cube 1	OPC	0.5	2400	1.18	9.5	------	-------	1:1:2	------	------
Cube 2						2.5 %	2.5%		0.06	0.06
Cube 3						7.5 %	7.5%		0.178	0.178
Cube 4						12.5 %	12.5 %		0.29	0.29
Beam 1		0.6		6.054	40.365	-----	------		-----	------
Beam 2						2.5 %	2.5%		0.252	0.252
Beam 3						7.5 %	7.5%		0.756	0.756

Note. Table 1 shows the proportion of the mix for cube and beams specimen according to British standards, that have fine aggregate, coarse aggregate water and cement including samples of palm ash and seashell powder.

3.3 Conducted Test

3.3.1 Workability Test

When the mixture was done, a slump test was conducted to determine the workability by using slump cone, by adding 5cm for each layer and tamping it 25 times to distribute the mixture evenly within the cone, and after that the cone was pulled and measured. In-fact, there are three types of concrete slump -true, shear and collapse slumps which can affect the workability according to water ratio in cement. After the cone test, the cement paste was transmitted to compressive and flexural test

IOP Conf. Series: Materials Science and Engineering 1282 (2023) 012010 doi:10.1088/1757-899X/1282/1/012010

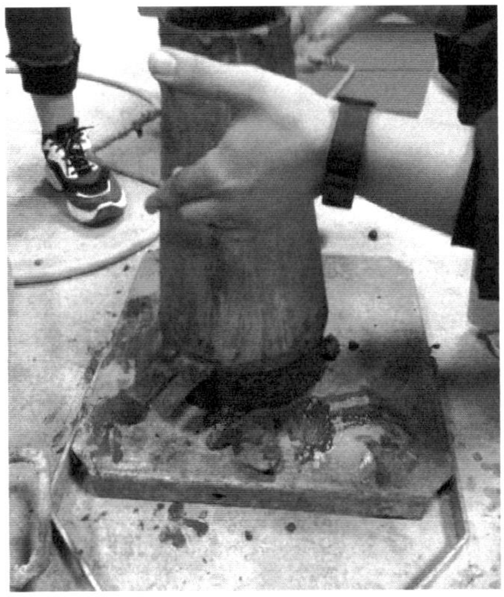

Figure 4. Slump test.

3.3.2 Compressive Test

0The Cement paste that was used in cone test was casted in 150mm cube and shaken for 30 times. Specimens were demolded after 24hours.

Figure 5. Compression Test.

IOP Conf. Series: Materials Science and Engineering 1282 (2023) 012010 doi:10.1088/1757-899X/1282/1/012010

3.3.3 Flexural Test

The flexural test proportions were 600mm*150mm*150mm which were considered to determine the tensile strength of concrete beams.

Figure 6. Flexural strength test.

3.3.4 Scanning electron microscope (SEM) inspection

This test was conducted in ALS Geochemistry to indicate chemical composition and surface for both samples by using Energy Dispersive X-Ray analysis (EDX) that is related to SEM. Using EXD has an advantage , as it can identify quality control, and the source of issues in the chain process[8].

3.3.5 Sieving analysis

This is conducted by arranging a set of sieve in the order of 4mm to 63micrometer according to the British standards, and then it was shaken for ten minutes by using a sieve shaker.

Figure 7. Sieving Analysis.

4. Results And Discussion

4.1 *Workability Test*

The following standard is European with rating of 175mm preferably, and 0.5 W/C ratio. However, table 2 shows that OPC has the highest workability by 134mm followed by 5% PA+SP with 115mm , with 15% PA+SP obtaining 90mm and 25% mixture with 62mm which is the lowest workability that was obtained. Though, the observed workability range was within high and medium [9] and according to the previous study, there is inverse relation between PA and SP as a Partial replacement of cement, and it can be observed in graph 1. The application of OPC, 5% and 15% replacement mixture can be used as slab, beam, columns and retaining walls, while 25% cement replacement can be utilized as bridge deck. Although there are factors that affect the workability as W/C ratio, size, and shape of aggregate, and mix proportion [10].

Table 2. Workability Results.

Specimen	PA%	SP%	Slump test values (mm)
OPC	0%	0%	134
5% PA+SP	2.5%	2.5%	115
15% PA+SP	7.5%	7.5%	90
25% PA+SP	12.5%	12.5%	62

Note. The main findings of this table indicate the workability decrease while increasing the percentageof PA&SP as a partial replacement of cement.

Graph 1. Relation between palm ash(PA) and seashell powder (SP) as a partial replacement ofcement.

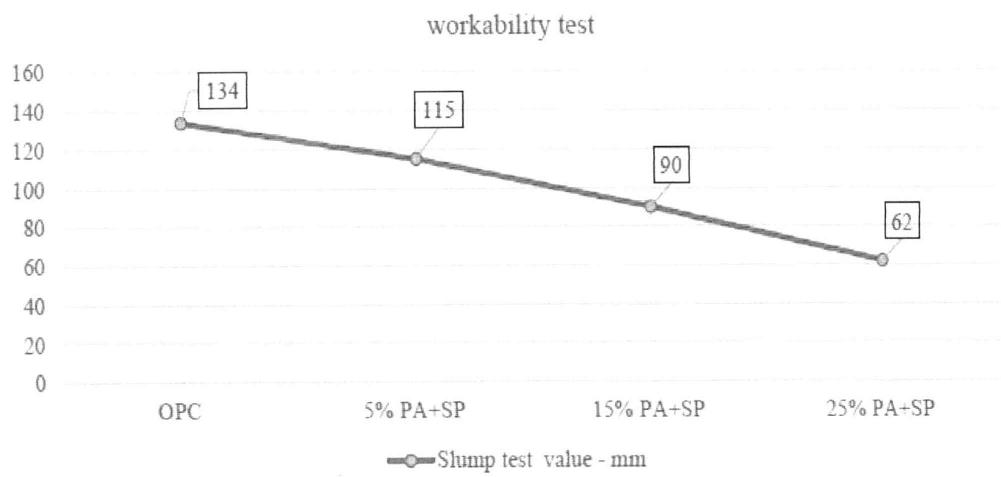

Note. There is an inverse relation between workability and cement partial replacement.

4.2 *Compression Test*

Table 3 shows the obtained result of compressive strength from 7 to 28 days of curing as the replacement percentage are 5%, 15%, and 25% which are equally distributed (12.5 palm ash and 12.5 seashell powder). As OPC gained 35.3MPa in 28days which was the highest value compared to 14 and7 days. 5% specimen was obtained 29.57MPa in 28 days which is the highest strength compared by 15%

specimen, although 25% showed 31.7Mpa in 28 days which was the highest strength compared with the partial replaced specimen. It was observed that there isa relation between strength and age of concrete which increased by 65% in 7 days, within 14 days the strength reached 90% and in 28 days the percentage of strength reached to the highest and the maximum strength which is 99% [11].

As Ofuyatan [6]claims the optimum percentage of palm ash is 10% , seashell powder is 4%[12] Our investigation showed that there is an inverse relation in palm ash and seashell powder when it mixes. According to Mishra [13] there are factors that may affect results such as air bubbles, water ratio, temperature and humidity of cube concrete. In table 4 the results of strength in palm ash was 33.24MPa, following seashell with 29.73MPa, and OPC with 24.5MPa; according to the findings, it was observed that palm ash has the highest strength compared with other specimens because it has the ability to produce C-S-H [14]. It was expected for seashell to have high silica content to form cementitious material [15]. As well as the characteristic of mortars is strength because it fills the void between sand particles, which will minimize the cracks by shrinking them[16].

Table 3. Compressive strength results for mortar at 7 days.

Percentage	Palm ash percentage (PA)	Seashell Percentage (SP)	Compressive Strength - MPa		
			7 days	14 days	28 days
0%	0%	0%	31.3	31.58	35.3
5%	2.5%	2.5%	17.9	27.3	29.57
15%	7.5%	7.5%	20.03	22.3	25.76
25%	12.5%	12.5%	16	18.4	31.7

Note. OPC has achieved the highest value in strength compared with other samples through the experimental period. However, there was an unexpected result in 25% replacement of cement in 28 days which is the highest value compared with the other replacement.

Table 4. Compressive strength results for mortar at 7 days.

Note. It was observed through the table that palm ash by itself has gained more strength than OPC and seashell replacement.

Sample	Added percentage of PA	Added percentage of SP	Compressive Strength-MPa
			7 days
1	0%	0%	24.5
2	2.5%	0%	33.24
3	0%	2.5	29.79

4.3 Flexural test:

The observed table 5 shows flexural strength in 7 days as OPC gained the lowest strength compared with the replacement specimen of 5% and 15% as the results showed 8 MPa , 12,56 MPa , and 13.19MPa respectively. According to Shahrior,[17] flexural strength represents 10 to 20% of compressive strength. As shown in the figure, the flexural strength of beams in 7 days indicates the relation between flexural strength and the added percentage of PA and SP replacement

Table 5. Flexural Strength Results for Beams In 7 Days.

Specimen percentage	palm ash - %	seashell powder - %	Flexural strength - MPa
			7 days
OPC	0%	0%	8
5% of SP+PA	2.5%	2.5%	12.56
15% of SP+PA	7.5%	7.5%	13.19

Graph 2. Flexural strength results of OPC and the replacement.

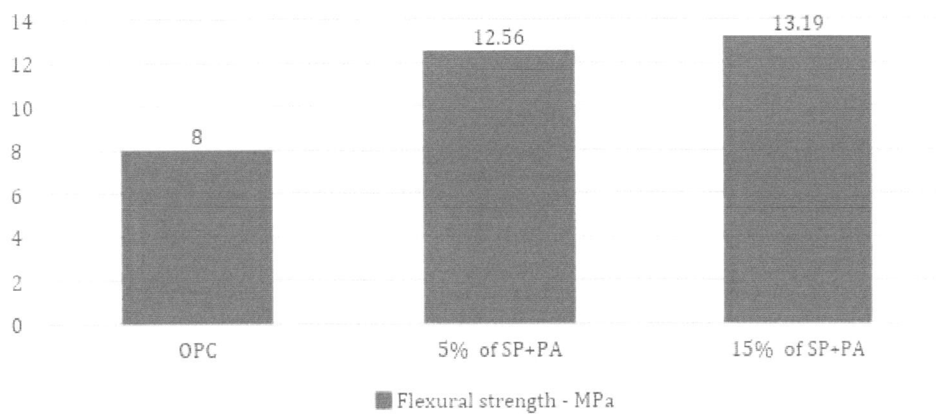

4.4 Scanning electron microscope (SEM) inspection

Table5 shows that the main chemical component of OPC are SiO_2, Al_2O_3 that have an effect on strength of the mixture [18].According to the obtained result, it was observed that PA gained more silicates 34.70%, 22.4% in silica, 24.43% in calcium carbonate; which led to an increase in setting time. Meanwhile, SP showed unexpected results as calcium carbonate gained 99.8% and silica was less than 0.01% due to exposing seashell in high temperature oven resulting in decreased strength.

As it is observed in figure 8, the texture of palm ash is smooth which indicates hydration that affects setting time and strength. While seashell showed rough texture as shown in fig. 9; when it is supposed to be smooth with less hydration.

Table 6. Chemical composition of OPC, Palm ash (PA), and seashell powder (SP).

Compound	Al_2O_3	$CaCO_3$	Fe_2O_3	SiO_2	SO_3	Mg	$CaCO_3$	$CaSO_4$	SiO4	NaCl	Cl	Na	FeS
OPC	3	60	0.5	17	1	0.1	------	------	------	------	-----	-----	-----
PA	------	------	------	22.40	----	8.93	24.43	3.87	34.70	2.38	2.97	-----	0.31
SP	------	------	------	0.01	----	0.06	99.8	------	------	------	-----	0.02	-----

Note.SiO_2, $Al_2O_3$are the main components that the effect strength. We noted that palm ash gainedSiO_2more than OPC which effects in strength during testing the mortar, also $CaCO_3$caused hydration leading to increase in the setting time.

Figure 8. SEM test on palm ash.

Figure 9. SEM test on Seashell Powder.

4.5 Sieving analysis

Table 6 shows the results of sieve analysis as passing percentage in fine aggregate, though the total weight of used fine aggregate is 1 Kg. There are four categorized zones for fine aggregate according to the passing percentage. zone one and two can be used with a concrete mix, zone three as a plaster, and zone four is rejected. Based on that 600 micro is in zone two with a passing percentage is 44.4, and300 micros consider as zone 1 since the passing percentage is 16.1, and the rest of sieving are rejected.

Table 7. Sieving analysis result for fine aggregate.

Sieving size	Wight Retained (gm)	Cumulative Weight Retained – gm	Cumulative percentage Retained	Passing Percentage
10mm	0	0	0	100
4.75mm	24	24	2.4	97.6
2.36mm	34	59	5.9	94.1
1.8mm	116	176	17.6	82.4
600micro	379	556	55.6	44.4
300micro	282	839	83.9	16.1
150micro	146	986	98.6	1.4
Pan	12	1000	100	0
		Total 1000		

Graph 3. Sieving Analysis Chart

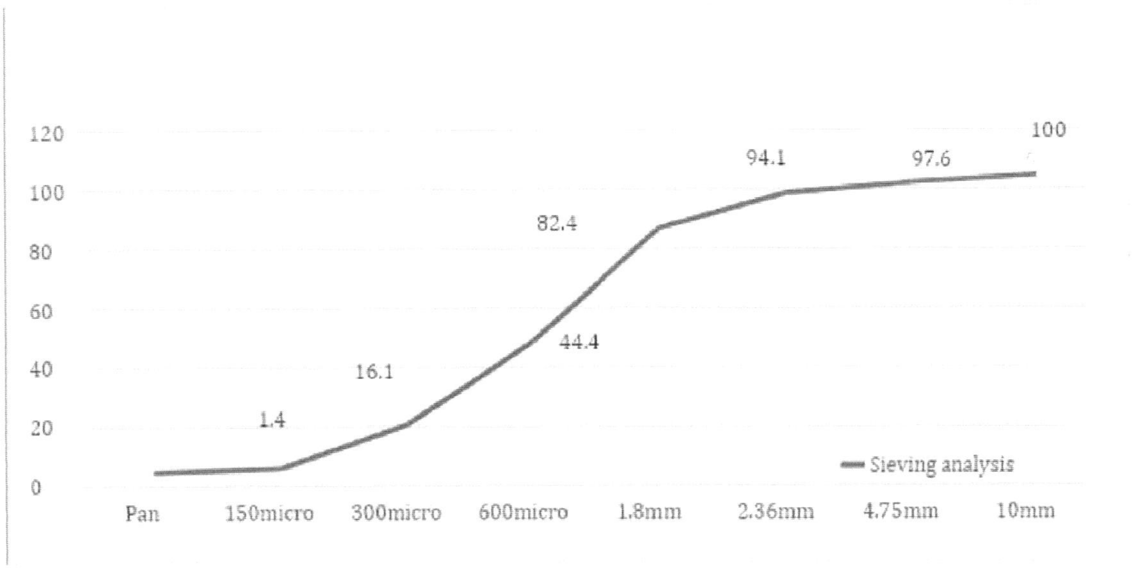

5. Conclusion

This article focuses on the behavior of palm and seashell powder as a partial cement replacement in term of mechanical and chemical properties. Following previous works it was found that palm ash(PA) and seashell powder (SP) as a partial replacement of cement workability decreases, compressive strength increases in 14 days curing, flexural strength decreases, according to Kuckian et al & Aaqib Dalvi [19] the experiment this could be used in interlocks pavers as the optimum percentage was found to be 15%.

As compared to 15 % of the seashell and palm powder similar results were obtained.

The main factors that effect on the results is the surface of palm ash and seashell powder, also during laboratory experiments such as humidity, air bubbles and temperature.

Acknowledgment
We appreciate lab staff and academic department at middle east college for giving us access to material and gaudiness also Albarrak laboratory for supporting us during this research.

References

[1] Tayeh, B. A., Hasaniyah, M. W., Zeyad, A., & Yusuf, M. O. (2019). Properties of concrete containingrecycled seashells as cement partial replacement: A review. *Journal of Cleaner Production*, **237**, 117723.

[2] Solomon, D., & Qin, M. (2007). *Overview of Greenhouse Gases | US EPA*. US EPA. https://www.epa.gov/ghgemissions/overview-greenhouse-gases.

*[3]*Dilna R D, BhagyasreeRaj and Dr. Anjana Santhakumar, 2020. Partial replacement of cement by glass andseashellpowder in concrete.*International Research Journal of Engineering and Technology, 7(6), pp.6902-6906.*

[4] Senthil Nathan, 2018.Increasing the compressive in concrete with partial replacement in cement by seashell powder. *Journal Of Engineering And Applied Sciences,* **13**(9), pp. 7111-7112

[5] Al-Kutti, W., Saiful Islam, A., & Nasir, M , 2019. Potential use of date palm ash in cement-based materials. *Journal of King Saud University - Engineering Sciences*, ***31***(1), pp. 26–31

[6] Ofuyatan, O. M., Olutoge, F., Omole, D., &Babafemi, A. (2021). Influence of palm ash onpropertiesof light weight self-compacting concrete. *Cleaner Engineering and Technology*, ***4***, 100233.

[7] Soltanzadeh, F., Emam-Jomeh, M., Edalat-Behbahani, A., &Soltan-Zadeh, Z. (2018). Development and characterization of blended cements containing seashell powder. *Construction and Building Materials*, ***161*,** pp.292–304.

[8] Intertek Group, 2020. *Energy Dispersive X-Ray Analysis(EDX)*.https://www.intertek.com/analysis/microscopy/edx/

[9] Backus , 2017. *What is workability of concrete?,* Global Gilson. https://www.globalgilson.com/blog/what-is-workability-of-concrete

[10] Mahajan, B. (2022). *Workability of concrete | factors affecting workability of concrete.* Civiconcepts. https://civiconcepts.com/blog/workability-of-concrete

[11] lceted, 2021. *Why do we choose 7, 14 and 28 days for concrete or compression tests?* LCETED INSTITUTE FOR CIVIL ENGINEERS.

[12] Wan Ahmad Soffian Bin Wan Mohammad et al 2017 IOP Conf. Ser.: Mater. Sci. Eng. **271** 012059

[13] Mishra , 2013. *Factors Affecting Strength of Concrete*. The Constructor.

[14] Al-Kutti, W., Nasir, M., Megat Johari, M. A., Islam, A. B. M. S., Manda, A. A., &Blaisi, N. I.(2018). An overview and experimental study on hybrid binders containing date palm ash, fly ash, OPC and activator composites. *Construction and Building Materials*, ***159***, pp.567–577

[15] Kim, T., &Olek, J., 2012. Influence of alkali silica reaction on the chemistry of pore solutions in mortars with and without lithium ions. *Brittle Matrix Composites* ***10***, pp. 11–20.

[16] Borlongan, J. (2011, June 18). *Properties of cement mortar.* eHow.https://www.ehow.com/info_8613013_properties-cement-mortar.html

[17] ShahriorAlam, (2022). Is there any relation between compressive strength and flexural strength forRHA concrete?. Research gate. https://www.researchgate.net/post/Is-there-any-relation-between-compressive-strength-and-flexural- strength-for-RHA-concrete

[18] Rahman, F. U. (2018, September 16). *Ordinary Portland Cement -Constituents, Properties, Typesand Uses*. The Constructor. https://theconstructor.org/concrete/ordinary-portland-cement/23181/

[19] Kuckian, S., & Dalvi, A. (2020). Strength Possibilities on Fly ash Based Interlock Pavers. In Applied Mechanics and Materials (Vol. 897, pp. 12-19). Trans Tech Publications Ltd.

Production and characteristics of stabilised interlocking mud blocks

H.G. Vivek Prasad[1], Mangala Keshava[2], P.S. Raghu Prasad[3], K.S. Jagadish[4]
[1,3]Dept. of Construction Technology and Management, Sri JayachamarajendraCollege of Engineering, JSSSTU, Mysuru, Karnataka
[2]Dept. of Civil Engineering, BMSCE, Bengaluru, Karnataka
[4]Retd. Professor, Dept. of Civil Engineering, IIsc, Bengaluru, Karnataka
Email: vivekprasad22@gmail.com

Abstract: This paper presents an ideology over physical and mechanical properties of interlocking and non interlocking stabilised mud blocks. These properties are useful to assess the performance and behaviour of load bearing masonry structures. This paper also compares the results of both non interlocking solid stabilised mud blocks (NSMB) of size 230 X 190 X 100 mm and interlocking solid stabilised mud blocks (ISMB) of sizes 230 X 190 X 80 and 230 X 190 X 70 mm. These blocks are tested for block density, Initial rate of water absorption, water absorption, wet compressive strength, flexural strength and Modulus of Elasticity. From the results obtained it has been found that shape and depth of interlocking has significant effect on properties of blocks. Flexural strength of 20mm (ISMB 20) and 30mm (ISMB 30) depth interlocking solid stabilised mud blocks produced are very much comparable with NSMB. Flexural strength of ISMB's has been tested in both the directions of interlocking effects of blocks which has given satisfactory results. This test shows that Shape considered for interlocking blocks do bond well with filler materials. Mechanical and physical properties of both ISMB and NSMB have exceeded the minimum requirements of blocks as per codal provisions.
Keywords: Masonry, Interlocking solid stabilised mud blocks, physical and mechanical properties, production

1. Introduction

Energy, Economy and Environment are the major areas to be carefully utilized by any nation. Increase in use of existing available natural materials leads to higher levels of pollution and ultimately results in exhaustion of these materials and also leading to many natural disasters [4]. At the same time natural resources need to be safeguarded and reusable materials to be produced at affordable cost to sustain economic growth. Hence it is high time to shift focus on usage of less or no carbon emitting materials for a sustainable growth [1]. A simple solution to cater the problems of Energy, Economy and Environment is to depend more on locally available materials to satisfy needs which are being used abundantly[6][8]. Even though bricks used for construction purpose is locally produced energy involved in production and transportation is high. Hence there is a need for sustainable alternative solution.

Human beings have used mud to build structures since the days of Egyptian and Mesopotamian cultures. Earliest examples of Soil-cement housing in India can be seen at Karnal in today's Haryana in year 1948. However there has been a drastic increase in construction of dwellings using soil-cement blocks in Bengaluru since 1988. [2][4]As compaction of soil improves its strength properties, utilisation of pressed soil block in construction of masonry structures has gained popularity. Development of pressed soil block construction in India advanced rapidly due to the research at CST (Centre for Sustainable Technologies) formerly ASTRA, in Indian Institute of Science[5]. A new machine was developed by ASTRA in 1979 called "ASTRAM" which was lighter and more flexible than other presses. Astram has been utilising for producing soil-cement blocks during 1980-

85.[3][6][7]Later an improved machine called "ITGE VOTH" was developed. Later a further improved machine was developed called "MARDINI PRESS " for better,easier compaction and handling. This machine is now widely used at many places in India. Blocks produced using this technology have performed well during the recent past hence people started experimenting with different sizes and shapes in order to produce the most pleasing effects on dwellings constructed[10].

There are various types of interlocking stabilised mud blocks being produced in different parts of the globe. [12]Interlocking size and shape varies from region to region. These interlocking blocks can be classified into two type's one with circular hollow grooves interlocking effect blocks and the other solid interlocking effect blocks. Solid interlocking block being produced in the market has an interlocking effect between 5 to 10mm in both horizontal and vertical directions[11]. This paper presents studies on size of interlocking effect of two different dimensions one with 20mm and another with 30 mm.[9][13][5] Interlocking effect has been provided only in horizontal direction since blocks can be produced with available existing mould rather than preparing a new mould to produce these blocks.

1.1 Objectives of the study

In the present construction scenario there are various structures which are being built by using stabilised mud blocks. It is reported by many researchers that stabilized mud blocks are less energy intensive as they are made from locally available soils and a small portion of stabilizer with a simple machine that operates using manual labour. This aspect of sustainability can be further improved by producing interlocking stabilized mud blocks. However there is very little information about interlocking effects in solid stabilised mud blocks. Thus an attempt is made in this direction to evaluate the mechanical characteristics of solid stabilised mud blocks with and without interlocking effects.

1.2 Literature review

Considerable amounts of research have happened in usage of interlocking effect in concrete blocks and burnt clay bricks but there are few articles which are completely dedicated for interlocking in stabilised mud blocks. Further the studies on interlocking in stabilised mud block technology is focussed on having interlocking effects by providing hollow effects in blocks which can also be filled with grout and/or steel rebars. Some of the earlier studies on interlocking effects are discussed below.

[3] have studied factors affecting non uniform properties of interlocking stabilised compressed earth blocks. These interlocking blocks were produced with groves within it ie hollow effect. They have concluded that usage of these interlocking blocks have reduced strength parameters such as compression, density, water absorption, flexural strength, and splitting tensile strength tests. However these blocks have satisfied minimum requirements for load bearing masonry constructions.

[4] have studied on axial compressive behaviour of rubberized interlocking masonry walls with hollow effect type interlocking blocks using crumb rubber. Results have revealed that hollow and grouted specimens, posses' strength which is 15 - 20% lower when compared with conventional interlocking masonry walls. However these blocks are more ductile because of utilisation of crumb rubber in its production process.

[5] have worked on understanding the characteristics of hollow effect type dry-stack interlocking compressed earth blocks by conducting compression and indirect tensile strength test on blocks. Produced samples were tested at an age interval of 7days till 28th day in order to determine the compressive strength characteristics of soil through cylindrical blocks. Results have ranged between 1.20-1.39 MPa for 5%, 2.16- 2.67MPa for 10% cement stabilised samples. Indirect tensile strength varied between 0.4 to 0.75MPa.

[6] studied the flexural strength of different types in compressed stabilized earth - bricks, solid

blocks, interlocking solid and hollow blocks and rammed earth. A very low level of pre compression was introduced for determination of flexural strength through masonry wallets. They have concluded by stating that all the types of bricks, blocks and rammed earth can be considered as safe and viable alternative solution for load bearing and internal walls up to 2 storied houses.

[7] understood compressive strength, stress-strain relationships, and elastic properties of stabilised masonry blocks. 6, 8 & 12% cement by weight of mix was used for block production and various combinations of mortars. They have concluded that masonry modulus increases as the block strength increases. The block strength will be increased with increase in cement content.

It can be observed through these literatures that a considerable amount of research is on-going in production of interlocking SMB. Interlocking stabilised mud blocks with circular hollow grooves have reduced strength carrying capacity when compared with plain stabilised mud blocks. This paper focuses on overcoming this issue and mainly emphasizes on understanding the characteristics of both non interlocking and solid interlocking stabilised mud blocks.

2. Methodology

Present study has two parts, Initial part as production and second part as characterization of stabilized solid mud blocks with and without interlocking effects. In order to produce blocks, an interlocking effect of 20mm and 30mm has been considered. These blocks have extensions in bottom and openings at top. These extensions and openings produce interlocking effects in masonry walls when placed one above another.120 x 190 mm extensions and openings are provided for both types of interlocking blocks produced. Extensions and openings are tapered with 1:1.5 leading to 10mm for 20mm depth and 15mm for 30mm depth interlocking blocks. In order to understand the characteristics of produced solid stabilised mud blocks with and without interlocking effects tests such as dimensionality, density, wet compression, water absorption, Initial rate of water absorption, flexural strength and modulus of Elasticity have been studied. Since modulus of elasticity is a material property it has been conducted only on stabilised mud blocks without interlocking effects. However the remaining tests are conducted on both types of blocks

Soil used in this investigation was sourced from soil excavated for hostel building in JSS Science and technology university campus located in Mysuru, Karnataka. Soil mixture had 78% of sand and silt & clay constituted to 22%. Liquid and plastic limits of soil are 36% and 15.25% respectively. Using this soil along with 7% cement as stabiliser blocks have been produced using ITGE VOTH. Blocks were cured for 28 days under wet burlap. Fresh density of 2.05 g/cc, has been maintained throughout production process for all three types of solid stabilised mud blocks. Non interlocking stabilised mud blocks with size 230 X 190 X 100 mm, interlocking stabilised mud blocks with an overall size of 230 X 190 X 80 mm and 230 X 190 X 70mm have been produced. Reduction in size of interlocking blocks is because of extensions and openings which have been introduced for production of special types of blocks. Interlocking effects has been provided by using wooden moulds. Produced blocks are shown in Fig 2.1.

Fig 2.1 Blocks produced with and without interlocking effects

3. Results and discussions

3.1. Dimensionality test:

Dimensionality test has been carried out to determine average dimensions of blocks produced. Twenty blocks are randomly selected and laid on level ground. Blocks were arranged on a surface which is free from any unevenness. Blocks were laid lengthwise, width wise and depth wise. Total length of all the blocks was measured using a steel tape in one stretch and then average measurements were determined. Table 3.1 shows different dimensions of blocks arrangement and average dimensions of blocks.

Table 3.1 Dimensionality test measurements for Stabilized Earth Blocks

Block type	Dimension	Length of blocks (mm)	Average dimensions of block (mm)
NSMB	Length	4609	230.45
	Width	3797	189.85
	Depth	2005	100.25
ISMB20	Length	4640	232
	Width	3830	191.5
	Depth	1620	81
ISMB30	Length	4640	232
	Width	3820	191
	Depth	1420	71

3.2 Block Density:

In order to determine block density, 12 samples were randomly selected out of the produced blocks aggregating to minimum 4 from each type. Blocks were free from loose particles and were oven dried in order to remove moisture. Results of the same are tabulated in Table 3.2. Average block density of all the three types of Stabilized earth blocks produced is found to be 1.86 g/cc which is greater than 1.750 g/cc.

Table 3.2: Block density test of solid stabilised blocks

Block type	Average density g/cc
NSMB	1.88
ISMB20	1.87
ISMB30	1.84

3.3. Initial rate of water absorption (IRA):

In order to determine IRA, 12 blocks were selected and were positioned lengthwise (Fig 3.1) so that immersion of either extensions or opening faces only is avoided for interlocking blocks. The blocks were immersed in water bath of 2.5cm depth for duration of 1 minute. Weight of blocks before and after immersion in water was noted and average IRA was determined. Average IRA value is found to be decreasing as depth of block is reduced as observed in Table 3.3.

Table 3.3 Initial rate of water absorption in solid stabilised blocks

Block type	Average IRA ($Kg/m^2/min$)
NSMB	6.81
ISMB20	5.29
ISMB30	4.76

Fig 3.1 Blocks being tested for IRA and positioned lengthwise

3.4 Water absorption

12 selected blocks were immersed in curing tank (cold water) for 24 hours. Wet and dry weights were noted down. Average water absorption of all the three types of blocks is tabulated in Table 3.4 which is in accordance with IS 3952-1998.

Table 3.4 Water absorption of stabilised solid blocks

Block type	Average water absorption %
NSMB	14
ISMB20	12
ISMB30	13

3.5 Wet Compressive Strength:

12 blocks were randomly selected to determine wet compressive strength of block. The test has been carried out in Compression Testing Machine. Prior testing- void spaces created because of extensions and openings in interlocking blocks and frogs of non interlocking stabilised solid blocks were filled with 1:2 rich cement mortar and were soaked in water for 48 hours. Load was applied axially at constant increasing rate at 14 N/mm^2/min till failure. Ultimate load at which unit failed was noted. Readings are tabulated in Table 3.5. Wet compressive strength of the NSMB is 3.10Mpa which clearly shows that blocks belong to grade 2.5. However the strength and grade increases with increase in depth of interlocking for same soil mix proportion and same quantity of stabiliser. This is due to apparent strength increase due to platen friction effects in testing machine. Web splitting, vertical cracking and face spalling were common modes of failures observed during testing.

Table 3.5 wet compressive strength of stabilised solid blocks

Block type	Avg wet compressive strength MPa
NSMB	3.10
ISMB20	4.10
ISMB30	4.21

Fig 3.2 Failed specimen under compression

Fig 3.3 Failed specimen under compression

3.6. Flexural strength:

In order to determine flexural strength of non interlocking and interlocking solid stabilised blocks 3 point loading arrangement was applied and tested in CBR apparatus with a rate of loading of 1.25mm/minute. Interlocking solid blocks were tested for both directions ie along direction of interlocking effect and perpendicular to direction of interlocking effects. Results in both directions

IOP Conf. Series: Materials Science and Engineering 1282 (2023) 012011 doi:10.1088/1757-899X/1282/1/012011

provided similar results but change in values can be observed in table 3.7 and table 3.8 because of change in moment of inertia value and not because of increase or decrease in load carrying capacity of interlocking solid stabilised mud blocks. All the blocks failed by bending. NSMB and ISMB30 behaved in similar fashion i.e. blocks failed by inclined shear as shown in Fig 3.6. Whereas ISMB20 were broken into two pieces which shows the brittleness of the blocks as shown in Fig 3.7

Table 3.6 Flexural strength test of stabilised solid blocks (along the direction of interlocking)

Block type	Avg flexural strength MPa
NSMB	0.668
ISMB20	1.160
ISMB30	1.212

Table 3.7 Flexural strength test of stabilised solid blocks (perpendicular the direction of interlocking)

Block type	Avg flexural strength MPa
ISMB20	0.760
ISMB30	0.806

Fig 3.4 Flexural strength test (along the directionof interlocking effect)

Fig 3.5 Flexural strength test (perpendicular tothe direction of interlocking effect)

Fig 3.6 Failed specimens of 30mm interlocking blocks

Fig 3.7 Failed specimens of 20mm interlockingblocks.

3.7 Modulus of Elasticity:

In order to determine modulus of elasticity three random samples were selected out of the produced blocks. Before testing, blocks were kept in moist condition under gunny bags for 24 hours. Test was conducted in 600kN capacity Universal testing machine. A demountable strain gauge having 100 mm gauge length was mounted on specimens exactly in central portion of blocks. The corresponding strain gauge readings were noted at regular intervals of 1kN. Strain gauge was removed after 20 readings and specimens were loaded until failure. These observations were noted and corresponding stress-strain graph was plotted in spreadsheet. All the sample readings were given as input in one graph and initial tangent modulus of the blocks have been arrived as shown in Fig 3.8. Average modulus of elasticity of blocks is 951.6 Mpa.

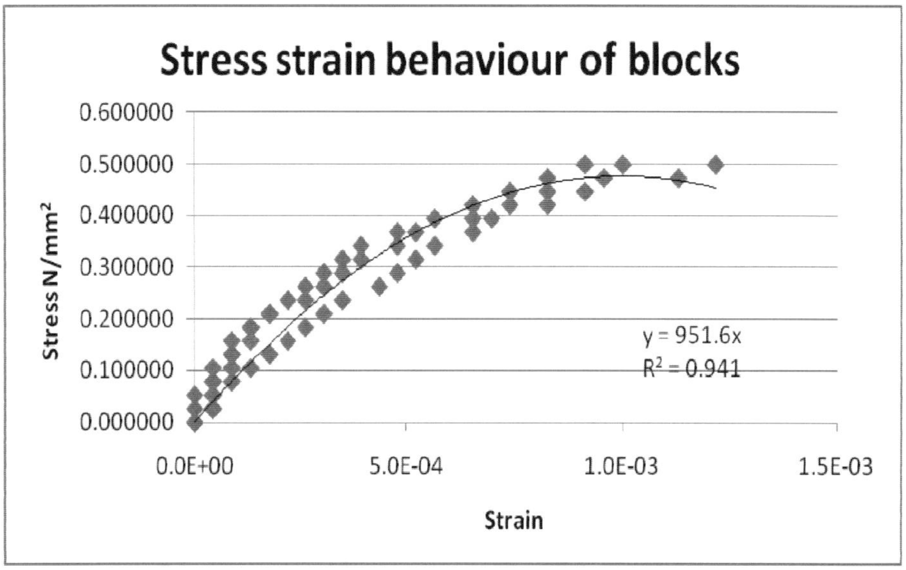

Fig 3.8 MOE of stabilised mud blocks.

Table 3.8 Summary of the results obtained.

Test results	NSMB	ISMB20	ISMB30
Dimensionality (mm)	230.35X189.85X100.25	232X191.5X81	232X191X71
Block Density (g/cc)	1.880	1.870	1.840
Initial rate of water absorption (Kg/m^2/min)	6.81	5.29	4.76
Water absorption (%)	14	12	13
Wet Compressive strength (MPa)	3.10	4.10	4.21
Flexural strength (along the direction of interlocking effect) (MPa)	0.668	1.160	1.212
Flexural strength (perpendicular to the direction of interlocking effect) (MPa)	-	0.760	0.806
Modulus of Elasticity (MPa)	951.6		

4. Conclusions:

With reference to investigations carried out following broad conclusions can be drawn: Size and shape of blocks:

- Blocks were produced by making slight modifications to existing mould.
- With the same mix and machine we were able to produce all the three types of blocks i.e. non-interlocking solid stabilised mud blocks, interlocking stabilised mud blocks with 20 and 30 mm interlocking effects respectively with ease and fine edges.

Strength Parameters:

- Wet compressive of ISMB 20 and ISMB 30 is around 4 MPa and hence can be considered for load bearing masonry construction. However the NSMB falls to Grade 2.5 which can be considered for G+ 1 storied structure.
- Blocks exhibited good flexural strength. Interlocking stabilised mud blocks too exhibited good flexural strength which concludes that the blocks will adjoin with the filling material i.e. mortar in both the directions.

This study has facilitated for further research in this area leading to production and usage of solid stabilised interlocking mud blocks. In non interlocking solid stabilised mud blocks, bond between mortar and block plays a major role in understanding bond strength. However, by providing interlocking effects in blocks this strength can be improved because of presence of interlocking effect along with mortar. Increase in bond strength will increase lateral load carrying capacity of a masonry wall. Enhancement in lateral load carrying capacity of masonry wall reduces the damages to a structure during a natural disaster. This leads to a sustainable infrastructure development as there will be less damages and casualties during a natural disaster.

References:

[1] Jagadish K S 2007 "Building with stabilised mud",I.K.International Publishing House Pvt.Ltd.,New Delhi.
[2] Jagadish K S, Reddy B V V, Rao K S N (2017) "Alternative building materials and technologies", New ageinternational Publishers, New Delhi.
[3] Saari S, et al (2021) "Factors of Non-Uniform Properties of Interlocking Compressed Earth Brick Units", Developments in the Built Environment
[4] Al-Fakih A, et al (2020) "Experimental study on axial compressive behavior of rubberized interlocking masonryWalls", Journal of Building Engineering.
[5] Sturm T, et al (2014) "Characterization of dry-stack interlocking compressed earth blocks" Materials andStructure.*PP 3059-3074*
[6] Jayasinghe C and Mallawarachchi R S (2008) "Flexural strength of compressed stabilized earth masonry materials"Materials and Design pages, 30(9) PP 3859-3868.
[7] Reddy B V V and gupta A (2006) "Strength and elastic properties of stabilised mud block masonry using cement-soilmortars"Journal of materials in Civil Engineering Vol. 18, No. 3, PP 472-476.
[8] Vivek Prasad H.G. (2009),"Interlocking in mud blocks for improved flexural strength and earthquake resistance",M.Tech. Thesis submitted to Department of Civil Engineering, RV College of Engineering, V.T.U. Belgaum.
[9] IS 1077: 1992 (reaffirmed 2007) "Common burnt clay building bricks – specification", Bureau of Indian standards,New Delhi.
[10] IS 3495-(Parts 1 to 4)-1992 (reaffirmed 2002) "Methods of tests of burnt clay building bricks", Bureau of Indianstandards, New Delhi.
[11] IS 1725 (1982): "Specifications for soil based blocks used in general building

construction" ,Bureau of IndianStandards, New Delhi.
[12] IS 5454:1978 :(reaffirmed 2010) "Methods for sampling of clay building bricks", Bureau of Indian Standards, NewDelhi.
[13] IS 3952-1998 (reaffirmed 2006) "Indian standard specification for burnt clay hollow bricks for walls and partitions",Bureau of Indian Standards, New Delhi.

Assessment of groundwater potential zones using an integration of Remote Sensing, GIS and 2D Electrical Resistivity imaging in the Echway watershed, Baro River Basin, Southwest Ethiopia

Jamuth Pajock, Jagadeshan Gunalan* Muralitharan Jothimani, Abel Abebe
Department of Applied Geology, College of Natural and computational Science, Arba Minch University, Ethiopia
Email:* jagaremote.1@gmail.com

Abstract: The objective of this study is to delineate and identify the groundwater potential zone of degraded land in the Echway watershed, southwest Ethiopia, employing a combined approach that includes 2D electrical resistivity, the analytical hierarchy process (AHP), geographic information systems (GIS) and remote sensing (RS). Using a geographic information system (GIS) and remote sensing data, groundwater potential zones were identified by taking into account the soil types, drainage density, geology, lineament density, Land use and land cover, rainfall and slope. Using the AHP method, calculate the weight of thematic layers and rank of subclasses based on the effects of various thematic layers on groundwater potential. A significant weight was provided on the geology of the research region, total annual rainfall, and lineament density. Due to significant effects on groundwater potential zones, only three of the seven theme levels were assigned significant weight in this analysis. The weighted overlay analysis was used to construct the diagram of the zones with groundwater potential. 2D electrical resistivity was utilized to find the groundwater aquifer, and four major electrical layers were discovered: clayey silt, sand, moderately weathered/fractured rock, and basement parent rock. The groundwater potential zones in the study area have been classified into five categories: very low (22.97%), low (13.43%), moderate (32.50%), high (25.12%), and very high (5.98%) covered, respectively. The groundwater potential zone image was created by combining GIS, remote sensing, AHP, and an electrical survey

Keywords: Groundwater potential zones, traverse line, thematic layers, resistivity, Echway watershed.

1. Introduction

Water is the most crucial component of life on earth and is essential to all living species' existence. Demand for water and management is becoming increasingly critical in the modern environment due to population growth, development and intensive farming [1]. Groundwater is the most significant vital supply on the earth, yet it is hard to identify since it is limited in form. Investigating groundwater potential zones were using various surface and subsurface geophysical methods, geomorphological, and remote sensing approaches capable of locate the depth of hydrogeological characteristics [2, 3]. The purpose of estimating groundwater potential is not only to meet requirements it primarily to safeguard and manage resources for a better future. Geographical information systems (GIS) can present one operating platform for rapidly and accurately analyzing geo-referenced collected information from multiple perspectives [4]. Remote sensing data is used to examine groundwater regulating elements such as geology, rainfall, slope, lineament density, drainage density, soil types,and land use/land cover [5]. As an outcome, the most effective and easy way to determine groundwater potential regions is the integration of Remote Sensing and Geographical Information systems. The analytical hierarchy process (AHP) use to construct similarity datasets for multi-criteria decision analysis by applying weights to thematic strata and grouping relevant subclasses [6, 7]. Integrating remote sensing, GIS, and AHP method is required for practical groundwater resource evaluation, observation, planning, prediction, and testing. In recent years, geophysical 2D resistivity surveys used in various applications, including environmental monitoring evaluation, mineral and oil exploration, civil engineering projects and hydrologic study [8]. The primary advantage of the 2D electrical resistivity concept over a 1D image, this method generates more detailed data on resistivity changes along a profile line in all vertical and lateral dimensions [9, 10, 11]. The two-dimensional electrical resistivity survey is called electrical resistivity tomography. It evaluated subsurface layers, porous and permeable zones, fractures/cracks, and potential water areas. Ethiopia is one of the countries experiencing a reduction in groundwater quantity and quality due to land degradation. According to current data, the average rate of deforestation in Ethiopia, owing to agricultural growth, is roughly 1,410 km^2[12]. The Gambella region's unique forest cover, vegetation, and rich biodiversity will be extinct by 2133 if current deforestation rates continue. It will have a significant impact on the LU/LC ecosystems all across the Gambella region. Groundwater potential in the Gambela Region has steadily declined due to fast land use and land cover changes, increased population, and agricultural activities in the Echway watershed. As a result of the gambela region's groundwater shortage in recent years as a result of soil degradation, identifying groundwater potential zones is essential to this region. The Echway catchment is one of the most densely inhabited places in Ethiopia's Gambela region, and as a result, there is a considerable demand for residential water supplies. The water treatment plant is inadequate to provide clean water to all areas of the study area, particularly Gambela municipality and the neighboring region inside this watershed, pushing inhabitants to rely on unsafe dug wells and adjacent rivers, which are unsuitable for household use. In order to identify groundwater potential zones, a combination of remote sensing, geographic information systems, AHP, and 2D electrical resistivity mapping was used. There hasn't been any additional research on this method of defining zones of groundwater potential in the Echway watershed.

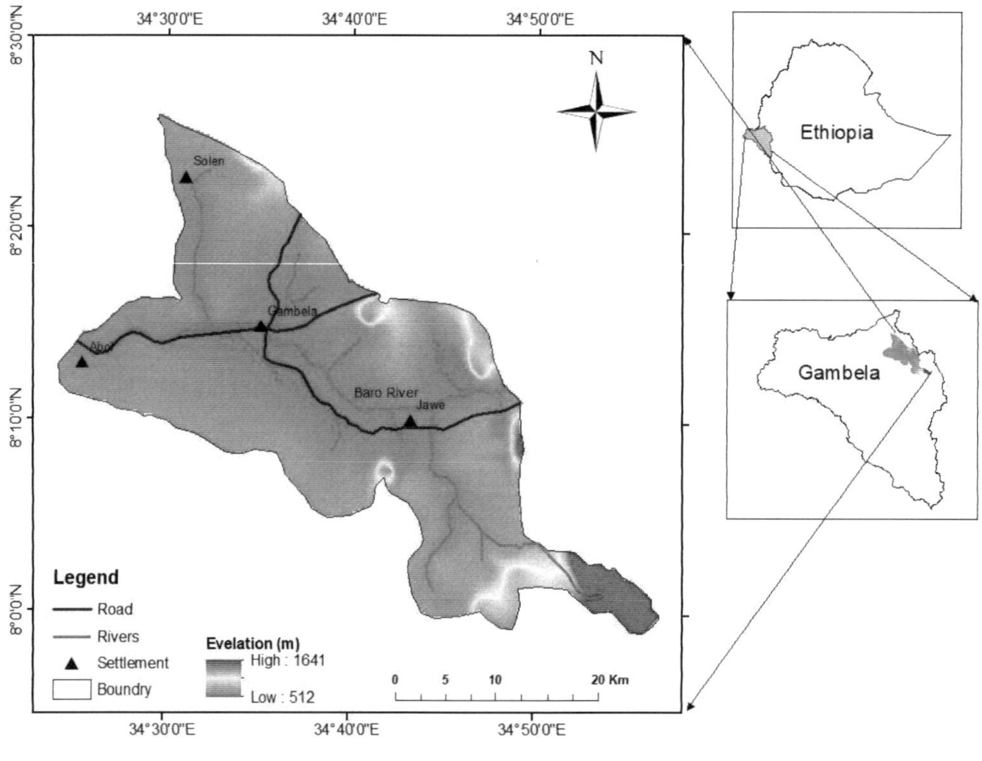

Figure 1. Location map of the Echway watershed basin

2. Materials and Methods

2.1 Description of the study area

The research region is located in the Baro river basin, Gambela region, southwest Ethiopia. The research region is roughly 1061 km^2 and is mainly delineated by watershed boundaries. It is approximately 766 kilometers from Addis Ababa, the capital city. Gambela town is part of the research area near to the Baro river bank (Figure 1). The natural ecosystem of the Gambella region has been dramatically affected by land use and land cover changes, with considerable sections of the land deforested [13]. The landscape in the research area is mainly flat to mildly undulating. In the periphery region of the southeast, a group of rugged hills occurs at an elevation of 1600 m, while the heart of the study area reveals a low elevation of 510m. The study topography variations are minimal near the river, most surface water flows as just a surface layer to the Baro River and minor rills erosion occurs. Many rivers flow east to west over the Gambela plain, eventually meeting the Sobat River on the South Sudan border. River discharge increases from May to September owing to the wet season and diminishes from October to April. During the dry season, the Baro River receives little discharge from the surrounding region's groundwater base flow and is characterized as a perennial river. The study region's maximum and lowest temperatures are around 45 ºC in March and 20 ºC in December. In the study region, annual rainfall ranges from 824 to 850mm.

2.2 Data collection and analysis

The presence, mobility and distribution of groundwater resources was established by combining GIS, remote sensing, AHP and 2D apparent resistivity techniques. The remote sensing and geographic information system data were collected from several sources, including the Ethiopian Ministry ofWater and Irrigation and the Ethiopian Geological Survey. In addition to assessing potential zones, these data were utilized to build seven different thematic maps, which included geology, soil, rainfall, lineaments density, drainage density, land use/cover and slope. The data from the Ethiopia Geological Survey was utilized to create the geology thematic map for this study using the ArcGIS clipping program. The Climatic Research Department provided rainfall data for the years 2011 through 2020. Data on land usage and land cover were collected from the Food and Agriculture Organization. Subsurface Soil information was acquired from the Ethiopian Ministries of Water and Energy, cropped and analysed with ArcGIS platform. The Schulte Radar Landscape Survey and a digital terrain map with a resolution of 30m were used to construct the slope, lineament density and drainage network. The ArcGIS clip tool extracted and manipulated the study's LULC. AHP is a well multi-stage decision-making methodology, and it is another essential strategy for stakeholders to manage complex choices in a simple manner by applying pair-wise correlation and assessing many elements based on its influence, particularly in aquifer delineation[14, 15, 16, 17]. The AHP approach was used in this research to distribute average weights per each parameter in the groundwater aquifer estimation. Using the AHP method, calculate the weight of thematic layers and rank of subclasses based on the effects of various thematic layers on groundwater potential. Significant weight is provided on the geology of the research region, total annual rainfall and lineament density. Only three thematic factors were given substantial weight in this study out of seven different thematic layers due to major effect characteristics on groundwater potential zones. The weighted sum software package in the ArcGIS 10.8 program was used to construct the aquifer areas map. Hydrogeological field surveys were conducted to visualize the research region's topographical plot and geological and hydrogeological formation. 2D electrical resistivity geophysical investigation was conducted in four different profile lines in different directions of the study region to determine the fracture length, level of weathering and crystalline basement rocks. Following the field study, the site inspection data and electrical resistivity values were analyzed, reviewed and processed with the Res2DInv tool. This survey was carried out in order to generate the 2D inverted model resistivity profile demonstrating underground geo-electrical layering of the region required for water-bearing structure characterization. The AMEB 400 resistivity device is 2D electrical resistivity equipment that was employed in this field investigation. Depending on the model system, the number of electrodes ranges from 25 to 100. This study utilised forty-two electrodes within 400m of each other, with an electrode spacing of 10m. The Wenner array arrangement was used to conduct the 2D resistivity study. When compared to other array configurations, the Wenner array had the greatest signal intensity and a significant depth of research in the survey [18]. The RES2DINV program was used to process and invert the acquired data. The RES2DINV inverts observed and estimated apparent resistivity values to produce an inverted design resistivity line corresponding to Earth's soil characteristics' true resistivity.

3. Results and Discussions

Groundwater storage factors vary according to location based on hydrogeological, meteorological and anthropocentric activities. The groundwater regulation factors land use/land contain, topsoil, lithology, drainage network, lineament, slope and rainfall were discovered to be crucial in determining groundwater recharge in this investigation. A 2D resistivity survey was done in the research region to determine the magnitude of weathering, fracture extent and foundation rock lithology to validate the groundwater carrying aquifer.

3.1 Identification of groundwater potential zone using remote sensing and GIS

3.1.1 Geology and lineament

The geological formation in the research region was classified as Precambrian and Pleistocene Holocene based on the geological time scale. The study area's Precambrian rocks consist primarily of metamorphic rocks such as syntectonic granitoid, post tectonic granitoid and lower complex basement rock. Such strata are poor prospectively and release little water. Only metamorphic rock faults and fractures may generate considerable volumes of water and function as porous material [19]. The Pleistocene Holocene strata comprise quaternary deposits of sand, silt, gravel and clay (Figure 2).These deposits were commonly categorised as water potential formations because of their considerable permeability and porosity [19, 5]. Lineament is the vertical characteristic that reflects subsurface structures such as a crack, fracture, join and fault. It facilitates substrate additional permeability and porosity [20, 21]. The spatial statistical unit density and categorisation program was used to determine lineament density from a lineament image. In the study field, lineament frequency was already divided into five categories: very low, low, moderate, high and very high (Figure 2). The substantial lineament density enhances infiltration and recharges efficiency due to the high secondary porous structure that represents a location with extensive groundwater occurrence. Poor lineament density influences recharge velocity and infiltration, suggesting a zone with limited groundwater capacity.

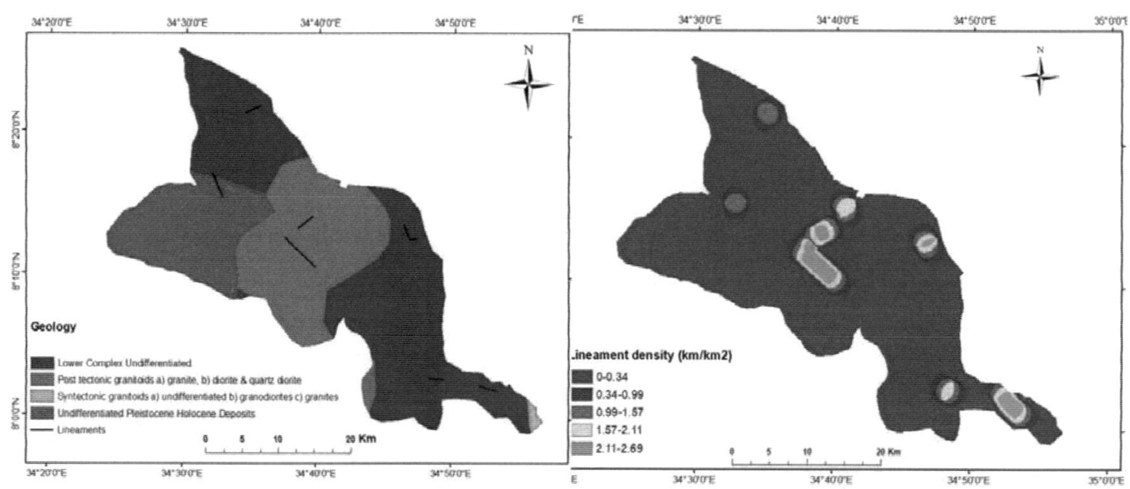

Figure 2. Geology and Lineaments density map of the study area

3.1.2 Slope and drainage density

The slope is one of the most critical components controlling groundwater storage and flow. Infiltration amount decreases when the terrain is steeper, and infiltration ratios are raised when the topography is plain. The amount of precipitation that percolates and reaches the aquifer highly depends on the slope's nature [22]. Most of the study area comprised plain to moderate slopes, with steep slopes representing only a small portion of the region. The catchment's slope was classified into five categories: very low, low, moderate, high, and very high (Figure 3). Drainages are surface and subsurface structures that control water flow above and below the earth [3, 23]. On the other hand, drainage density is the sum of the distances including all tributaries within a specific area. The research region's drainage density was divided into five groups: very high, high, moderate, low, and very low (Figure 3). More drainage density structure shows that the formation rocks are well resistant to erosion and weathering owing to very high surface water runoff and limited groundwater infiltration capacity, but low drainage density gives little surface runoff and considerable groundwater potential.

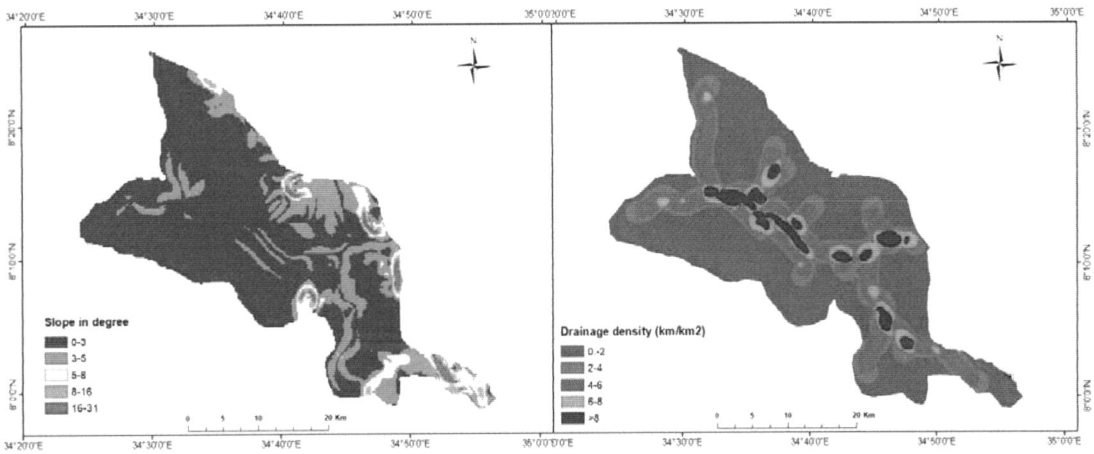

Figure 3. Map of the research area's slope and drainage percentage

3.1.3 Soil and Land Use and Land Covers

Groundwater potential zones can be strongly influenced by soil properties such as porosity, soil composition, texture and permeability [24]. Soil types were classified into five category, eutric cambisols, eutric fluvisols, eutric lithosols, gleysols, and eutric vertisols. The soils mentioned above are categorized depending upon size distribution as coarse-grained, moderate-grained and small-grained. As a result, eutric gleysols and eutric vertisols are small-grained soils with low to shallow potential groundwater zones. Eutric cambisols are intermediate grain soils with reasonable groundwater potential. Eutric fluvisols and eutric lithosols are coarse-grained soils with high and very high likely groundwater (Figure 4). The surface water runoff, infiltration and circulation of groundwater are all strongly affected by LULC. Anthropogenic changes in land use/land cover can affect recharge percentages, therefore, may impact a detrimental effect on groundwater chemistry [25]. During precipitation, the roots of plants may absorb and retain rainwater, resulting in high infiltration and little surface runoff. However, urban areas increase surface runoff and decrease the percentage of groundwater recharge [5]. This research region's five types of LULC include annual crop, open grass, dense forest, habitation, and water body. An annual yield is a small unit in the region, with open grass and lush vegetation occupying most of the landscape (Figure 4). As a result, groundwater occurrence in the Echway watershed, the water body, and the forests are highly valued, annual crops and open grass is reasonably valued, and settlement is low valued [26].

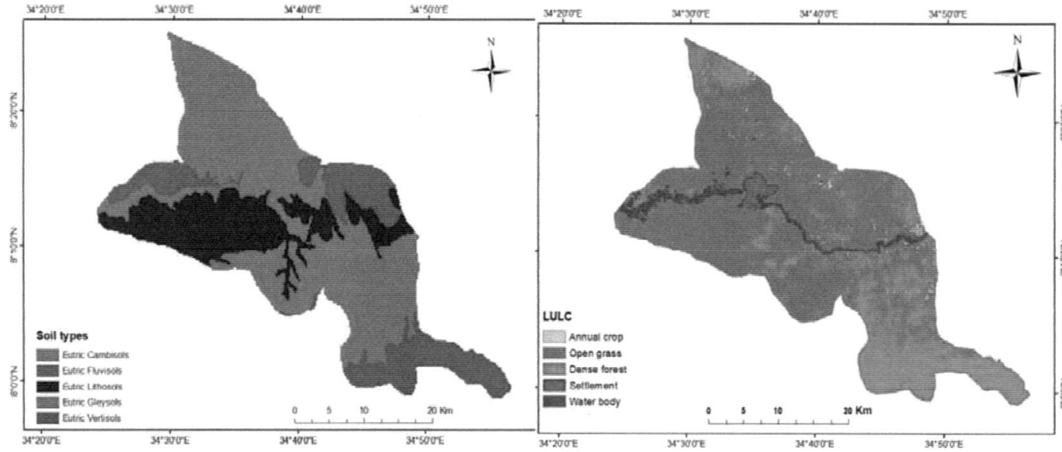

Figure 4. Soil and Land use & Land cover map of the study area

3.1.4 Rainfall

Rainfall influences subsurface recharging and plays a vital role in the water balance cycle. Continuous long-period precipitation much more reflects good recharge than low precipitation, which indicates poor groundwater recharge [27]. The rain in the research region was divided into five categories: very low, low, moderate, high, and very high. The rainfall image shows significant precipitation in the centre of the research zone, however, the research region's southeast trend indicates that the amount of precipitation there is quite low (Figure 5).

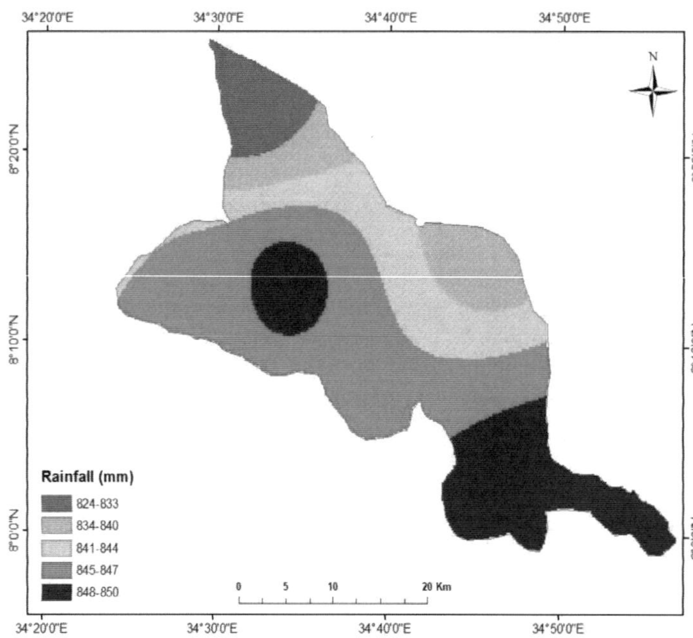

Figure 5. Rainfall map of the study area

4. Groundwater potential zones mapping using AHP approach

The potential groundwater zone image for the research region was created by combining the scores of seven parameters and associated subdivision score using the weighted overlay approach. (Table 1) [28, 29, 30].

Table 1. The weight of criteria and their subcategory ranking

Criteria	Subcriteria	Groundwater potential	Class rank	Normalized weight		Final weight
				Criteria	Subcriteria	
Geology	Undeferential Pleistocene	Very high	5	0.35	0.53	0.1855
	Holocene deposited	Moderate	3		0.24	0.084
	Post tectonic granitoid	Moderate	3		0.15	0.0525
	Syntectonic granitoid	Very low	1		0.07	0.0245
	Lowe Complex Undeferential					
Rainfall (mm)	850-848	Very high	5	0.24	0.49	0.1176
	848-847	high	4		0.31	0.0744
	847-841	Moderate	3		0.18	0.0432
	841-833	Low	2		0.11	0.0264
	833-824	Very low	1		0.08	0.0192
Lineament density (km/km2)	2.69-2.11	Very high	5	0.16	0.41	0.0656
	2.11-1.57	High	4		0.26	0.0416
	1.57-0.99	Moderate	3		0.16	0.0256
	0.99-0.34	Low	2		0.11	0.0176
	0.34-0	Very low	1		0.06	0.0096
Soil types	Lithosols	Very high	5	0.11	0.41	0.0451
	Fluvisols	High	5		0.29	0.0319
	Cambisols	Moderate	3		0.16	0.0176
	Gleysols	Low	2		0.08	0.0088
	Vertisols	Very low	1		0.06	0.0066
LULC	Water body	Very high	5	0.07	0.36	0.0252
	Dense forest	Very high	5		0.31	0.0217
	Annual crop	Moderate	4		0.17	0.0119
	Open grass	Low	2		0.12	0.0084
	Settlement	Very low	1		0.02	0.0014
Slope in degree	0-2	Very high	5	0.05	0.34	0.017
	2-4	High	4		0.31	0.0155
	4-6	Moderate	3		0.19	0.0095
	6-8	Low	2		0.11	0.0055
	>8	Very low	1		0.06	0.003
Drainage density Km/km2	0-3	very high	5	0.02	0.33	0.0066
	3-6	high	4		0.30	0.006
	6-16	moderate	3		0.19	0.0038
	16-31	low	2		0.12	0.0024
		very low	1		0.07	0.0014

The potential groundwater zone of this study was classified into five categories, namely, very low, low, moderate, high, and very high, with an area covered of 229.44km2 (22.97%), 134.08km2 (13.43%), 324.88km2 (32.50%), 25.52km2 (25.12%), and 59.76km2 (5.98%), respectively (see table 5.8). Due to the complicated basement rock and lower rainfall within study site, the aquifer diagram of the area reveals shallow water capability regions in the northwest to east. High and very high groundwater potential occupied the study area's west, central, and southeast parts. These areas have coincided with quaternary sediment, highly weathered post-tectonic lineament, water body, and a relatively high amount of rainfall.

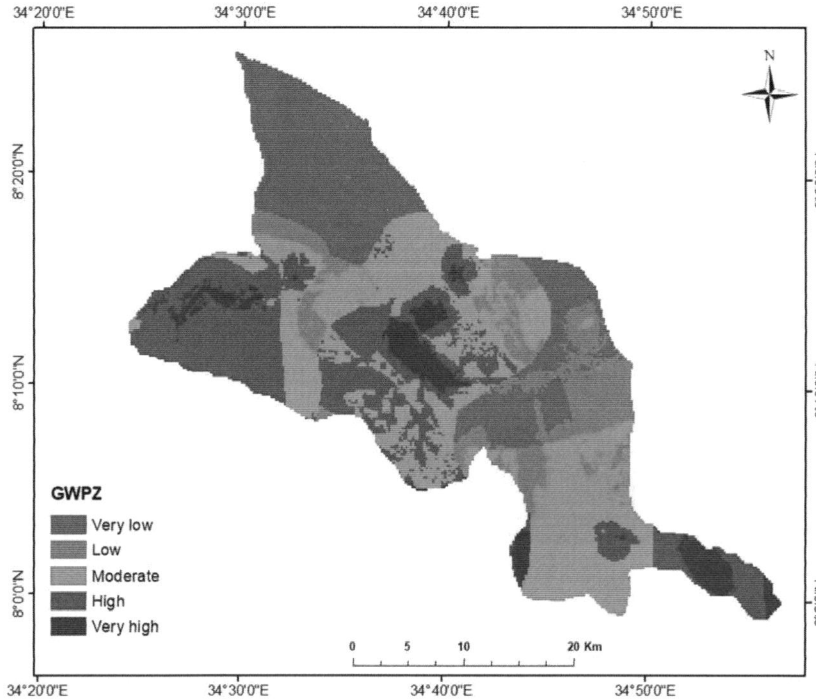

Figure 6. Groundwater potential map of the study area

Table 2. Groundwater potential zones classification

Class	Area (km2)	Area (%)
Very low	229.44	22.97
Low	134.08	13.43
Moderate	324.52	32.50
High	250.88	25.12
Very high	59.76	5.98

5. Identification of groundwater potential zone using 2D Electrical Resistivity

According to the RS, GIS, and AHP methods groundwater potential map, Compared to the northwest and southeast, the research region's middle has a greater water possibility region. Four resistivity traverses were carried out in the central portion of the research area. To determine subsurface lithology, number of aquifers and the depth of groundwater bearing formation the inverted model's pseudo-traverse data were used[31, 32, 33]. The figure's x-axis measures the length above the surface in meters, while the y-axis shows the depth of penetration of the two-dimensional inverted traverse in meters (Figure 7). The traverse resistivity value is evaluated using the research area's local geology and hydrogeology.

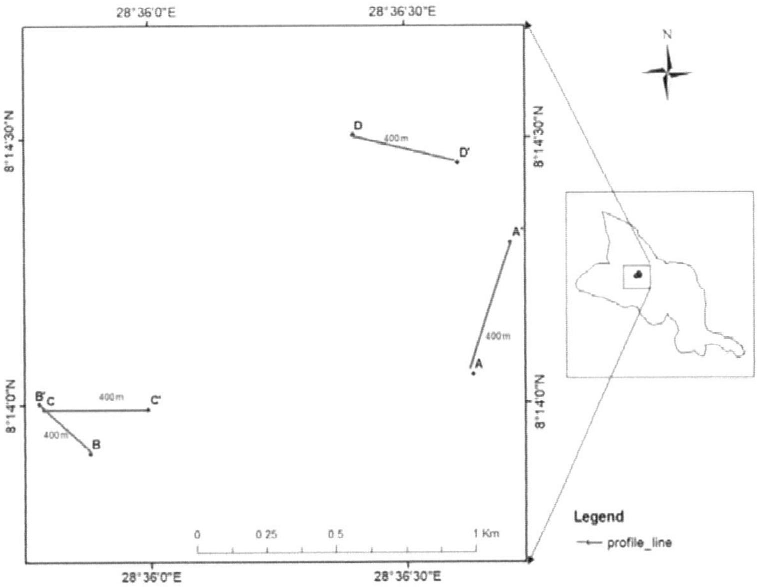

Figure 7. Location of 2D electrical resistivity traverse

5.1 Traverse one (AA')

The first traverse line (AA') was run perpendicular to the Baro River for 400m, running south-northeast (Figure 8). Due to the compact sandy silt that was located on the surface top layer on the left side of the traverse are high resistance value and its resistivity values vary from 200 to 270Ωm. The second layer, which occurs at the end of the inverted model's right wing, was interpreted as wet clayey soil, with a low resistivity value range of 20 to 85Ωm and thickness of 10m. The third subsurface layer, referred to as the sand layer, denotes a water-bearing zone with resistivity values ranging from 95 to 150 Ωm. The thickness of this water-bearing saturated sand layer is thin on the right side of the model and progressively increases toward the left side of the traverse. The last layer was characterized as a slightly weathered rock with a resistivity of 160 to 245Ωm.

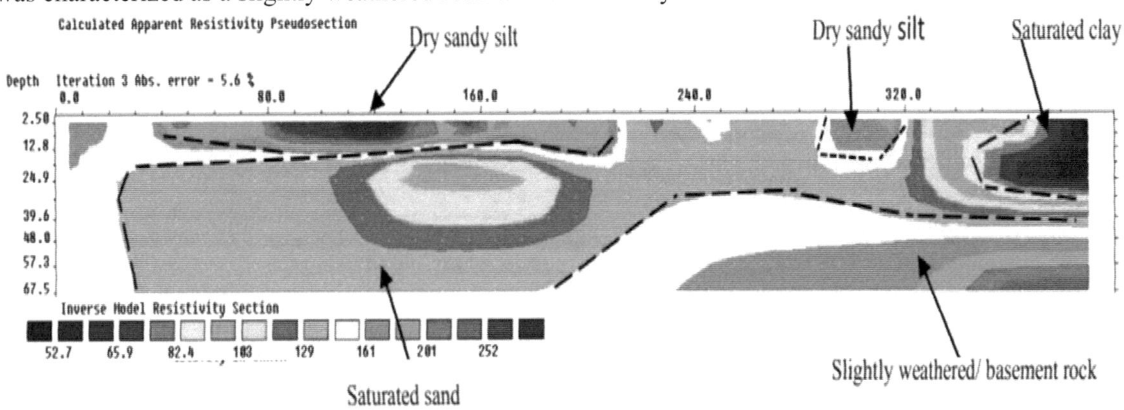

Figure 8. Pseudo section of resistivity traverse line 1

5.2 Traverse two (BB')

The second traverse line (BB') was conducted parallel to the Baro River in the southwest direction

(Figure 9). This traverse's data points were 145m long with a maximum length of 400m. This traverse showed four distinct geological layers: clayey silt, sand, weathered/fractured layer and massive parent basement rock. The top layer of the traverse was made up of moist clay soils, with a thickness of 3 to 5m and a resistivity range of 10 to 30 Ωm. The second layer, which lies beneath the top layer of clay, was characterized as saturated sand and had a resistivity range of 38-85Ωm. The thickness of this layer is relatively thin in the middle to left part of the traverse because of the occurrence of hard granitoidbasement rock, but this hard granitoid rock is absent on the right side of the traverse. The thickness of this second sand layer ranges from 30 to 45m, and it was assumed to be a suitable groundwater-bearing zone. The third layer was identified to be slightly weathered granitoid parent rock with a resistivity of 100 to 200 Ωm and a thickness of 6 m. The last layer is a considerable depth of solid granite that occurs at resistivity levels of 300 Ωm and higher.

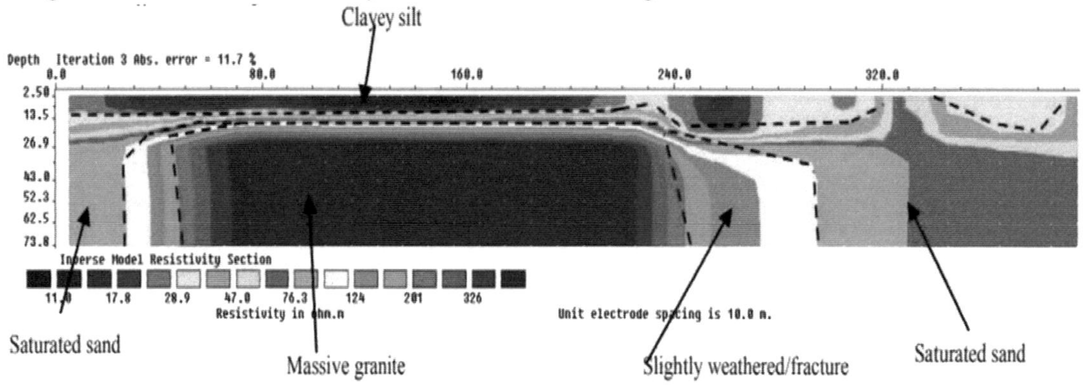

Figure 9. Pseudo section of resistivity traverse line 2

5.3 Traverse third (CC')
This third traverse line (CC') was carried out at the study area's southwest corner and ran west to east (Figure 10). The upper surface of the traverse has a 30m thick layer of clayey silt with a resistivity value range of 10 to 35 Ωm. The left side of the traverse had a 5m thickness of saturated sand, while the right side had a 30m thickness. The thickness of the saturation formation gradually increases from the traverse's left to right direction. This second layer is considered a water-bearing zone. The third layer, which resistivity values are more than 200 Ωm and shows slightly weathered granitoid rock on the left and middle of the traverse.

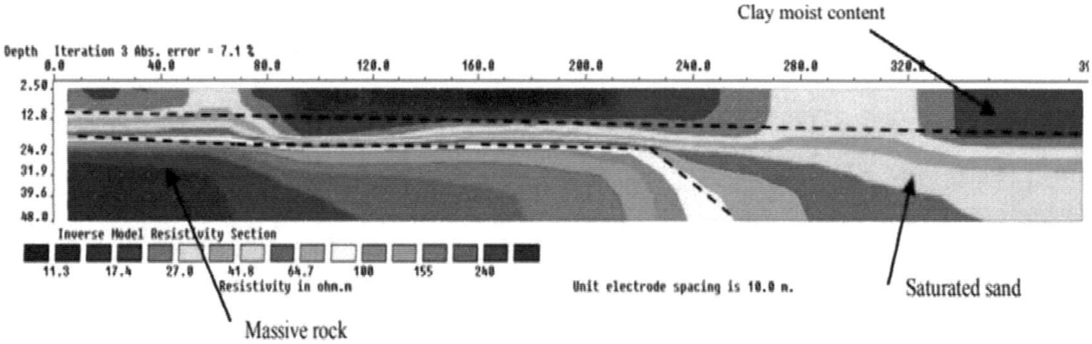

Figure 10. Pseudo section of resistivity traverse line 3

5.4 Traverse Fourth (DD')
This profile was carried out parallel to the new Gambela University campus road (Figure 11). This

traverse model depicts the broad range of underground resistivity regions with values ranging from 20 to 6950 Ωm. From left to right, 10 to 20m thick first layer traverse is the low resistivity value region of 20 to 170Ωm made up of conductive clay. High resistivity values of 2574 to 2682 Ωm dry or compacted clay have been detected in only small areas of 5m thickness in the right corner of the traverse. The highly saturated sand layer with a resistivity value of 150 to 320 Ωm occurs on both the left and right sides of the inverted traverse model beneath the clay layers. The middle part of the traverse comprises granitoid rock. With a thickness of 30 to 45m, this second layer shows suitablegroundwater bearing zones indicating the aquifer system on both sides of the model. The last layer of Precambrian rock in the middle of the inverted model is fractured and has a resistivity value of 268 to 6568 Ωm.

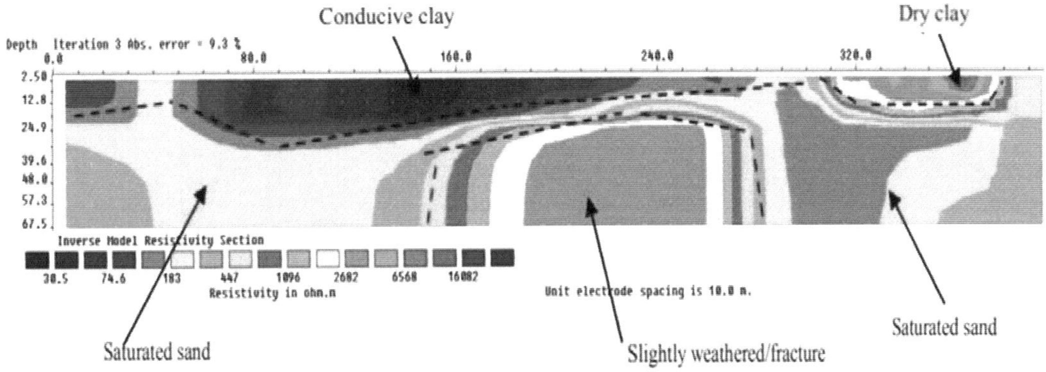

Figure 11. Pseudo section of resistivity traverse line 4

6. Integration of Remote Sensing, GIS, AHP and 2D Electrical Resistivity

The 2D resistivity traverse result was integrated with groundwater potential zone mapping to improve accuracy, identify shallow groundwater potential zones, and recommend the best location for the community to drill the water well [34, 33]. The area's geo-electrical layers were divided into clayey silt, sand, slightly weathered/fractured rock and massive basement rock. Sand layers occurrence of Pleistocene Holocene region represents high groundwater saturated zone and thickness vary from 12 to 70 m. The southern part of the study area traverses BB' and CC' is excellent groundwater carrying site with a shallow depth of 12m. Although clayey silt layers constitute the moderately water-bearing zone, they are not a suitable location for drilling the water well. Poor and very poor potential zones are indicated by slightly weathered and massive hard granitoid rocks. Based on the overall interpretation results, the middle portion of the second layer of sand deposits in the Pleistocene Holocene region is a good groundwater potential zone with shallow water depth (Figure 12). This location is ideal for drilling water well to supply the water needs of Ethiopia's Gambella town.

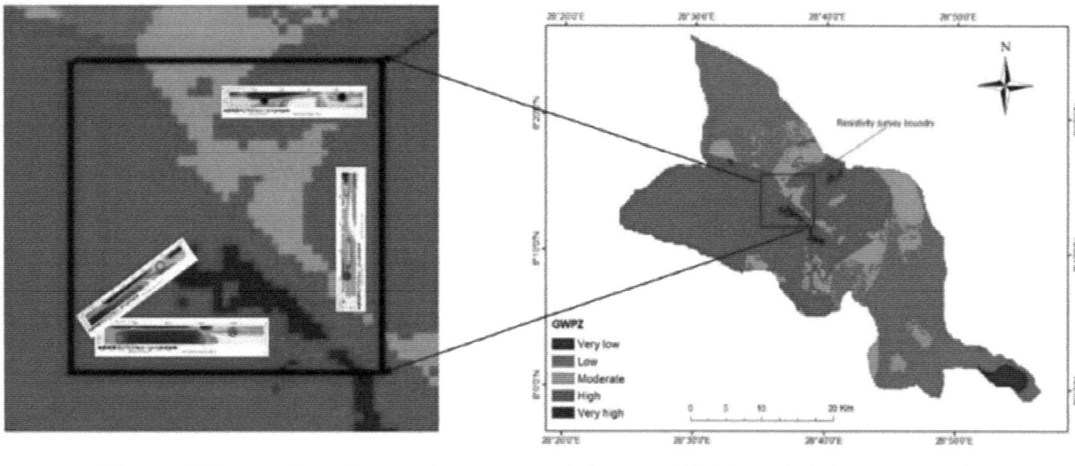

Figure 12. Integration of groundwater potential zone with 2D resistivity traverse line

7. Validation for the probable locations for groundwater

Applying data from shallow and open dug wells, the GWPZs have been validated [34, 35]. The wells data were obtained from the Gambela water and irrigation bureau. The wells data, including wells, yield data,static water level, and transmissivity. There are forty-four wells in the study area. Out of 44 wells, only 20 wells have yield and transmissivity. The model's validity was examined using these 20 wells. The pump well yields vary between 0.01-14l/s and transmissivity range from 1 to 100 m^2/d above in the study area. These bore wells yield and transmissivity have been classification as very low 0.01-0.5l/s and 1-10m^2/d, low 0.5-2l/s and 10-50m^2/d moderate 2-5l/s and 50- 100m^2/d, High 5-10l/s and100m^2/d, and very high 10-14l/s and 100 m^2/d above (Figure 13). Five wells fall on very high potential zones, seven wells on high potential zones, four wells on moderate potential zones, two wells on low potential zones and two wells on shallow potential zones. According to the validation findings, 16 out of the 20 pump wells are located in aquifer zones with a medium to high capacity. This revealed that 80% of the wells' yield and transmissivity correspond to potential groundwater zones. The identification of the aquifer potential regions matched well inventory information that was already available. The overall borehole wells drilled in the study area are shallow wells, with a depth range of 45-65m, exceptional of 2 open dug wells with a depth of 10m. Regional trends of rock type and rainfall can be seen in the spatial variability of the several aquifer potential regions that were determined by the results.

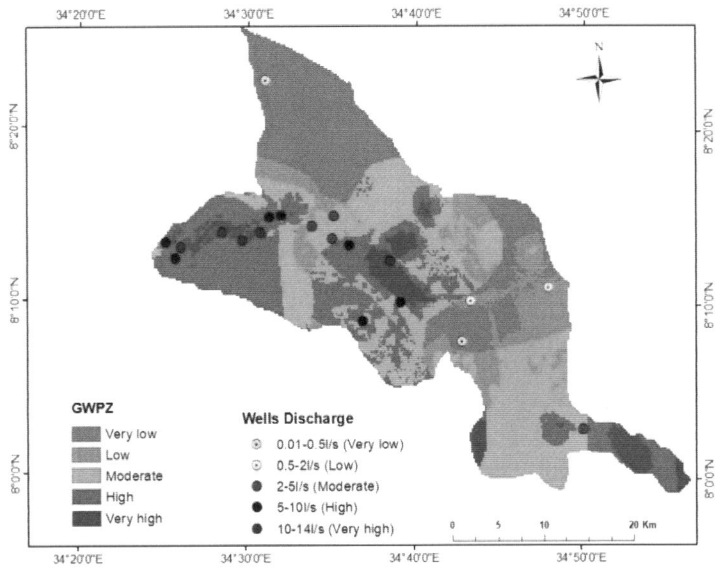

Figure 13. Validated Groundwater potential map

8. Conclusions

Groundwater is a reliable source which can be successfully controlled and managed to satisfy the country's demands. The current work studied an integrated combination of RS, GIS, AHP and 2D electrical resistivity to assess and identify significant groundwater potential zones. Water potential zones are considered using seven distinct thematic maps, including geology, soil, rainfall, lineaments density, drainage density, land use/cover and slope. The study area's Precambrian rocks consist primarily of metamorphic rocks such as syntectonic granitoid, post tectonic granitoid and lower complex basement rock. The Pleistocene Holocene strata comprise quaternary sand, silt, gravel and clay deposits. The geology of the research region, annual total rainfall and lineament density are all given significant weight. This study's groundwater potential zone was divided into five groups: very low (22.97%), low (13.43%), moderate (32.50%), high (25.12%), and very high (5.98%)covered, respectively. Based on RS, GIS, and AHP groundwater potential map, the research's west, north, and centre portions have moderate, high and very high groundwater potential. Because of the region's Quaternary sediments, flat to gentle slope, high lineament density and good precipitation. Fourresistivity traverses were carried out in the central portion of the research area to determine subsurface lithology, number of aquifers, and depth of groundwater bearing formation. The area's geo-electrical layers were divided into clayey silt, sand, slightly weathered/fractured rock and massive basement rock. Sand layers occurrence of Pleistocene Holocene region represents high groundwater saturated zone and thickness vary from 12 to 70 m. The southern part of the study area traverses BB' and CC' is excellent groundwater carrying location with a shallow depth of 12m. Poor and very poor potential zones are indicated by slightly weathered and massive hard granitoid rocks.

9. Acknowledgements

The authors acknowledge the Geology Department, College of Natural and Computational Science, and Arba Minch University in Ethiopia for their contributions.

10. References

[1] Farrag, A, Riad, S, Ahmed, MH, Groundwater situation and evaluation in the Nile basin east

of Assiut, Egypt,*Journal of Engineering Sciences*, 30 (2):517–542, 2002..
[2] SkMafizul, H, Suresh, K, Ajay, KT, Dipika, K, Rakesh, KB, Prashant, KC Prakash, C, Identification of groundwater resources zones in the active tectonic region of Himalaya through earth observatory technique,*Groundwater for Sustainable Development* 10-1000337, 2020
[3] Omolaiye, GE, Oladapo, IM, Ayolabi, AE, Akinwale, RP, Akinola, AA, Omolaye, KL, Sanuade, OA. Integration of remote sensing, GIS and 2D resistivity methods in groundwater development, *Applied Water Science* 10(6) doi:10.1007/s13201-020-01219-x, 2020
[4] Umit, Y, Identification of Groundwater Potential Zones Using GIS and Multi-Criteria Decision-Making Techniques: A Case Study Upper Coruh River Basin NE Turkey,*International Journal of Geo-Information* 10(6)- 396; doi.org/10.3390/ijgi10060396, 2021
[5] Kibrit GB and Samuel DH, Identification of Groundwater Potential Zones Using Proxy Data: Case study of Megech Watershed, Ethiopia,*Journal of hydrology* 28-100676, doi.org/10.1016/j.ejrh.2020.100676, 2020
[6] Kaliraj, S, Chandrasekar, NMagesh, N, Identification of potential groundwater recharge zones in Vaigai upper basin, Tamil Nadu, using GIS-based analytical hierarchical process (AHP) technique,*Arabian Journal of Geosciences* 7:1385–1401, DOI:10.1007/s12517-013-0849-x, 2014.
[7] Kumar, VA, Mondal, NC, Ahmed, S, Identification of groundwater potential zones using RS, GIS and AHP techniques: a case study in a part of Deccan volcanic province (DVP), Maharashtra, India,*Journal of the Indian Society of Remote Sensing* 48(1), DOI:10.1007/s12524-019-01086-3, 2020
[8] Sharma SP and Baranwal, VC, Delineation of groundwater-bearing fracture zone in a hard rock area integrating Very Low Frequency electromagnetic and resistivity data, *Journal of Applied Geophysics* 57(2):155-166, doi.org/10.1016/j.jappgeo.2004.10.003, 2005.
[9] Mahmoud, M, Hossam, ES, Amr, H, Combined application of electrical resistivity and GIS for subsurface mapping and groundwater exploration at El-Themed, Southeast Sinai, Egypt,*Egyptian Journal of Aquatic Research* 42(4), DOI:10.1016/j.ejar.2016.10.007, 2016
[10] Thanasamy JR and Dushiyanthan C, Evalauation of Groundwater Potential Zones using Electrical Resistivity Response and Lineament Pattern in Uppodai Sub Basin, Tambaraparani,*Journal of Geology & Geosciences* 05(02), DOI:10.4172/2329-6755.1000241, 2016
[11] Shivendra, NR, Thiagarajan, S, Dewashish, K, Kanchan, MD, Rai, PK, Ramachandran, A and Nithya, B, Electrical resistivity tomography for groundwater exploration in a granitic terrain in NGRI campus,*Current Science* 105(10):1410-1418, 2013
[12] Muke M and Haile B, Land-use/cover change analysis using remote sensing techniques in the landscape of Majang Zone of Gambella Region, Ethiopia, African Journal of. *Environmental Science and Technology* 12(4):141–149, 2018.
[13] Othow, OO, Gebre, SL, Gemeda, DO, Analyzing the rate of land use and land cover change and determining the causes of forest cover change in Gog District, Gambella Regional State, Ethiopia,*Journal of Remote Sensing & GIS* 6(218):2, 2017
[14] Abdul, AH,Vanum, G Amare, GM, Evaluation of groundwater potential using geospatial techniques. *Applied Water Science* 7:2447–2461, DOI:10.1007/s13201-016-0433-0, 2016
[15] Ibrahim, K. and Ahmed, S.A. 2016. Geospatial technology for delineating groundwater potential zones in Doddahalla watershed of Chitradurga district, India. *The Egyptian Journal of Remote Sensing & Space Sciences* 19(2): 223–234, doi.org/10.1016/j.ejrs.2016.06.002
[16] Arulbalaji, P, Padmalal, D, Sreelash, K, GIS and AHP Techniques Based Delineation of Groundwater Potential Zones: A case study from Southern Western Ghats, India,*Scientific Reports* 9:2082, 2019
[17] Preetilata, M, Rose, RS, Krishnan, N, Spatial analysis of groundwater potential using remote sensing and GIS in the Kanyakumari and Nambiyar basins, India,*The Journal of the Indian Society of Remote Sensing* 37(4):681–692, 2009
[18] Ayodeji, Jand Michael AO, 2-D Electrical Resistivity Tomography for Groundwater Exploration in Hard Rock Terrain,*International Journal of Science and Research*4(4):156-163, 2015
[19] Seifu, K, Groundwater in Ethiopia: Features, Numbers and Opportunities. Heidelberg,

Germany. Springer,*Hydrogeology*, doi:10.1007/978-3-642-30391-3, 2013
[20] Nezar H, Ali EN, Mohammed, B,An integrated approach to groundwater exploration using remote sensing and geographic information system,*Journal of Water Resource and Protection* 4(9), DOI: 10.4236/jwarp.2012.49081, 2012
[21] Nasir, KA Groundwater detection in basement complex of Northwestern Nigeria using 2D electrical resistivity and offset wenner techniques,*International Journal of Science and Technology* 2(7):529-535, 2013
[22] George ZN and Yali EW, Integrated Assessment of Groundwater Potential Using Geospatial Techniques in Southern Africa: A Case Study in the Zambezi River Basin,*Water* 13(19), 2610, doi.org/10.3390/w13192610, 2021
[23] Thomas R and Duraisamy V, Hydrological delineation of groundwater vulnerability of droughts in semi-arid areas of western Ahmednagar district,*The Egyptian Journal of Remote Sensing & Space Sciences* 21 (2) 121-137, doi.org/10.1016/j.ejrs.2016.11.008, 2017
[24] Kumar, P, Herath, S, Avtar, R, Takeuchi, K, Mapping of groundwater potential zones in Killinochi area, Sri Lanka, using GIS and remote sensing techniques. Sustain,*Water Resources Management* 2, 419–430, doi.org/10.1007/s40899-016-0072-5, 2016
[25] Zhenfeng, S, Huqa, E, Bowen, C, Orhan, A, Yan, L, Integrated remote sensing and GIS approach using Fuzzy-AHP to delineate and identify groundwater potential zones in semi-arid Shanxi Province, China,*Environmental Modelling & Software* 134, 104868, 2020
[26] Fenta, AA, Kifle, A, Gebreyohannes, T, Hailu, Spatial analysis of groundwater potential using remote sensing and GIS-based multi-criteria evaluation in Raya Valley, northern Ethiopia. *Hydrogeology Journal* 23:195–206. doi.org/10.1007/s10040-014-1198-x. 2015.
[27] Kotchoni, DO, Vouillamoz, JM, Lawson, FM Relationships between rainfall and groundwater recharge in seasonally humid Benin: a comparative analysis of long- term hydrographs in sedimentary and crystalline aquifers,*Hydrogeology Journal* 27:447–457, doi.org/10.1007/s10040-018-1806-2, 2019.
[28] Suganthi, S, Elango, L and Subramanian, SK, Groundwater potential zonation by Remote Sensing and GIS techniques and its relation to the Groundwater level in the Coastal part of the Arani and Koratalai River Basin, Southern India,*Earth Sciences Research Journal* 17(2):87-95, 2013.
[29] Kumar A and Pandey AC Geoinformatics based groundwater potential assessment in hard rock terrain of Ranchi urban environment, Jharkhand state (India) using MCDM-AHP techniques,*Groundwater for Sustainable Development*, 2–3:27–41. doi.org/10.1016/j.gsd.2016.05.001, 2016
[30] Savita, RS, Mittal, HK, Satishkumar, U, Singh, PK, Yadav, KK, Jain, HK, Mathur, SH, Sham, D, Delineation of Groundwater Potential Zones using Remote Sensing and GIS Techniques in Kanakanala Reservoir Subwatershed, Karnataka, India,*Applied Sciences* 7(1):273-288, doi.org/10.20546/ijcmas.2018.701.030, 2017
[31] Ratnakumari, Y Rai SN, Thiagarajan S and Dewashish K, 2D Electrical resistivity imaging for delineation of deeper aquifers in a part of the Chandrabhaga river basin, Nagpur District, Maharashtra, India, *Current Science*, 102 (1), 2012
[32] Alagbe O.A., Sanusi S.O, Bamigboye A.A, 2D Electrical Resistivity Prospecting For Groundwater At The Premises Of The Staff Quarters Of Elizade University, Ilara Mokin, Ondo State, Nigeria, *World Journal of Innovative Research*, 6(2):100-108, 2019
[33] Jagadeshan G, Dessalegn, B, Abel A,Muralitharan, J, Evaluation of Groundwater Potential Zones by Resistivity Survey in a part of Masgerdo River Basin, Borena, Southern Ethiopia, *Indian Journal of Environmental Protection* 41(10):1160-1165, 2021
[34] Gabriel, EO, Ilesanmi, MO, Adebowale, EA, Rotimi, PA, Akinpelumi, A, Akinola, Kayode LO, Oluseun, AS, Integration of remote sensing, GIS and 2D resistivity methodsin groundwater development, *Applied Water Science*, 10:129doi.org/10.1007/s13201-020-01219, 2020
[35] Akwasi, AA, Charles, KS, Patrick, AM, Reginald, MN, 2D Electrical Resistivity Imaging for Delineation of Crystalline BasementAquifer in Northern Ghana, *Journal of Geoscience and Environment Protection*, 7:102-115 2019

Evaluation of the Strength properties of Sarooj based concrete design

Laila Hamis Nasser Al Gharbi[1], Sachin Kuckian[2], Kiran Kumar[3], Dr. Hussin Yahia[4], Dr. C. Venkata Siva Rama Prasad[5]

[1]Student, Middle East College, Muscat, Oman
[2]Lecturer, Middle East College, Muscat, Oman
[3]Lecturer, Middle East College, Muscat, Oman
[4]Assistant Professor, Middle East College, Muscat, Oman
[5]Assistant Professor, Peter's Engineering College, Telangana, India.

E-mail: 16f16125@mec.edu.om
skuckian@mec.edu.om
kpoloju@mec.edu.om
hyahia@mec.edu.om
cvsrprasad90@gmail.com

Abstract. Despite having several negative effects on human life and the environment, cement is the main component of concrete. Carbon dioxide is released into the atmosphere as a byproduct of cement production. As a result, cement production must be curtailed to lessen environmental impact. Therefore, it was crucial to find a material that could replace cement but wouldn't harm people or the environment in the process. When compared to cement, burnt clay (or sarooj) is seen as more eco-friendly because of its similar properties and uses. It was investigated in this study whether sarooj could be used to replace cement in concrete at different percentages (0%, 10%, 20%, 30%, and 40%). Several tests, including a sarooj chemical test, a slump test, a water absorption test, a rebound hammer test, a compressive strength test, and a flexural strength test, were performed on the concrete to ascertain its mechanical qualities. Based on the results, 10% sarooj is the best option because it outperforms 20%, 30%, and 40% in compressive and flexural strength, respectively.

Keywords: Sarooj, compressive strength, flexural strength.

1. Introduction

The Earth's natural materials can be used as a sustainable alternate to manufacture ones in construction. In the past, cement wasn't readily available in Middle Eastern countries, so Omani architects had to find alternatives like sarooj. Cement was crucial in the construction of the ancient world. Mortar was made from sand and lime by the Romans and Greeks, while cement was made from calcined gypsum by the Egyptians [1].

There are problems and significant challenges in the construction industry everywhere [2]; the earliest use of cement dates back to the Egyptian civilization. This includes issues such as the number of workers, the quality of their work, the cleanliness of the environment, the reliability of transportation and water supplies, the quality of raw materials, the longevity of construction products, their resistance to chemicals, and much more. Gypsum and limestone cement was used to build Egypt's massive pyramids. The durability of the cement used is evidenced by the fact that these pyramids have survived to the present day. However, it is important to note that modern cement is not the same as Egyptian cement. Later, during the Roman era, the idea of cement advanced. Pozzolana and slaked lime were used by the Romans in the construction of many noteworthy buildings [3]. When it comes to binders in modern construction, cement is widely regarded as indispensable. Marble, shells, chalk, slate, clay, and shale are some of the other materials that may be included in cement along with sand (silicon), limestone (calcium), iron ore, and bauxite (aluminium). Cement and water, through a series of chemical reactions, form a paste that is then used to secure the building's components together [4].

The increase in environmental pollution can be directly attributed to the building industry. New technologies, however, like the use of PET in concrete mixtures, have shown to be an effective answer to the problem of environmental pollution [5]. Saroj, a calcined clay pozzolana, was used as a cementing material in ancient times by being combined with water and lime. Houses, bridges, dams, and other man- made structures have all benefited from the use of sarooj in engineering. For thousands of years, sarooj was a staple in Omani forts, castles, and homes. However, these days the sarooj is only employed in the restoration of forts and castles [6]. In the traditional Omani method of making sarooj, the raw materials are gathered from fields and farms. These ingredients go through a sieve to get rid of any debris or large pieces. The soil is then wetted down and left to sit for at least three days so that it can absorb as much moisture as possible. The dry leaves are placed between the dry logs and help to facilitate the burning process. A bedform is used to compact the soil on the third layer, and a fire is lit within the container. Sarooj, or soil that has been burned, should be left in the same spot for about two months before being collected and used [7].

More time and money must be invested in the manufacturing of concrete in order to ensure that the material has the desired long-term durability, strength, and structural sustainability. Concrete is also the most important material used in construction. Cement is the backbone of concrete, but making it causes a lot of pollution and climate change. Pozzolana is a silicate that, when combined with calcium hydroxide, hydrates the cement and creates additional cementitious materials. Since more than 25% of hydrated Portland cement is calcium hydroxide, and since this component does not add to the durability and strength of concrete, its consumption is viewed favourably. Pozzolana reacts with calcium hydroxide to produce calcium silicate hydrate, which is crucial for the durability of the concrete. Pozzolana's chemical make-up and particle size determine concrete's tensile strength. The use of clay in place of cement results in chemical activation and a decrease in greenhouse gas emissions [8].

Cement has a long history of use in Iraqi construction and is therefore widely used there. The ancient structure, made of bricks and cement, represents one of the most pivotal figures in Iraq's rich history. Because of its advanced age, this architectural style necessitates modernizations and repairs, such as the installation of sarooj to reinforce the brick and cement columns. Cement mortar (organic binder) is used in place of epoxy... sarooj mortar (compressive strength 50MPa and 0.76MPa, respectively) [9].

Cement is frequently used as a road filler, despite the fact that it is extremely hazardous to the environment due to the presence of toxic metals and other inorganic elements it contains. This cement can then be used in a variety of primary applications after metal values have been recovered using an appropriate leaching processing strategy. After being processed into useful goods or used as a raw material substitute in the production of useful goods, sarooj becomes an effective product and garners interest as a potential raw material with less hazardous, reduced toxicity [10].

Stone buildings are typical in the Dhofar region of southern Oman. Due to modernization efforts initiated during Sultan Qaboos's rule beginning in the 1970s, however, industrial construction materials have largely replaced traditional ones. Due to the lack of research into conventional production methods and product use, many buildings are in jeopardy of being lost forever. Traditional masons in the area have told me that sarooj is used for joints and finishing touches on walls, while nurah is used for the construction of floors and walls [11].

Cement prices are a major issue currently. As a result, it needs to be sturdy enough. Because of this criticality, new approaches have been developed, some of which are discussed in research and make use of materials like Sarooj, Lime, Bagasse Ash, Rice Hush Ashpowder, and base Graduated Blast Furnace to reduce the swelling potential of cement and increase its strength. Concrete is responsible for the most notable increase in fluid cutoff expansion. It's worth noting that using ten percent lime, ten percent cement, ten percent Sarooj, and four percent cement plus lime all improved quality as much as was practically possible. To take advantage of this setting, we first augmented the swell weight with expansions of 4% sarooj, 4% lime+4% cement, and 6% lime+6% cement [12].

Modern construction sustainability rests on the development of new, low-emission building materials, making life cycle analysis an essential part of this process (LCA). Traditional cement, mortar, plasters, and other concrete-based products are increasingly used, contributing to urban heat and discomfort.

Sarooj is a man-made cementitious and pozzolana material obtained by the traditional calcination of unprocessed clayed soil. Although sarooj had been widely used in Oman and the neighboring countries for centuries in a variety of architectural and defensive contexts, its application in these areas has become much more restricted in recent decades. The primary purpose of this research is to enhance the physical mechanical properties of the ancient manufactured Sarooj and develop it into a binding product while also reducing its impact on the environment [13].

The primary goal of this research is to determine whether carbon-dioxide-causing carbon-cement substitution is feasible. While cement is essential to the making of concrete, it does have some unfavorable effects on both human and environmental health. Carbon dioxide is released into the atmosphere during the cement manufacturing process. Decreased cement production is therefore necessary to lessen environmental damage. Because of this, finding an alternative to cement that doesn't harm people or the planet was of the utmost importance. One environmentally friendly material that serves the same purpose and has similar properties to cement is burnt clay (sarooj). Experimental results from a range of cement replacement percentages (0-40%) in concrete using sarooj are presented. Numerous tests, such as a sarooj chemical analysis, a slump test, a water absorption test, a rebound hammer test, a compressive strength test, and a flexural strength test, were carried out to evaluate the mechanical properties of the concrete. From what we can tell, 10% sarooj has greater compressive and flexural strength than 20%, 30%, and 40% sarooj, so it's the one we recommend using.

These are our goals: -

1. Create a method for precisely mixing concrete

Create a concrete recipe using sarooj as a partial replacement for cement (10%, 20%, 30%, and 40%, respectively).

Third, to compare mechanical properties fairly across developed mix designs. This paper investigates the effects of sarooj cement replacement on concrete properties. Concrete mixtures with and without sarooj additions of 10%, 20%, 30%, and 40% were tested. A variety of tests, including those for chemical composition, slump, water absorption, rebound hammer strength, compressive strength, and flexural strength, were carried out to determine the concrete's qualities.

2. Literature review

Five different methods were tested to determine if rice husk, a common form of agricultural waste, could be used to make a highly reactive ash that could be used in the production of durable concrete- based products. The methods included optimising rice husk ash with refined combustion methods, testing the pozzolanic activity of the prepared husk ash, boosting mortar strength with the material, conducting a sulphate effect experiment to probe the substance's potential longevity, and assessing the impact of the material on the environment. There has been an increase in the percentage of rice husk ash in the mortar mix, but this has not resulted in increased compressive strength. [14].

Concrete and cement are two of the most widely used man-made materials, and their applications are currently under scrutiny. Cement and concrete production places a strain on limited resources like water while also generating a large amount of greenhouse gas emissions. We zero in on tweaks that can be made all the way through the value chain, like increasing the clinker content of cement and incorporating additional cementitious ingredients. If everyone gets on board, the sum of these small gains can have a huge impact on reducing emissions of greenhouse gases by as much as half. A slight binding of the cement mixture does occur, however, when moving up the value chain. [15]. Graphene oxide (GO) nanosheets and other 2D nanomaterials have demonstrated promising results in reinforcing and enhancing the functionality of composites based on ordinary Portland cement (OPC).

These researchers used tetraethyl orthosilicate to create a nanoscale sheet of graphene oxide (GO) (TEOS). To better understand graphene oxide (GO), we first defined it and then compared its distribution in cement pore solution to that of graphene oxide spheres (GOS). The thin silica layer served as a spacer, resulting in low aggregation [16].

There have been recent innovations in how organic waste fibres are put to use. Egypt has a large number of landfills where a wide variety of organic waste is dumped. Both human and environmental health are put at risk by these actions. Recycling and reusing the solid wastes, especially organic waste fibres, is encouraged by a variety of international waste management policies, and the decision- making community is well aware that doing so is an efficient way to get rid of these materials. It takes some finesse to recycle in this manner [17].

Studies have been conducted on the carbonating of concrete-based materials for carbon dioxide sequence, including their carbon dioxide sequestration capacity, the effects of carbonating on the mechanical and physical behaviour of concrete and recycling aggregate, methods for increasing efficiency, and ways to speed up the sequence. The effectiveness of cement-based materials at sequestering carbon dioxide is highly dependent on factors such as the amount of liquid water present, the size of the sample, the surrounding pressure, the concentration of carbon dioxide, and the temperature [18].

The cement matrix, formed when cement is combined with water, is the essential component of the composite. The remaining parts are made up of inorganic, metallic, and polymeric substances. Cement composites, of which concrete is the most common, are crucial to the building of safe and comfortable homes for people all over the world. Different types of fibre reinforced cement composites (FRCCs) with varying levels of mechanical strength can improve functionality. Cement-based composites, despite their widespread use, have poor tensile strength and fracture resilience [19].

Cement companies are using their enormous processing capacity to deal with the rising tide of municipal trash (MSW). The growing number of incineration units has raised the question of whether or not cement kiln co-processing (CKC) should be used as an additional method of managing municipal solid waste. The MSW treatment pressure and its temporal and geographical evolution in the provinces were statistically evaluated, and the CKC promotion and emission reduction potential for adjusting to regional MSW demand were estimated, using a large amount of recent data and information [20].

Carbon emissions from the concrete sector have emerged as a major environmental issue in recent years. Partial cement alternatives, nanotechnology, and carbon capture and storage are all being used to lower the concrete industry's carbon footprint. To make the cement industry more environmentally friendly, a suite of measures is proposed. incorporates three distinct methods for making eco-friendly cement (CCS, SCMs, and nanotechnology). [21] In doing so, it finds a happy medium between short- term gain and long-term sustainability.

The degradation of cementitious materials is typically caused by sulphate attack. In contrast to XRD, SEM-EDS, and 29Si NMR, we investigated the effects of pH value on the hydration phase, morphology, and microstructure of concrete hydration products under sulphate assault. Lowering the pH of the sulphate solution causes the concrete paste to precipitate gypsum crystals with platelet and columnar shapes, while ettringite growth is stifled. Research [22] investigated the full hydration and corrosion progression mechanism of concrete when sulphate was present.

While Portland cement's widespread use in construction is unquestionable, the manufacturing process is extremely energy intensive and contributes significantly to greenhouse gas emissions. This Expansion's alkali activators are responsible for nearly eighty-one percent of the carbon dioxide (CO2) emissions. By switching out the cement, we were able to cut emissions by 388.85 kgCO2- eq/m3. It was more likely that countries like Europe and the United States would take steps to reduce their carbon emissions. The environmental impact of transporting locally sourced materials by rail is lower than that of driving. When larger ships are used, carbon emissions may be reduced [23].

3. Materials Used

The materials that used in this project are:

3.1 *Sarooj*:

Sarooj is used as cementing material and it is produced by burning a specific type of clay using date trees as fuel. Sarooj is a mixture of water, lime, and calcined clay.

3.2 Cement:

Cement is made by burning clay and limestone at temperature 1300 C° and 1400 C°. Cement is the main substance in concrete, and it is used to bind the building materials together.

3.3 Aggregates:

Aggregates strongly affect the concrete properties, mixture proportions, and economy because the aggregates occupy from 60% to 75% of concrete. Therefore, choosing the proper type and quality of aggregates to use in concrete is significant because the aggregates occupy a high percentage. The types of aggregates that were used are coarse aggregate and fine aggregate. Coarse aggregates are crushed stones with a 5 mm size or gravel. Fine aggregates are sand or crushed stones less than 5 mm in size.

3.4 Water:

The water quantity controls several properties in concrete, such as permeability, durability, workability, compressive strength, and weathering. Therefore, controlling the water quantity in concrete is crucial for constructability and service life.

4. Methodology

To study the possibility of replacing the cement with sarooj, several tests must be done to check the properties of conventional concrete and 10%, 20%, 30%, and 40%. Slump tests, water absorption tests, rebound hammer tests, compressive strength tests, and flexural strength tests were performed on fresh and hardened concrete. These tests were performed after 7 and 28 days of curing, respectively.

5. Discussion andfindings

5.1. The Chemical Test of Sarooj

Cement's chemical composition may be analysed to learn a great deal about how concrete fails. Chemical tests are the ones used on concrete constructions the most commonly. Based on the chemical test results, sarooj consists of several compounds such as Sulfur Trioxide (SO3), Aluminum Oxide (Al2O3), Magnesium Oxide (MgO), Calcium Oxide (CaO) and Silicon Dioxide (SiO2), and the percentage of each component is 6.04%, 7.91%, 11.19%, 20.02%, and 26.56% respectively.

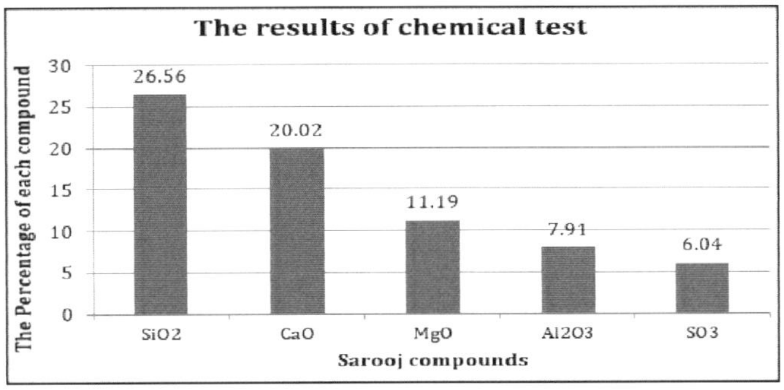

Figure 1. The Chemical Test Results of Sarooj

5.2. The Effect of Sarooj Percentage in Workability of Concrete

A concrete slump test measures the consistency of a mixture to establish how easily it will flow. The test not only ensures that each batch is consistent, but it also identifies any defects in the mixture so that the operator can correct them before pouring.

According to the slump test results, the result of conventional cement was 40 mm, while the results of 10%, 20%, 30%, and 40% of sarooj in concrete were 40mm, 35mm, 35mm, and 25mm respectively. It is observed that when the percentage of sarooj increased the workability of concrete decreased.

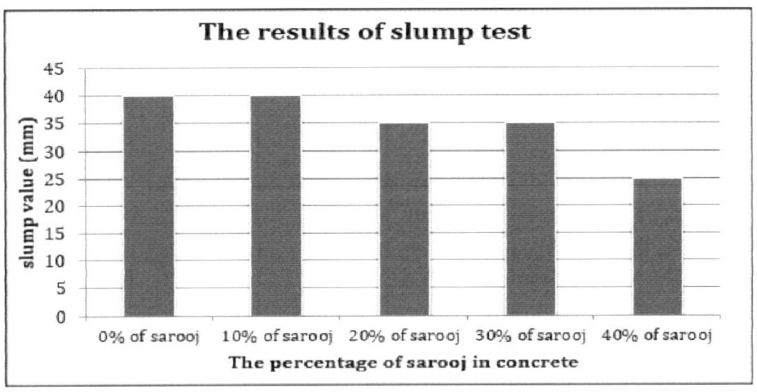

Figure 2. The slump Test Results

5.3. The Effect of Sarooj Percentage in Moisture content of Concrete

The water absorption test reveals how quickly the inner and outer surfaces of the concrete absorb water (sorptivity). Submerging concrete in water can give an estimate of its total (reachable) pore volume, but it won't reveal the concrete's permeability, which is more important for durability. Water absorption tests conducted after seven days of curing showed that regular concrete had a result of 13%, while concrete containing 10%, 20%, 30%, or 40% sarooj had results of 28%, 65%, 46%, or 37%.

Water absorption tests revealed that standard concrete, after 28 days of curing, absorbed 18% of the water used in the test, while concrete containing 10%, 20%, 30%, or 40% sarooj absorbed 19%, 26%, 22%, or 39% of the water used in the test, respectively. This is because sarooj concrete has a high pore content, which allows for a greater absorption of water.

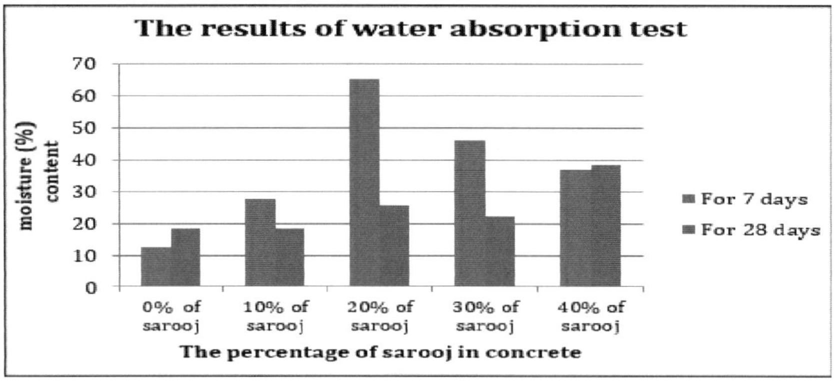

Figure 3. The Water Absorption Test Results

5.4. The Effect of Sarooj Percentage in Compressive Strength of Concrete by Using Rebound Hammer Test

Testing with a rebound hammer, which is nondestructive and provides a rapid and simple assessment of concrete's compressive strength, is becoming increasingly popular. The plunger in the rebound hammer, also called a Schmidt hammer, is operated by a strong spring, and moves through a tube housing. After seven days of curing, the results of a rebound hammer test showed that conventional concrete had a result of 26 N/mm^2, whereas concrete with 10, 20, 30, and 40 percent sarooj in it had values of 24, 23, and 21 N/mm2 respectively. In addition, the rebound hammer test results for standard concrete after 28 days of curing were 30 N/mm^2, while the results for concrete with 10, 20, and 40% sarooj were 26, 25, and 23 N/mm^2 correspondingly.

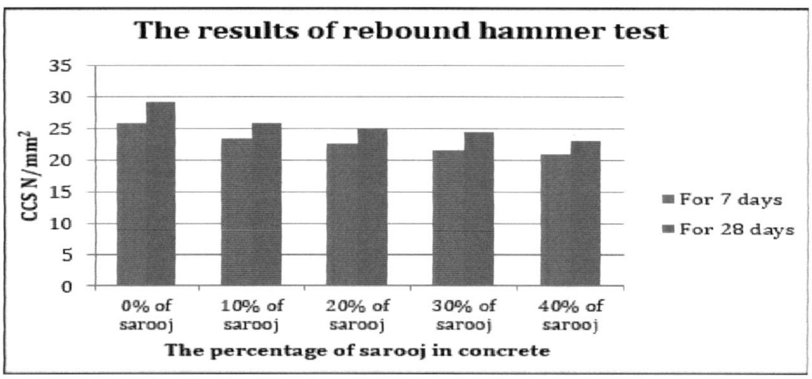

Figure 4. The Rebound Hammer Tests Results

5.5. The Effect of Sarooj Percentage in Compressive Strength of Concrete by Using Compressive Strength Test

The ability of a substance to withstand cracking and splitting defines its compressive strength. In this test, we determined the maximum compression that could be applied to the cement specimen without breaking it, as well as the impact force applied to both sides of the mortar specimen made of cement. Following seven days of curing, the compressive strength of standard concrete was determined to be 31 MPa, whereas the values for 10%, 20%, 30%, and 40% were 22 MPa, 21 MPa, 19 MPa, and 15 MPa, respectively. The compressive strength of conventional concrete after 28 days of curing was 33 MPa, while the results for 10%, 20%, 30%, and 40% sarooj in concrete were 28 MPa, 26 MPa, 24 MPa, and 23 MPa, respectively. The compressive strength of conventional concrete is greater than that of sarooj concretes after 7 and 28 days of curing. As the sarooj proportion in concrete increased, the compressive strength fell, but overall, sarooj concrete got stronger over time. Early strength in concrete is a result of the presence of tricalcium silicate (C3S), which is present in cement. The cement content of traditional concrete is higher than that of sarooj concrete. Because of this, the reaction time for sarooj is much longer than that of cement. The sarooj thus functions as a retardant, altering the rate of hydration at the outset, and thus the rate at which strength in built [24].

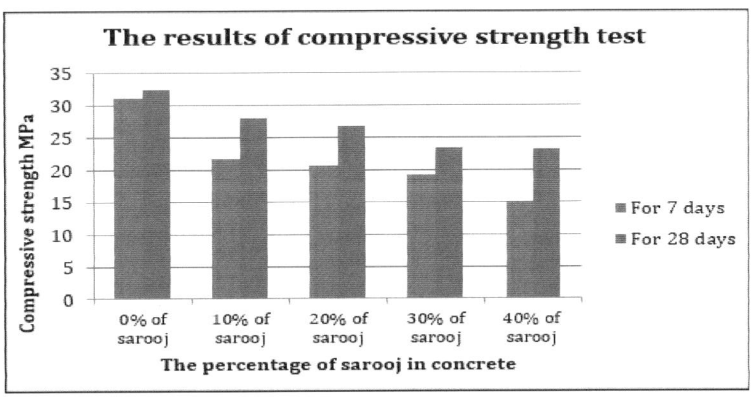

Figure 5. The Compressive Strength Tests Results

5.6. The Effect of Sarooj Percentage in Flexural Strength of Concrete by Using Flexural Strength Test

Concrete's tensile strength can be inferred indirectly through the flexural test. This technique is used to calculate the bending strength of a concrete beam or slab. Concrete's flexural test results are reported in terms of the material's rupture modulus, which is abbreviated as (MR) and is given in either megapascals (MPa) or pounds per square inch (psi).

After curing for 28 days, the flexural strength test revealed that regular concrete had a result of 11 MPa, while the results for 10%, 20%, 30%, and 40% were 7, 4, 3, and 2 MPa, respectively. Test results were obtained because the standard concrete had cured for 28 days.

As a result, it appears that conventional concrete has a higher flexural strength than sarooj does. Furthermore, it demonstrates that sarooj proportion is inversely related to flexural strength.

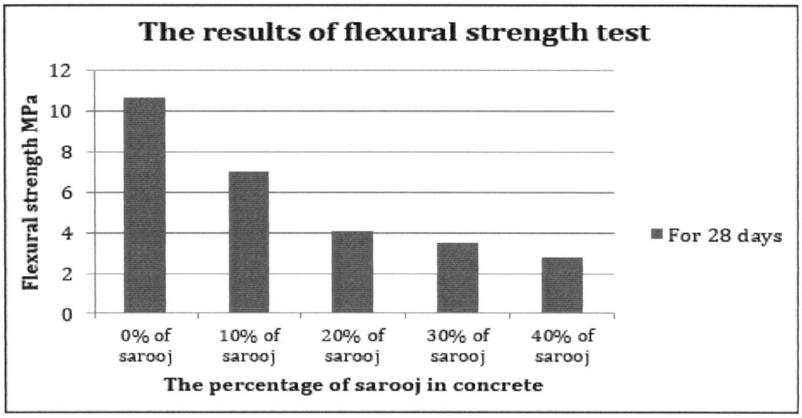

Figure 6. The Flexural Strength Tests Results

6. Conclusion

- The percentages of the various ingredients in sarooj are as follows: 6.04% sulphur trioxide (SO3), 7.91% aluminium oxide (Al2O3), 11.19% magnesium oxide (MgO), 20.02% calcium oxide (CaO), and 26.56% silicon dioxide (SiO2).
- Compared to sarooj concrete, regular concrete is easier to work with. The workability of concrete ranged from 40 mm for 10% sarooj to 25 mm for 40% sarooj. What this means is that adding more sarooj to concrete makes it less workable.
- In comparison to standard concrete, karoo concrete has a higher moisture content. The high pore content of sarooj concrete is responsible for its high-water absorption [25].
- When comparing the compressive strength of conventional concrete and sarooj concrete, the former has a higher value after 7 days of curing and after 28 days of curing, as measured by the rebound hammer test and the compressive strength test. After seven days of curing, the compressive strength of both regular concrete (26 N/mm2) and concrete with 10% sarooj (24 N/mm2) was measured using the rebound hammer test. The compressive strength test revealed that after seven days of curing, conventional concrete had a compressive strength of 31 MPa, while concrete containing 10% sarooj had a compressive strength of 22 MPa. A rise in compressive strength was observed after 28 days of curing for both traditional and sarooj concrete. Compressive strength also decreased with increasing sarooj content.
- After 28 days of curing, conventional concrete had a flexural strength of 11 MPa, while sarooj at 10% only had a flexural strength of 7 MPa. The flexural strength also rose along with the sarooj content. For comparison, the flexural strength of 10% sarooj was 7 MPa, while that of 40% was only 2 MPa.
- Since 10% sarooj has higher compressive and flexural strength than 20%, 30%, and 40% sarooj, it can be concluded that this percentage can be used.

7. Acknowledgement

We would like to extend our gratitude to the management and lab staffs of the Middle East College for their assistance in the successful completion of the studies. We would also like to thank those members of the faculty who participated either directly or indirectly to the project.

8. References

[1] Al-Rawas, A A., Hago A W, Corcoran T C, and Al-Ghafri, K. M. (1998). Properties of Omani artificial pozzolana (sarooj). Applied Clay Science, 13(4), 275-292.
[2] Kuckian S, and Dalvi, A. (2020). Strength Possibilities on Fly ash Based Interlock Pavers. In Applied Mechanics and Materials (Vol. 897, pp. 12-19). Trans Tech Publications Ltd.
[3] Al-Rawas, A. A., Hago, A. W., and Al-Sarmi, H. (2005). Effect of lime, cement and Sarooj (artificial pozzolan) on the swelling potential of an expansive soil from Oman. Building and environment, 40(5), 681-687.
[4] Hago A W, Al-Rawas A A., and Al-Riyami, A. (2002). Effect of varying cement content and curing conditions on the properties of sarooj (artificial pozzolana). Building and environment, 37(1), 45-53.
[5] Salim O J, and Kuckian S (2020). Solidification of Recycled Concrete using Plastic waste. Solid State Technology, 63(6), 19506-19529.
[6] Al-Rawas A A., Hago A W, Corcoran T C, and Al-Ghafri K M (1998). Properties of Omani artificial pozzolana (sarooj). Applied Clay Science, 13(4), 275-292.
[7] Meddah M S, Benkari, N, and Al-Busaidi M (2019, February). Potential Use of Locally and Traditionally Produced Bending Construction Material. In IOP Conference Series: Materials Science and Engineering (Vol. 471, No. 4, p. 042008). IOP Publishing.
[8] Hago A W, and Al-Rawas, A A (2008). Design of a Rotary Kiln for Production of Sarooj. The Journal of Engineering Research [TJER], 5(1), 55-61.
[9] Dakhil, A, Naji, Z M, and Al Jasim, S (2022). Improving clay brick Column's compression capacity using CFRP sheets and Reinforced concrete jacketing. Iraqi Journal of Civil Engineering, 15(2), 29-40.
[10] Dan, A. K., Bhattacharjee, D., Ghosh, S., Behera, S. S., Bindhani, B. K., Das, D., and Parhi, P. K. (2021). Prospective utilization of coal fly ash for making advanced materials. In Clean Coal Technologies (pp. 511-531). Springer, Cham.
[11] Takubo Y, Matsumoto N, Otsubo M, Hayashi K, Koshihara M, Benkari N. 2022 Traditional building materials in coastal Dhofar: A sustainability perspective. In: KEM ;913. Trans Tech Publications Ltd.
[12] Mishra, S., and Prajapati, B. EFFECT OF CORN HUSK FIBER ON THE CONSISTANCY OF EXPANSIVE SOIL.
[13] Meddah, M. S., Benkari, N., Al-Saadi, S. N., and Al Maktoumi, Y. (2020). Sarooj mortar: From a traditional building material to an engineered pozzolan-mechanical and thermal properties study. *Journal of Building Engineering*, *32*, 101754.
[14] Hu, L., He, Z., and Zhang, S. (2020). Sustainable use of rice husk ash in cement-based materials: Environmental evaluation and performance improvement. Journal of Cleaner Production, 264, 121744.
[15] Habert, G., Miller, S. A., John, V. M., Provis, J. L., Favier, A., Horvath, A., and Scrivener, K. L. (2020). Environmental impacts and decarbonization strategies in the cement and concrete industries. Nature Reviews Earth and Environment, 1(11), 559-573.
[16] Lin, J., and Shamsaei, E. (2020). Basquiroto de Souza F.; Sagoe-Crentsil K.; Duan WH Dispersion of graphene oxide–silica nanohybrids in alkaline environment for improving ordinary Portland cement composites. Cem. Concr. Compos, 106(10348810.1016).
[17] Guirguis, M. N., and Moussa, R. R. (2019, November). Investigation on utilizing garbage as a resource for a sustainable neighbourhood: Case study of a neighbourhood in New Cairo, Egypt. In IOP Conference Series: Earth and Environmental Science (Vol. 397, No. 1, p. 012018). IOP Publishing.
[18] Wang, D., Xiao, J., and Duan, Z. (2022). Strategies to accelerate CO2 sequestration of cement-based materials and their application prospects. Construction and Building Materials, 314, 125646.
[19] Moosazadeh, M. K., and Zandi, Y. (2019). Comparative Evaluation of Sarooj Mortar Mix Design according to Various Sources and Best Mix Design Determination.
[20] Kosajan, V., Wen, Z., Zheng, K., Fei, F., Wang, Z., and Tian, H. (2021). Municipal solid waste (MSW) co-processing in cement kiln to relieve China's Msw treatment capacity pressure. Resources, Conservation and Recycling, 167, 105384.
[21] Poudyal, L., and Adhikari, K. (2021). Environmental sustainability in cement industry: An integrated approach for green and economical cement production. Resources, Environment and Sustainability, 4, 100024.
[22] Zhang, G., Wu, C., Hou, D., Yang, J., Sun, D., and Zhang, X. (2021). Effect of environmental pH values on phase composition and microstructure of Portland cement paste under sulfate attack. Composites Part B: Engineering, 216, 108862.
[23] Sandanayake, M., Law, D., and Sargent, P. (2022). A new framework for assessing the environmental impacts of circular economy friendly soil waste-based geopolymer cements. Building and Environment, 210, 108702.
[24] Moosazadeh, M. K., and Zandi, Y. (2019). Comparative Evaluation of Sarooj Mortar Mix Design according to Various Sources and Best Mix Design Determination.
[25] Hago, A. W., Al-Rawas, A. A., and Al-Sidairi, A. (2002). Effect of the Fineness of Artificial Pozzolana (Sarooj) on the Properties of Lime-Pozzolana Mixes. Sultan Qaboos University Journal for Science [SQUJS], 7(2), 251-258.

IOP Conf. Series: Materials Science and Engineering 1282 (2023) 012014 doi:10.1088/1757-899X/1282/1/012014

Experimental study on concrete by partial replacement of fine aggregate with steel slag and cement with metakaolin and glass powder

E Sandhyarani[1] and Dr. Ch Srinivasa Rao[2]
[1] Student, CMR Technical campus, Medchal, Telangana
[2] HOD, CMR Technical campus, Medchal, TelanganaE-mail: sandhyaranierrabelli@gmail.com

Abstract. The present study will describe the characteristics and the effects of concrete properties when fine aggregate is partially substituted by steel slag and cement partially substituted by metakaolin and glass powder at 0%, 10%, 20% and 30. Compressive, tensile, flexural strength, and durability tests were conducted on the cube and cylinder specimens. The test results were compared with conventional concrete specimens. The tests are conducted on specimens for 7days and 28days. Based on the outcomes of the study, it was concluded that replacing cement and fine aggregate with glass powder, metakaolin, and steel slag increases the strength of concrete and reduces attack at 30% replacement. Hopefully, the results have clearly demonstrated the concrete constructions can also be made from eco-friendly alternative materials which are cost-effective and sustainable for longtime.

Keywords: metakaolin, glass powder, conventional, eco-friendly, cost-effective, sustainable.

1. Introduction

The world uses a great deal of concrete for construction. Water, fine aggregate, and cement are all mixed to create the mix. In the present day, the construction industry is growing rapidly and requires an enormous amount of Concrete [3]. The demand for concrete leads to greater utilization of cement and aggregates[5]. As we know, cement has outstanding binding properties, but the manufacturing of cement leads to greater emission of CO_2 [7]. Also, the increasing demand for fine aggregates/natural sand and becoming more expensive which leads to the reduction of natural resources [1]. To reduce those risks, some of the alternative materials are introduced in the construction industry. Some of the alternative materials are steel slag, metakaolin and glass powder, rice husk. The present study utilizes some of alternative materials like steel slag, metakaolin, and glass powder as partial replacements with different percentage replacements to make concrete[2].

.

1.1. Research significance

Most researchers have conducted tests on concrete with one or two alternative materials added as partial replacements[6]. However, this study aims to make concrete specimens with three alternative materials, Cement can be substituted partially with metakaolin, glass powder, and fine aggregate can be substituted partially with steel slag [4].

2. Materials

A concrete mix is made up of fine aggregates, coarse aggregates, cement, and water. In the present study, concrete is prepared with Metakaolin and glass powder are used to replace part of the fine aggregate and cement with steel slag.

- Coarse aggregate is an aggregate particle which retain on a 4.75mm sieve is called as. It is a larger size particle. A coarse aggregate of 12.5mm is used in the present study.
- Fine aggregate is an aggregate particle which passes through a 4.75mm sieve is called as fine aggregate. It is smaller in size
- Metakaolin is a high quality pozzolanic material. It is made by calcining kaolin clay at a temperature between 650°C to 850°C.
- Glass powder is a powder made from waste ground glass. When the particle size of glass is less than 75m, it is amorphous and may be pozzolanic. Because of its silica content, ground glass is considered a pozzolanic material and as such exhibit's properties like other pozzolanic materials.
- Steel slag is a by-product of steel. When impurities are separated from molten metal, it is formed. Slag as an alternative material for concrete reduces the need for natural rock, thus reduces the preserving our natural rock resources.

Fig.1 Metakaolin

Fig.2 Glass powder

Fig.3 Steel slag

3. Methodology

Literature review is studied in the first phase of research. Identifying the gap by analyzing the literature reviews. Then objectives are taken from the gap of literature review. Next is to find the materials and collect them from the source/ locally available store.

After that, materials properties are noted. Then, mixing of materials are done by selecting proper grading of materials and mix them as per proportions. First Conventional Concrete is made without any percentage of replacement. Then alternative materials like steel slag, metakaolin and glass powder are added in percentage wise of 10%, 20% and 30% and then mix is made as per Mix design of M30 Grade. For both concrete mixes cubes and cylinders were cast. Then, all type of specimens which are cast were placed in water tank for curing for 7days and 28 days. For 7th day, specimens were removed from tank and compressive, tensile, and flexural test was performed. Same tests are applied for 28th day of curing. Durability tests like acid attack and alkali attack test were also conducted.

4. Tests

Various tests are conducted on concrete specimens made with alternative materials like steel slag, glass powder and metakaolin along with ordinary concrete specimens.

4.1 Compression strength test

The specimen is compressed and deformed at various loads is recorded. Compression Test is conducted using standard Compression Testing Machine.

$$F=P/A$$

where, F is Compressive strength kn/m^2, P is the force applied to the sample in kn, and area of the sample(A) in m^2.

4.2 Tensile strength test

Effect of alternative materials on tensile strength of hardened concrete with alternative resources is proposed to be studied at 7 days and 28 days by conducting a tensile test determines the behaviour of materials under crushing loads. The maximum load applied to the specimen when first crack appears is then noted.

$$T= 2P/\Pi l d^2$$

Where, tensile strength(T) in KN/m^2, load(P) in KN, Length(L) in meters, and Diameter(d) in meters.

4.3 Flexural strength test

Specimens were kept in water for 7 and 28 days for curing, and then tested in dry condition.

$$F=PL/bd^2$$

Where, F is flexural strength in kn/m^2, applied load(P) on the sample in KN, Length(L) in meters, Breadth(b) in meters, and Diameter(d) in meters.

4.4 Durability test

Effect of alternative materials on durability properties of hardened concrete with alternative resources is proposed to be studied at 28 days by conducting an alkali and acid attack test determines the behavior of materials. The present study uses H_2SO_4 and NaOH as acid and alkali, respectively. An alkali attack test was conducted by immersing specimens in 4% of NaOH solution.

5. Result and discussion

Test results are obtained by testing the specimens with UTM in accordance with IS: 516 - 1959.

Table 1: Compression strength results for 7, 28 days

Mix designation	Compressive strength after 7 days (N/mm^2)	Compressive strength after 28 days (N/mm^2)
MGS_0	35.11	38.03
MGS_{10}	36.32	31.45
MGS_{20}	38.2	40.65
MGS_{30}	37.54	39.23

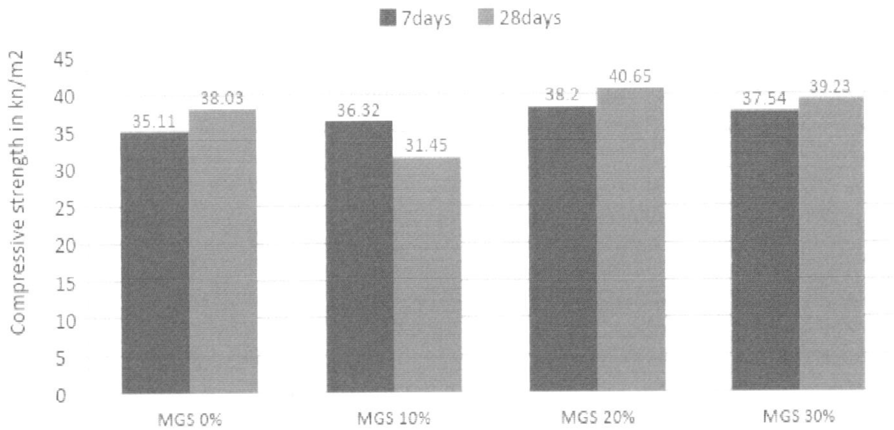

Graph 1: 7 days & 28 days compressive strength

Based on the above test results, we can conclude that the compressive Strength of the alternative materials Mix concrete is increasing from $35.11N/mm^2$ for MGS_0 to $37.54N/mm^2$ for MGS_{30} within seven days. The strength is slightly increased from $38.03N/mm^2$ for MGS_0 to $39.23N/mm^2$ for MGS_{30} at 28 days. However, the compressive strength results for different Mixes show higher values than MGS_0

Table 2: Tensile strength results for 7, 28 days

Mix designation	Tensile strength after 7 days (N/mm^2)	Tensile strength after 28 days (N/mm^2)
MGS_0	3.4	4.2
MGS_{10}	3.8	4.7
MGS_{20}	4.2	5.2
MGS_{30}	5.5	5.91

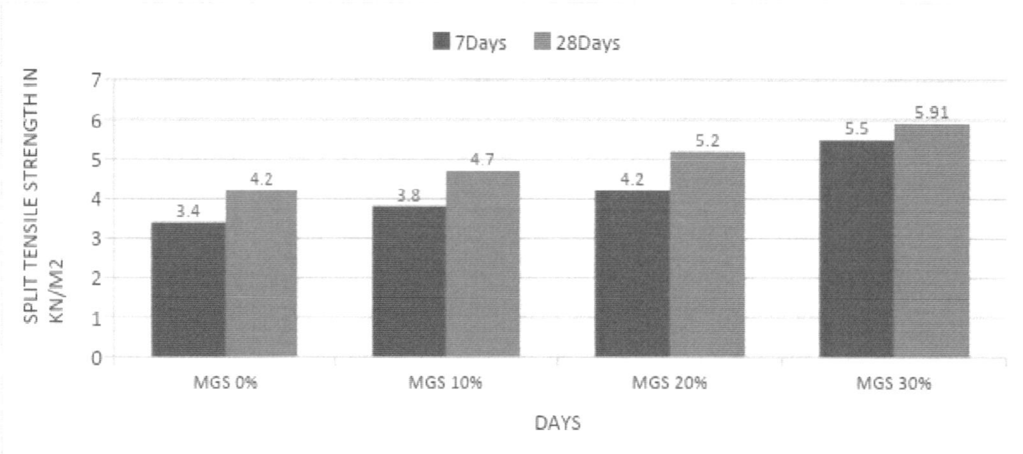

Graph 2 : 7days & 28 days split tensile strength

Based on the above test results, we can conclude that the split tensile Strength of the alternative materials Mix concrete is increasing from 3.4N/mm^2 for MGS_0 to 5.5N/mm^2 for MGS_{30} after seven days. The strength is slightly increased from 4.2N/mm^2 for MGS_0 to 5.91N/mm^2 for MGS_{30} at 28 days The Flexural strength of the alternative materials Mix concrete is increasing from 2.4N/mm^2 for MGS_0 to 2.5N/mm^2 for MGS_{30} after seven days. The strength is slightly increased from 3.5N/mm^2 for MGS_0 to 3.6N/mm^2 for MGS_{30} at 28 days.

Table 3: Flexural strength results for 7 and 28 days

Mix designation	Flexural strength after 7 days (N/mm^2)	Flexural strength after 28 days (N/mm^2)
MGS_0	2.4	3.5
MGS_{10}	2.1	3
MGS_{20}	2.8	3.8
MGS_{30}	2.5	3.6

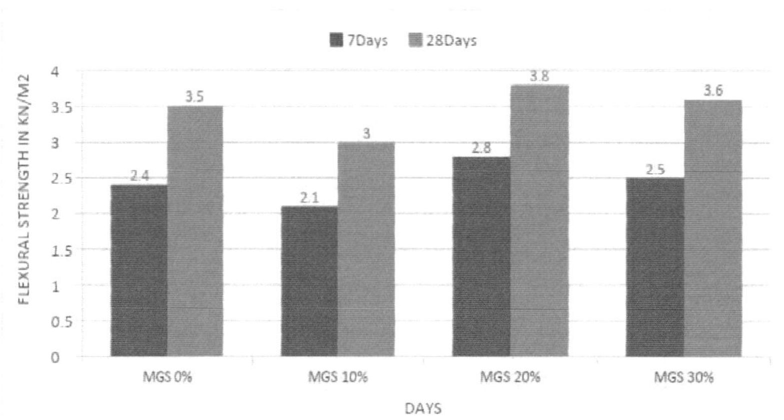

Graph 3: 7days & 28 days flexural strength

Table 4: Durability test for 28 days

Mix designation	Acid attack test at 28 days (%)	Alkali attack test at 28 days (%)
MGS_0	1.8	2.2
MGS_{10}	1.7	1.9
MGS_{20}	2.1	1.75
MGS_{30}	1.1	0.95

The durability test of the alternative materials mix concrete, shows the acid and alkali attack test gives minimum attack at 30% replacement.

6. Conclusion

The results obtained from the tests concluded the following

- By increasing the percentage of MGS, there is an increasing in the strengths of concrete. But for compression strength at 10% replacement of MGS, there is slight decrease in strength compared to conventional concrete strength.
- The optimum replacement percentage of Cement containing slag and fine aggregates containing metakaolin and glass powder is at 20%.
- Concrete durability properties are less affected by alternative materials at 30% replacement.

References

[1]. S. Rahman, M.N. Uddin "Experimental Investigation of Concrete with Glass Powder as Partial Replacement of Cement" Civil Engineering and Architecture 6(3): 149-154, 2018, DOI: 10.13189/cea.2018.060304.

[2]. R. Padmapriya, V.K. Bupesh Raja, V.G. Kumar and Baalamurugan.J "Effect on Properties of Concrete In Partial Replacement Of Fine Aggregate By Steel Slag And Cement By Metakaolin" Vol. 12 | No. 4 |1744 - 1751| October - December | 2019 ISSN: 0974-1496 | e-ISSN: 0976-0083.

[3]. Abhijeet B, Mayur Singi, Ajay Bidare "Experimental Study on Concrete as Partial Replacement of Cement by Glass Powder" International Research Journal of Engineering and Technology (IRJET) e-ISSN: 2395-0056 www.irjet.net p-ISSN: 2395-0072 Volume: 05 | Issue: 07 | Page 243.

[4]. Dilip Kumar Roy, Amitava Sil "Effect of Partial Replacement of Cement by Glass Powder on Hardened Concrete" International Journal of Emerging Technology and Advanced Engineering (ISSN 2250-2459, Volume 2, Issue 8, August 2012).

[5]. S.Deepak, Vijay.G, Vinoth kumar.R, and Gokulakannan. P "An Experimental Study on Concrete by Partial Replacement of Glass Powder and Quartz Powder in Cement with Steel Fiber" ISSN:0974-4290, Vol.10 No.8, pp 259-262.

[6]. Subramani.T and S.B.Sankar "Experimental Study on Properties of Concrete using Cement with Glass Powder "ISSN (e): 2250-3021, ISSN (p): 2278-8719 Vol. 05, Issue-05 (May. 2015), ||V3||PP 43-53.

[7]. M. Mallesh, Suresh "Experimental study on concrete with partial replacement of cement with Ggbs and fine aggregate with steel slag" International Research Journal of Engineering and Technology (IRJET) e-ISSN: 2395-0056 Volume: 04 | issue: 09 | Sep – 2017 p-ISSN: 2395-0072.

Comparative Study on Seismic Response of Regular and Plan Irregular Base isolated structures

Dr. Kavitha B[1] **Ajay Kumar Kore** [2]
[1]Assistant Professor, Department of Civil Engineering, National Institute of Technology, Warangal, Telangana, India-506004
[2], MTech student, Department of Civil Engineering, National Institute of Technology, Warangal, Telangana, India-506004. Email:[1]kavithab@nitw.ac.in, [2] ajaykumarkore01@gmail.com, Contact: [1]+91-9944139989

Abstract: Earthquake is one of most destructing threat which cause severe damage to life and property. Recently, due to increasing demand and need for the present generations has made the designers and engineers to plan for structures having irregularities. It is impossible to avoid designing uneven buildings in reality. It is important to research how this irregular structure behaves in earthquake-prone areas, especially for structures with base isolation. When compared to regular structures with base isolation, structures with plan irregularity react differently. When a reinforced concrete building with varied plan configurations, such as an H, U, or L shape, is subjected to seismic loading, the impact of plan irregularity is explored. In the current study, a regular structure is also considered to compare the responses of the isolated irregular base structure. Building response spectrum analysis for seismic analysis was done in seismic zone III using IS 1893-2002 code [5]. The structural analysis software tool ETABS 18 was used to conduct the analysis. Among different types of isolators present, in the present study LRB isolator and friction isolators was used as a Nonlinear Support. The dynamic response of the structure such as base shear, time period and story displacement are used to compare the regular and irregular structures. The effectiveness of the base isolators in the structures in earthquake prone area is evaluated in the present study.

Key words: Base isolation, irregular, LRB isolator, Friction isolator, Seismic Analysis, Response Spectrum Analysis, Etabs

1. INTRODUCTION:

One of nature's biggest dangers, earthquakes have historically resulted in a substantial number of fatalities and significant property damage, particularly in areas that are more vulnerable to them. Because of these earthquakes, engineers and architects developed a variety of crucial design criteria to get beyond the standard design procedure. The seismic base isolation can be utilised to effectively address a variety of seismic design issues thanks to well-established methodologies that have been examined by several researchers.

One of the most successful strategies and one that has gained relevance around the world is the employment of base isolation techniques to safeguard buildings against damage brought on by seismic strikes. This is due to base isolation techniques, flexible bases that largely decouple the structure from ground motion, and the fact that structural reaction accelerations are often lower than ground accelerations. In general, the structure's displacements and acceleration reactions may be reduced as extra damping is added. It was observed from the literature reviews that the researchers had done a lot of work regarding the use of base isolation technique. They have modelled different story RC framed structures and are analyzed using different base isolators, using different seismic analysis methods and found the effectiveness of the different isolators by comparing the base shear, displacements, time period, story drifts etc (Ashok et al., (2012) [1], Reddy and Ramesh (2015) [2], Darshale and Shelke (2016) [3], Babu and Jose, (2018) [4]). This study's primary goal is to assess how plan irregularity affects a base isolated structural system's response. The current study investigates if a regular base isolated structure performs better during an earthquake than an irregular base isolated structure. It also investigates whether a low rise building performs better than a medium rise structure with base isolator and imperfections. In the current study, G+5 and G+15 story reinforced concrete moment-resisting frame buildings with and without base isolation are taken into consideration in order to observe the impact of base isolation.

2. MODELLING AND ANALYSIS OF THE STRUCTURE

This study includes the modeling and analysis of the structures using ETABS 18 structural analysis software. The following design steps are performed:

- Modelling of the G+5 & G+15 storey RC framed building for both regular & plan irregular structure
  and assigning all the material and section properties for all the members of the building such as beams,columns, slabs etc.
- Assigning the joint restraints properties as fixed at the bottom of the column.
- Assigning the loads on the members and also the lateral loads.
- Analyzing the fixed model for given loads for response spectrum for zone III and type II soil.
- Assigning properties of base isolators as LRB isolator and Friction isolator using N-link spring at baseof the structure.
- Analyzing the base isolated structure for response spectrum for zone III.
- Comparing the results of earthquake zone III for fixed base and base isolated RC buildings.

2.1. Material, load and section Properties:

Grade of concrete = M30

Grade of steel = Fe415

Floor Live load = 3 kN/m^2

Live Roof load = 1.5 kN/m^2

Floor finish for all floors = 1.5 kN/m^2

External wall =230 mm thick and load is 11.7 kN/m

Internal wall = 115 mm thick and load is 5.8 kN/m

Parapet height = 1 m and load is 4.6 kN/m

Density of wall masonry = 20 kN/m^2

G+5 model	**G+15 model**
Beam size = 230 mm x 450 mm Column size= 400 mm x 400 mm Slab thickness = 150 mm	Beam size = 230 mm x 450 mm Column size for bottom 6 storeys = 600 mm x 600 mm Column size for top 10 storeys =400 mm x 400 mm Slab thickness = 150 mm

2.2. Isolator Properties:

Table 2.1: Properties of LRB Isolator

U1 Linear Effective Stiffness	1500000 kN/m
U2 and U3 Linear Effective Stiffness	800 kN/m
U2 and U3 Nonlinear Stiffness	2500 kN/m
U2 and U3 Yield Strength	80 kN
U2 and U3 Post Yield Stiffness Ratio	0.1
U1 Post Yield Stiffness Ratio	0.1

Table 2.2: Properties of Friction Isolator

U1 Linear Effective Stiffness	15000000 kN/m
U2 and U3 Linear Effective Stiffness	750 kN/m
U2 and U3 Nonlinear Stiffness	15000 kN/m
U2 and U3 friction coefficient slow	0.03
U2 and U3 friction coefficient fast	0.05
Rate parameter	40
Radius of sliding surface	2.23

Table 2.3: Models and their representation:

S.NO	Model Name	Model Representation
1.	Regular G+5 plan model	G+5 R
2.	'H' shaped G+5 plan irregular model	G+5 H
3.	'U' shaped G+5 plan irregular model	G+5 U
4.	'L' shaped G+5 plan irregular model	G+5 L

5.	Regular G+15 plan model	G+15 R
6.	'H' shaped G+15 plan irregular model	G+15 H
7.	'U' shaped G+15 plan irregular model	G+15 U
8.	'L' shaped G+15 plan irregular model	G+15 L

2.3 Plan view of all the models:

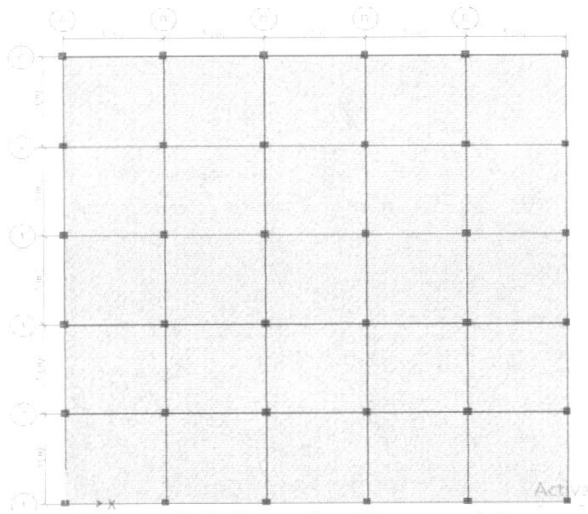

Figure 2.1: Regular Plan model

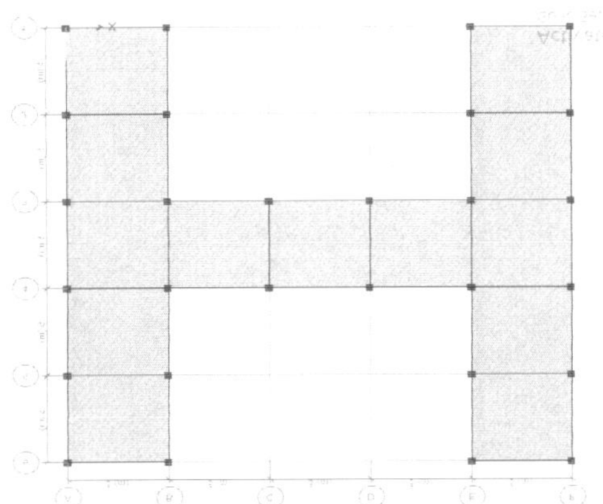

Figure 2.2:'H'-shaped irregular Plan model

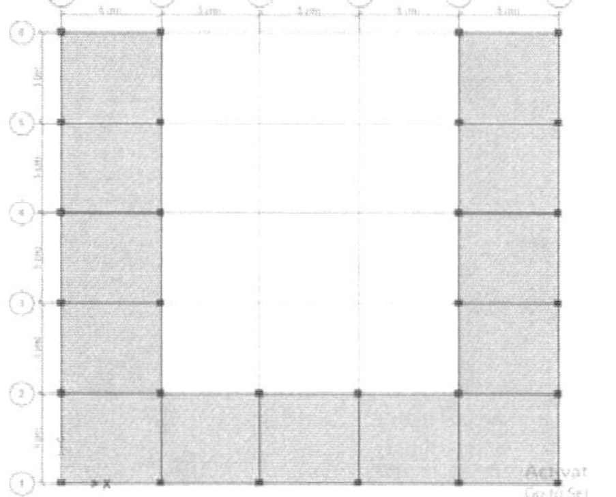

Figure 2.3:'U'-shaped irregular Plan model

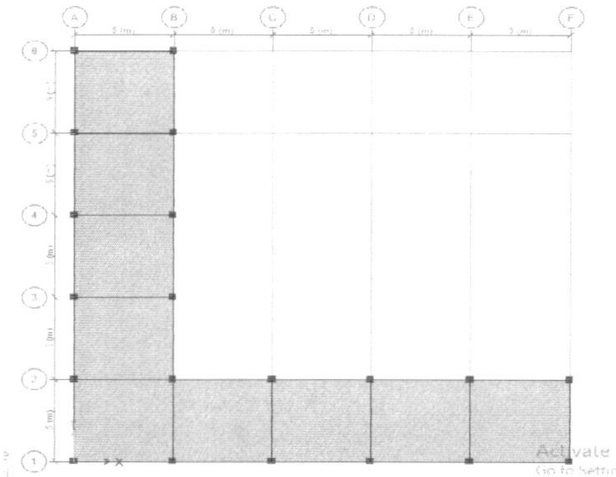

Figure 2.4:'L'-shaped irregular Plan model

3. RESULTS AND DISCUSSION

The dynamic response of a regular and irregular structures was evaluated for isolated and fixed case to show the effect of isolation for both G+5 and G+15 story structures. The fixed and isolated structures' base shear, t time period and maximum story displacement values characterize the seismic performance of the structures. The maximum values at each story were evaluated and compared for the analysis.

Table: 3.1 Base shear for G+5 regular and plan irregular models

S.no	Models	Fixed Base (kN)		LRB isolator (kN)		Friction isolator (kN)	
		X direction	Y direction	X direction	Y direction	X direction	Y direction
1	G+5 R	799.56	799.56	370.95	370.95	361.98	361.98

2	G+5 H	523.25	549.52	260.19	263.54	255.22	258.24
3	G+5 U	521.48	549.94	261.40	263.61	255.86	258.32
4	G+5 L	376.86	376.86	195.27	195.27	191.50	191.50

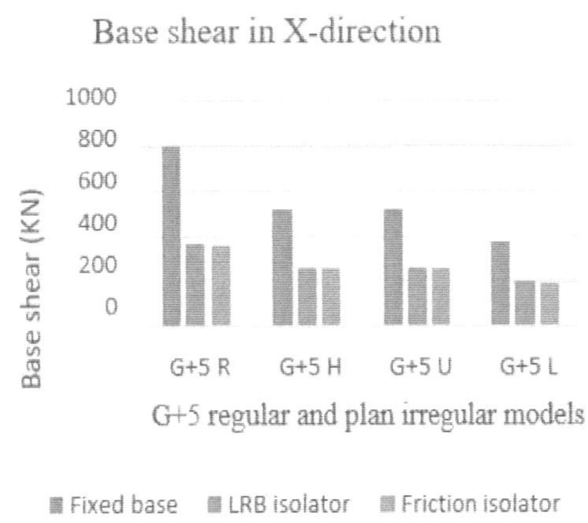

Figure 3.1: Base shear for G+5 models inX-direction

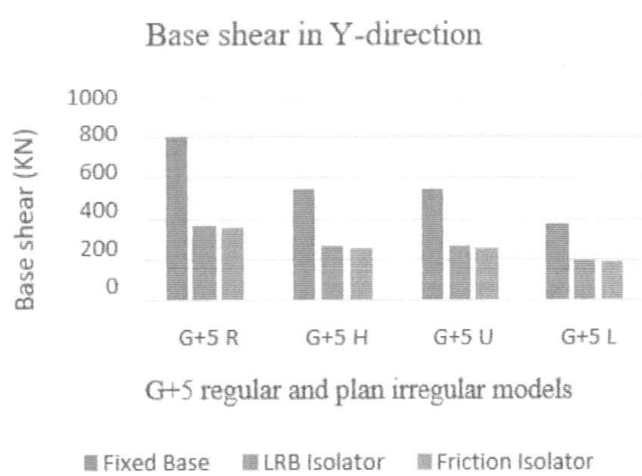

Figure 3.2: Base shear for G+5 models in Y-direction

It has been observed that base isolators are very efficient in reducing the base shear when compared to fixed base structure. From the results it observed that the maximum decrease in base shear is 53.60 % and 54.72 % with LRB isolator and Friction isolator respectively in case of G+5 R model when compared with other models in case low story structures.

Table: 3.2 Base shear for G+15 regular and plan irregular models

S.no	Models	Fixed Base (kN)		LRB isolator (kN)		Friction isolator(kN)	
		X direction	Y direction	X direction	Y direction	X direction	Y direction
1	G+15 R	926.26	926.26	711.19	711.19	712.57	712.57
2	G+15 H	581.57	630.30	463.27	466.45	464.82	467.47
3	G+15 U	571.46	629.84	483.88	473.75	486.97	474.82
4	G+15 L	422.87	422.87	364.61	364.61	369.85	369.85

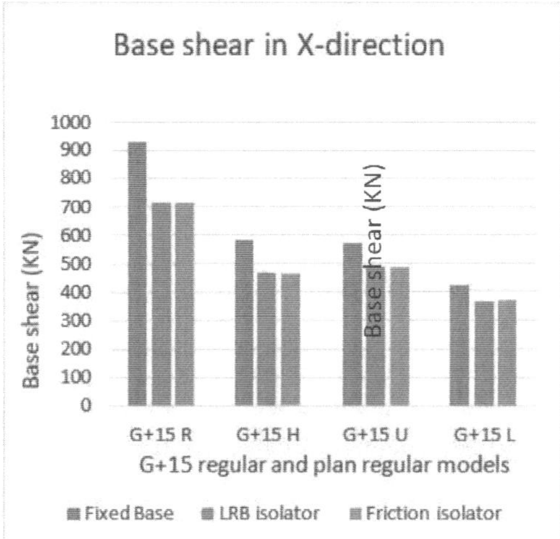

Figure 3.3: Base shear for G+15 models in X-direction

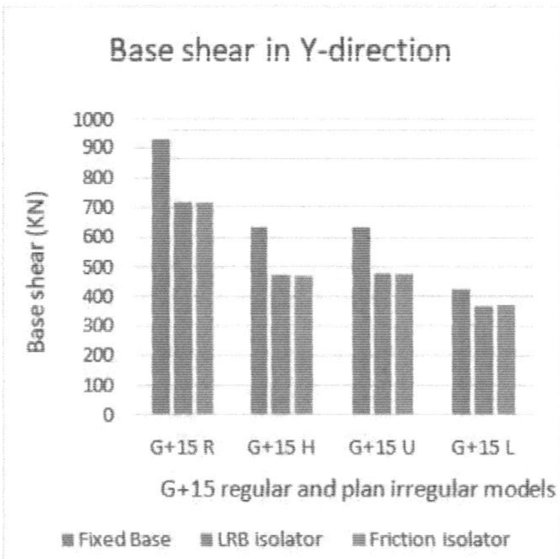

Figure 3.4: Base shear for G+15 models inY-direction

From the above results it is clear that the maximum decrease in the base shear is 23.21 % and 23.07% for LRB isolator and Friction isolator respectively in case of G+15 R model when compared with other models.

Time Period:

Table: 3.3 Time period for G+5 regular and plan irregular models

S.no	Models	Fixed Base (sec)	LRB isolator(sec)	Friction isolator(sec)
1	G+5 R	1.11	2.79	2.86
2	G+5 H	1.11	2.62	2.67
3	G+5 U	1.12	2.62	2.68
4	G+5 L	1.11	2.58	2.64

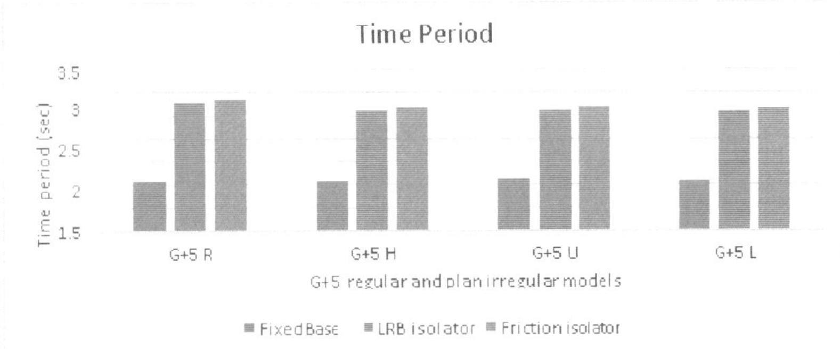

Figure 3.5: Time period for G+5 regular and plan irregular models

Base isolation shifts the main period of the earthquake and the fundamental period of the structure. When compared to other models, the fundamental period of the fixed structure for the G+5 R model is raised by 60.21% for the LRB isolator and 61.18% for the friction isolator, respectively. This lessens the accelerations of the floor and prevents the building from experiencing severe shaking.

Table: 3.4 Time period for G+15 regular and plan irregular models:

S.no	Models	Fixed Base (sec)	LRB isolator(sec)	Friction isolator(sec)
1	G+15 R	2.72	4.93	5.0 2
2	G+15 H	2.86	4.79	4.8 5
3	G+15 U	2.91	4.81	4.8 6
4	G+15 L	2.94	4.75	4.8 0

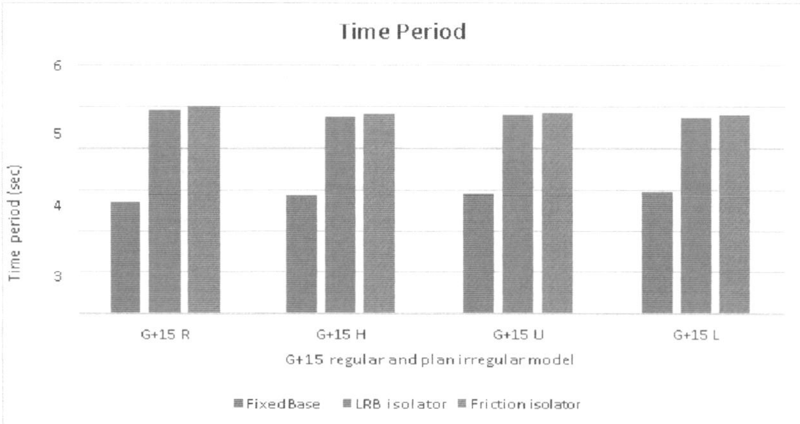

Figure 3.6: Time period for G+15 regular and plan irregular models

The fundamental period of the fixed structure is increased maximum for G+15 R model by 44.82% and 45.81 % for LRB isolator and Friction isolator respectively when compared with other models, which helps the structure to avoid severe shaking due to period shift provided by isolators.

Story Displacement:

Table: 3.5 Maximum Story Displacement in X-direction for Fixed and isolated G+5 R model

S.no	Story	G+5 R(mm) Fixed base	G+5 R (mm) LRB isolator	G+5 R (mm) Friction isolator
1	6th story	9.04	18.72	19.07
2	5th story	8.48	18.53	18.89
3	4th story	7.39	18.14	18.52
4	3rd story	5.79	17.54	17.94
5	2nd story	3.79	16.72	17.14
6	1st story	1.58	15.53	15.98
7	Base	0.0	12.51	13.05

Table: 3.6 Story Displacement in X-direction for Fixed Base and isolated base condition for G+5 H model

S.no	Story	G+5 H(mm) Fixed base	G+5 H (mm) LRB isolator	G+5 H (mm) Friction isolator
1	6th story	9.20	17.93	18.26
2	5th story	8.56	17.68	18.03
3	4th story	7.40	17.23	17.59
4	3rd story	5.75	16.54	16.94

5	2nd story	3.70	15.61	16.04
6	1st story	1.48	14.26	14.73
7	Base	0.0	11.42	11.96

Table: 3.7 Story Displacement in X-direction for Fixed Base and isolated base condition for G+5 U model

S.no	Story	G+5 U (mm) Fixed Base	G+5 U (mm) LRB isolator	G+5 U (mm) Friction isolator
1	6th story	10.39	17.25	17.43
2	5th story	9.63	16.95	17.17
3	4th story	8.28	16.46	16.71
4	3rd story	6.39	15.74	16.03
5	2nd story	4.07	14.76	15.09
6	1st story	1.58	13.35	13.74
7	Base	0.0	10.62	11.09

Table: 3.8 Story Displacement in X-direction for Fixed Base and isolated base condition for G+5 L model

S.no	Story	G+5 L(mm) Fixed Base	G+5 L (mm) LRB isolator	G+5 L(mm) Friction isolator
1	6th story	11.03	18.69	18.85
2	5th story	10.21	18.38	18.57
3	4th story	8.79	17.85	18.08
4	3rd story	6.78	17.07	17.34
5	2nd story	4.32	16.00	16.33
6	1st story	1.69	14.48	14.86
7	Base	0.0	11.51	12.00

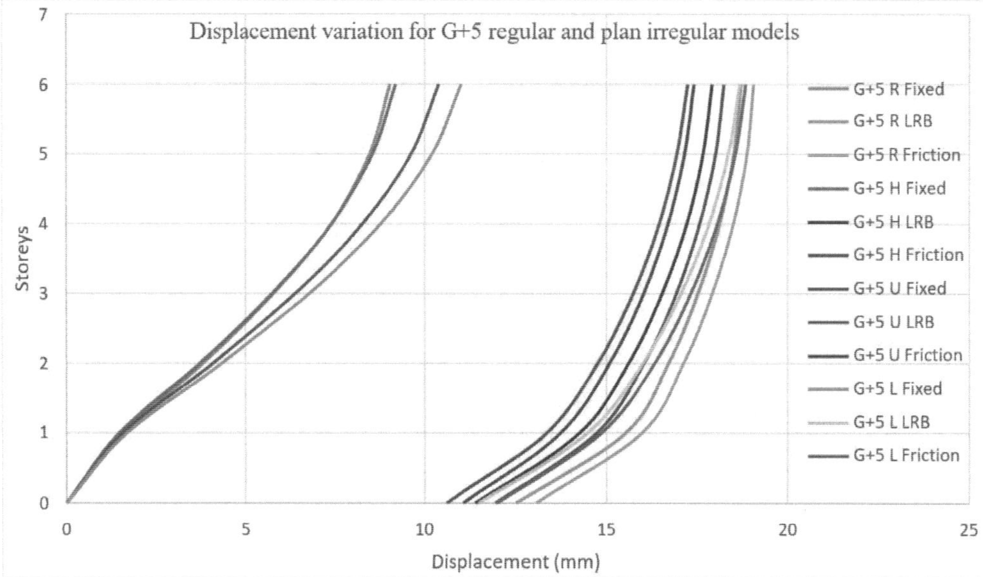

Figure 3.7: Displacement variation for G+5 regular and plan irregular models

Table: 3.9 Storey Displacement in X-direction for Fixed and isolated G+15 R model

S.no	Story	G+15 R (mm) Fixed Base	G+15 R (mm) LRB isolator	G+15 R (mm) Friction isolator
1	16th story	22.77	43.56	44.95
2	15th story	22.41	43.34	44.74
3	14th story	21.77	42.97	44.39
4	13th story	20.88	42.43	43.88
5	12th story	19.76	41.74	43.21
6	11th story	18.44	40.89	42.38
7	10th story	16.94	39.89	41.40
8	9th story	15.26	38.74	40.27
9	8th story	13.43	37.46	39.01
10	7th story	11.45	36.03	37.61
11	6th story	9.34	34.48	36.08
12	5th story	7.70	33.26	34.87
13	4th story	5.99	31.95	33.60.
14	3rd story	4.20	30.55	32.20
15	2nd story	2.40	29.0	30.67
16	1st story	0.81	27.15	28.84
17	Base	0.0	24.44	26.14

Table: 3.10 Storey Displacement in X-direction for Fixed and isolated G+15 H model

S.no	Story	G+15 H (mm) Fixed Base	G+15 H (mm) LRB isolator	G+15 H (mm) Friction isolator
1	16th story	24.26	42.67	43.26
2	15th story	23.67	42.24	42.88
3	14th story	22.85	41.68	42.37
4	13th story	21.79	40.94	41.69
5	12th story	20.52	40.04	40.85
6	11th story	19.07	38.99	39.85
7	10th story	17.45	37.79	38.71
8	9th story	15.69	36.44	37.42
9	8th story	13.79	34.97	36.00
10	7th story	11.76	33.37	34.46
11	6th story	9.65	31.67	32.82
12	5th story	7.97	30.21	31.41
13	4th story	6.04	28.69	29.94
14	3rd story	4.17	27.07	28.38
15	2nd story	2.33	25.30	26.66
16	1st story	0.76	23.20	24.60
17	Base	0.0	20.44	21.90

Table: 3.11 Storey Displacement in X-direction for Fixed and isolated base G+15 U model

S.no	Story	G+15 U (mm) Fixed Base	G+15 U (mm) LRB isolator	G+15 U (mm) Friction isolator
1	16th story	31.31	48.53	47.57
2	15th story	30.41	47.83	47.00
3	14th story	29.23	47.01	46.27
4	13th story	27.77	46.00	45.38
5	12th story	26.07	44.81	44.30
6	11th story	24.15	43.44	43.06
7	10th story	22.04	41.92	41.66
8	9th story	19.76	40.24	40.12
9	8th story	17.33	38.44	38.44
10	7th story	14.78	36.51	36.65
11	6th story	12.11	34.48	34.75
12	5th story	9.82	32.70	33.08
13	4th story	7.49	30.84	31.35
14	3rd story	5.11	28.86	29.50
15	2nd story	2.83	26.69	27.46
16	1st story	0.91	24.18	25.08
17	Base	0.0	21.04	22.09

IOP Conf. Series: Materials Science and Engineering 1282 (2023) 012015 doi:10.1088/1757-899X/1282/1/012015

Table: 3.12 Storey Displacement in X-direction for Fixed and isolated base G+15 L model

S.no	Story	G+15 L (mm) Fixed Base	G+15 L (mm) LRB isolator	G+15 L (mm) Friction isolator
1	16th story	32.04	55.29	54.45
2	15th story	31.11	54.52	53.80
3	14th story	29.90	53.58	52.98
4	13th story	28.40	52.44	51.97
5	12th story	26.66	51.10	50.76
6	11th story	24.70	49.56	49.35
7	10th story	22.54	47.83	47.76
8	9th story	20.21	45.93	46.00
9	8th story	17.74	43.88	44.09
10	7th story	15.13	41.70	42.05
11	6th story	12.41	39.38	39.88
12	5th story	10.08	37.35	37.98
13	4th story	7.69	35.23	35.99
14	3rd story	5.26	32.97	33.87
15	2nd story	2.91	30.49	31.53
16	1st story	0.93	27.63	28.81
17	Base	0.0	24.05	25.38

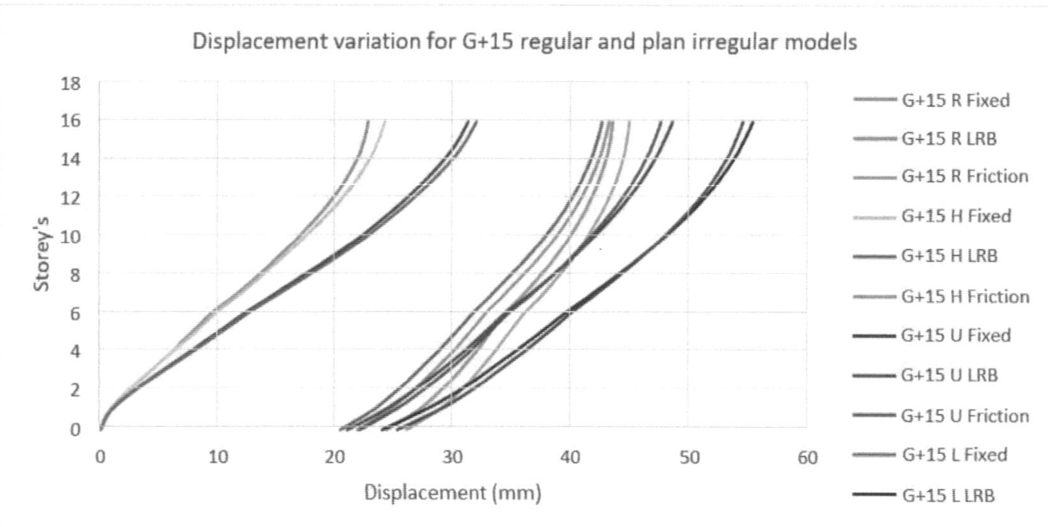

Figure 3.8: Displacement variation for G+15 regular and plan irregular models

From the above Tables, it is observed that the displacement is zero at fixed base and maximum displacement at top story but in case of base isolated structure there is displacement at base of the structure that is due flexibility of the structure due to isolator. From the Displacement tables, it is observed that the top story displacement is maximum for G+5 L model with a value of 11.03 mm, for G+5 U model with a value of 10.39 mm, for G+5 H model with a value of 9.20 mm and for G+5 R model with a value of 9.04 mm in case of fixed base for low story structure. It is observed that the top story displacement is maximum for G+15 L model with a value of 32.04 mm, for G+15 U model with a value of 31.31 mm, for G+15 H model with a value of 24.26 mm and for G+15 R model with a value of 22.77 mm in case of fixed base for medium story structure. From top story displacement values, it is observed that as irregularity increases the top story displacement values are high. So irregular structures are more prone to damages dueearthquake when compared with regular structure.

4. CONCLUSION:

The results from the analysis of the G+5 story and G+15 story structure shows how the main effect of base isolation is to greatly reduce the damage due to base shear and the effects of natural period.

It is observed that the maximum decrease in base shear is 53.60 % and 54.72% with LRB isolator and Friction isolator respectively in case of G+5R model when compared with other models in case of low story structures. From the results it is clear that the maximum decrease in the base shear is 23.21 % and 23.07 % for LRB isolator and Friction isolator respectively in case of G+15 R model when compared with other models in case of medium story structures. It is observed that fixed base buildings have higher base shear and isolated buildings has low base shear for both low story and medium story buildings. Therefore, it is clear that base isolation reduces the earthquake forces acting on the structure and thereby makes the structure stable. It is observed that the reduction in base shear is more in case of low story structure when compared with medium story structures. Therefore, it is best suitable for low story structures. It is also observed that the both isolators that is LRB isolator and Friction isolator shows same amount of reduction in base shear. Therefore, both the isolators are effective to use in the structure vulnerable to earthquake. The fundamental period of the fixed base structure is increased maximum for G+5R model by 60.21 % and 61.18 % for LRB isolator and Friction isolator respectively when compared with other models in case of low story structure. The fundamental period of the fixed structure is increased maximum for G+15 R model by 44.82 % and 45.81 % for LRB isolator and Friction isolator respectively when compared with other models in case of medium rise structure. It is observed that the period of the base isolated structure increased when compared with fixed base. Therefore, with increase in the fundamental time period of the structure there is reduction in the maximum acceleration and earthquake forces on the structure. From the maximum story displacement results, it is observed that displacement is zero at fixed base but in case of isolated structure there is displacement at the base of the structure. Thus, makes the structure flexible and has less damage due to earthquake.

From the maximum story displacement results, it is observed that irregular structures have maximum displacement when compared to regular structures. Therefore, irregular structures are more vulnerable to earthquake when compared with regular structures.

REFERENCES

1. Ashok, Shirule Pravin, et al. "Response spectrum analysis of multi storeyed base-isolated building." *International Journal of Civil, Structural Environmental and Infrastructure Engineering Research and Development, ISSN* (2012): 2249-6866.
2. Reddy, Anusha R., and V. Ramesh. "Seismic Analysis of Base Isolated Building in RC Framed Structures." *International Journal of Civil and Structural Engineering Research* 3.1 (2015): 170-176.
3. Darshale, S. D., and N. L. Shelke. "Seismic Response Control of Vertically Irregular RCC Structure using Base Isolation." *International Journal of Engineering Research* (2016).
4. Babu, Suni Susan, and Manish Jose. "Seismic assessment of irregular reinforced concrete frame structures using triple friction pendulum bearing." *IOP Conference Series: Materials Science and Engineering*. Vol. 396. No. 1. IOP Publishing, 2018.
5. IS-1893:2002 Criteria for Earthquake Resistant Design of Structures: (Part 1) General Provisions and Buildings, New Delhi, India.

Analysis and Parametric Assessment of Reinforced Soil Wall

S S Yashmin[1] and A Chakraborty[2]
[1] PhD student, AcSIR-CRRI Delhi.
[2] Assistant Professor, Tezpur University, Tezpur, Assam.
E-mail: sabina.yash1310@gmail.com

Abstract. Due to rapid urbanization and growth of countries, the rate of building infrastructures is also on the higher side. Owing to a lack of adequate high bearing capacity soil, one is forced to select regions with lower bearing capacity, which can cause long-term instability problems. Various improvement techniques available in the field of Civil Engineering or Geotechnical Engineering are to be adopted in such cases to avoid instability issues. One such well known method is Reinforced Soil wall that has been investigated and focused here. In this paper, an effort is made to compare numerical and analytical analysis, as well as to investigate their consistency and understand the effect of various parameters on its performance. The numerical analysis, which is based on Lee et al.'s lateral earth pressure theory, has been paired with an analytical one using the Finite element method (FEM) to evaluate its performance and efficiency in the field. Furthermore, various parameters such as reinforcing element length, soil friction angle, spacing, height of the wall, and maximum tensile forces produced have been considered, with their variations affecting the overall efficiency of the system being investigated. Further, an attempt has been made to understand the behavioural pattern by comparing the results using Regression Analysis. The findings from both analyses are very consistent and complement one another.
Keywords: Reinforced soil wall, numerical analysis, lateral earth pressure concept, Finite element method, Regression Analysis.

1 Introduction

Slope instability is a troublesome issue and has a great significance in the fields of geotechnical and geological engineering as it is known to have claimed substantial loss of lives and properties over the time. Again, slope instability is not just limited to natural slopes but also man-made slopes as in the case of railway embankments, road embankments, road cuts, etc. Various improvement techniques such as earthworks, retaining walls, and reinforced earth mechanisms have also been suggested in the fields to stabilise the slope and achieve satisfactory performance [1]. The field technique that is to be focussed in this paper is the mechanical stabilisation of earth (MSE) wall system, whereby reinforcements are provided within the earth mass and has been proved to be better over the other traditional retaining walls. MSE walls have developed as a solution to restrained right of way, complex sub-surface situations and the projects that were delayed in the past be accelerated with the use of MSE wall system. Over the past few decades, the mechanism of MSE wall has gained prominent utilisation in the field of civil engineering due to its design resiliency, relative flexibility in construction, and cost efficiency compared to other conventional methods [2-3]. It can tolerate much higher settlements (differential) being more flexible and cost-effective than the conventional methods of the past. It is an improvement technique whereby mechanically stronger materials such as metallic or polymeric strips, sheets, geogrid, etc. are provided layer by layer with the compacted backfill in between them. It could have been accomplished because of the interaction that occurs through the inclusion of reinforcement in the compacted earth mass between the soil and the reinforcing component. The design and behaviour of reinforced earth wall is a complex system with a combination of various elements including facing, backfill soil, foundation soil, and reinforcements which requires a lot of experimentations and case studies of the existing ones, which is not always economically feasible. But have been made possible due to the development of analytical approaches using LEM or FEM models which allows analyzing the response and behaviour of reinforced soil wall corresponding to various parameters and loading conditions with great certainty that holds well with the field results [4-5]. Further, the type of reinforcements used in reinforced soil structures have a significant effect on the length and lateral earth pressures endured by the walls [6]. The lateral earth pressures endured by the walls can be largely affected by end conditions of the reinforcements, whether freely suspended or fixed to the walls. The length of the reinforcements (L) to height of the wall (H) ratio i.e., L/H influences the variation of resultant earth pressure and resultant overturning moments [7]. In practical conditions, the optimum length (L) of the reinforcing strips to the height (H) of the wall was determined to be approximately 0.6-0.8 [8-10]. Again, the development of non-dimensional design charts has aided in the determination of the lateral earth pressure formed on the wall and its point of application over the wall foundation. The point of load implementation has been proven to affect the behaviour of the wall that varies along the backfill with the pattern of reinforcement distribution [11]. The reinforcement length to wall height (L/H) ratio is the most significant geometric parameter that influences the behaviour of reinforced soil wall structures, and evenly spaced truncated reinforcement with L/H=0.7 provides a reasonably efficient reinforcement force distribution [12-13]. The overall characteristics of reinforced soil are influenced by the mechanical properties of the soil and the reinforcement, as well as the relative proportions and geometrical structure [14]. When examining the lateral deformation pattern of the walls, it was found that the deformation pattern depends on multiple factors, such as loading conditions, structural geometry, type of facing, etc. The deformation of the front face at the base is minimal, but the deformation at the top is substantial, and the deformed shape is almost parabolic in deviation from base to top [15-16]. The retaining wall system's maximum vertical settlement, as determined by FE study and in the field, is very similar to the wall facing, and settlements in the backfill soil can be managed by using higher reinforcement and wall facing stiffness [17]. Further, if the desired backfill material is not available nearby the site, then one can utilize bottom ash which is a waste product of thermal power plants and would result in combatting waste management as well as cost-efficacy [18-19]. Reinforced soil technology is not just limited to civil engineering fields but also extends to military fields as well as to safeguard staff and property against unintended detonation of concealed bombs, munitions, and ammunition plants [20].

The goal of the present study is to carry out stability analysis of a reinforced soil wall specifically on a mechanically stabilised earth (MSE) wall using a purely numerical approach and an analytical one to understand the influence of the various parameters of the wall system. From the parametric assessment of the wall system, the importance of each parameter has been analysed using SPSS software whose results have been compared with the 10% incremental and decremental method suggested by Pandey et al. [21]. Furthermore, the data used for developing the models have been analysed statistically using Linear Regression Analysis in SPSS software and then validated. The numerical approach adopted in the study is founded on the concept of lateral earth pressure which has been investigated and tested by Lee et al [22] for the development of reinforced soil walls. The analytical analysis has been carried out using FEM approach employing PLAXIS 2D computer program. The study was conducted using various combinations of soil and reinforcement parameters and the results obtained have been compared from both the approaches.

2 Methodology

The methodology adopted for the present study is based on two approaches- numerical and an analytical using FEM approach, which are being described in detailed in the following sections. A conventional soil nail wall of varying heights and other parameters as mentioned in succeeding sections, with vertical face and horizontal backfill is considered along with homogeneous distribution of reinforcements in both horizontal and vertical direction for the analysis for exemplification and better understanding.

2.1 Numerical Analysis

For reinforced soil walls, there are primarily three design approaches: lateral earth pressure of Rankine, lateral earth pressure of Coulomb and concept theories of Equivalent Confining Stress. Compared to Coulomb's, Rankine's approach is usually easy; it is therefore more commonly used than any other design approaches. The lateral earth pressure concepts, as developed for the design of traditional retaining structures, can be applied to the design of reinforced earth walls. Lee et al [22] have studied and investigated this design process for their applicability in the fields of reinforced soil walls. Similar design proposals were subsequently issued by Bell [23] and Koerner [24] for the design of mechanically reinforced earth walls using geogrids as material reinforcement.

Tie length calculation: Two lengths are included in the total tie length to be provided: length inside the Rankine failure zone (L_a) and length outside the Rankine failure zone, i.e., the effective length (L_e) (Fig 1). In order to resist tie pull-out failure, the length of the tie at any stage should be such that it must establish adequate bond strength with the surrounding medium (granular soil).

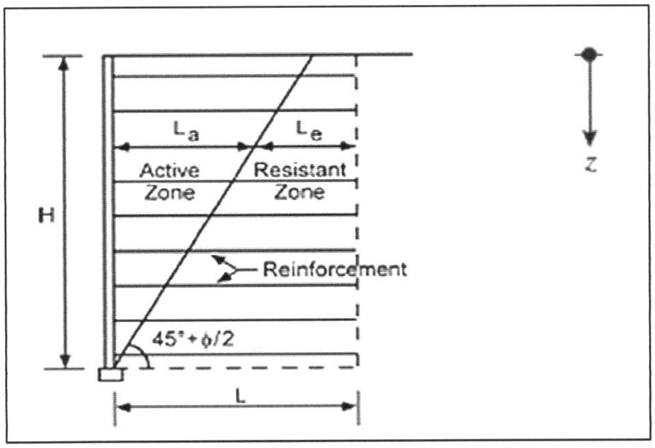

Fig 1. Representation of failure surface plane [22]

The length within the Rankine failure zone or active zone, La has been calculated using the following expression

$$L_a = \frac{(H - z)}{\tan\left(45 + \frac{\varnothing}{2}\right)} \quad (1)$$

The tie length required beyond the Rankine failure zone, i.e., the effective length, L_e has been determined using the following expression

$$L_e = \frac{(F_{SP}.K_a.\gamma.z.h.s)}{(2.w.\gamma.z.\tan\delta)} \quad (2)$$

Further, the stress induced, σ_i at each reinforcement level using the lateral earth pressure concept has been worked out using the following expression

$$\sigma_i = K_a.z_i.\gamma \quad (3)$$

Finally, the tensile force, T_i at each reinforcement level has been calculated using the following equation

$$T_i = \sigma_i.s.h \quad (4)$$

The above steps have been followed for numerical analysis of reinforced soil wall for various friction angles, wall heights and spacing of the reinforcement elements employed in the system.

2.2 Analytical Method using PLAXIS 2D

The Finite Element Method (FEM) is a computational technique for computing approximate numerical solutions for various engineering problems. It enables accurate visualization that displays the distribution of pressures and strains within a structure. PLAXIS is a geotechnical application finite element software in which soil simulations are used to predict the behaviour of the soil. For different types of geotechnical applications, PLAXIS is a special purpose two-dimensional finite element computer software used to perform deformation and stability analysis. This uses a convenient graphical user interface that helps users to easily create a geometry model and a finite element mesh depending on the condition at hand in a representative vertical cross-section. In this study, PLAXIS is used to simulate the reinforced soil wall using finite element dependent models under drained conditions and as a plane strain case. A finite element mesh with a sufficient fineness is generated using 15-noded triangular elements. The model of reinforced soil wall considered along with metallic strips as reinforcement elements has been represented in figure 2. The input parameters that have been utilised in modelling are represented in table 1.

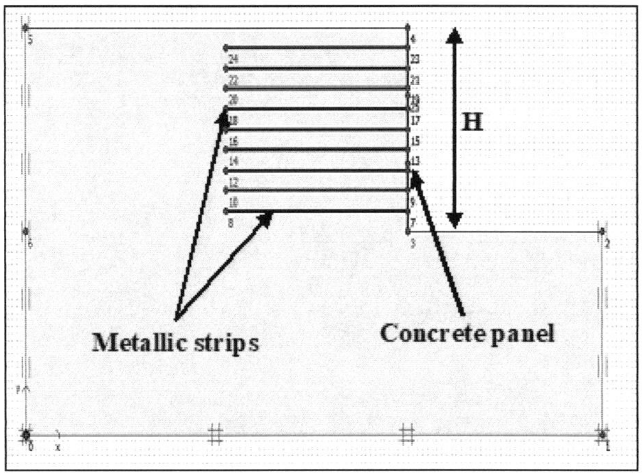

Fig 2. Representation of reinforced soil wall model considered for the analysis.

Table 1. Input parameters for model analysis

Parameter	Value
Model	Plane Strain
Element	15 Node
Angle of internal friction (ø)	30-42 degree
Height of the wall (H)	5-15 m
Spacing (S)	0.5-1.5 m
Cohesion (c)	5 KN/m^2
Material Model	Mohr- Coulomb
Material Type	Drained
Unit Weight of soil, (γ)	17 - 20 KN/m^3
Poisson's Ratio, (v_s)	0.300
Stiffness modulus	13000 KN/m^2
Metallic Strips	
Material Type	Elastic
Young's Modulus	2.98 x 10^6 KN/m
Rigidity Modulus	1.136 x 10^5 KNm^2/m
Weight (each strip)	20 KN
Poisson's Ratio, (v_{ms})	0.35
Concrete Panel	
Material Type	Elastic

Young's Modulus	3×10^6 KN/m
Rigidity Modulus	2.25×10^4 KNm^2/m
Weight (each panel)	8.4 KN
Poisson's Ratio, (v_{cp})	0.15
Dimension of panel	5 m

3 Results and Discussions

Several physical properties of the soil play a major role on the consistency and structural behaviour of the reinforced soil wall. The results obtained shall be discussed in this section with numerical results succeeded by the analytical ones then sensitivity analysis, Multilayer Perceptron and Linear Regression Analysis in SPSS software.

3.1 Numerical Analysis

3.1.1 Variation of Tmax w.r.t Spacing

Following the numerical analysis methodology, the maximum tensile force developed in the metallic strips from the concept of lateral earth pressure at various depths have been computed. The variation of Tmax with respect to spacing has been represented graphically for better understanding in figure 3. It can be concluded from the graph that as spacing increases, the area enclosed by the strips also increases along with the increase in embedment depth of the reinforcements which influences the tensile forces induced in the wall. Again, the highest tensile force can be experienced by the reinforced soil wall whose reinforced soil has lower angle of internal friction, i.e., 30 degree and lowest by the one with higher friction angle, i.e., 42 degree in this analysis. The internal friction angle of the adjacent soil must also be considered when constructing reinforced soil wall structures. This is due to the fact that the distribution of ground pressure is mostly determined by the soil's internal friction angle. For example, the term K, which denotes the soil's lateral earth pressure coefficient, can be measured using the formula, $K = 1 - \sin \phi$ (i.e., for at-rest earth pressures). A higher value of ϕ results in a lower K, lowering the amount of stress the soil exerts on the retaining walls.

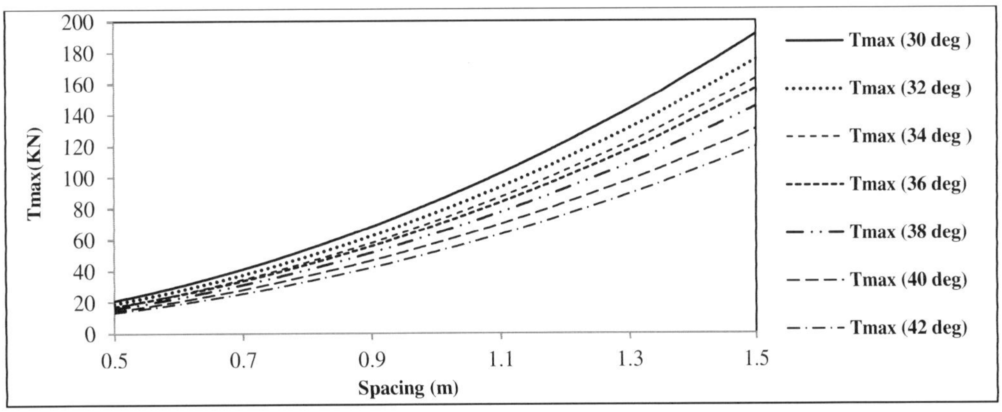

Fig 3. Graphical representation of Spacing v/s Tmax for different friction angles.

3.1.2 Variation of length w.r.t Spacing

The variation of lengths w.r.t spacing for various angle of internal friction taken into consideration is represented graphically in figure 4. It can be concluded that the length (L) required for developing sufficient bond strength as well as to prevent pull-out failure have decreased significantly upon increasing the values of internal friction angle from 30 degree to 42 degree. Thus, the length of the metallic strips is directly dependent on the type of soil to be employed for construction of reinforced fill. Further, the lengths of the metallic strips have indeed showed significant variations with spacing, implying that the lengths of the strips are to be increased to cope up with the increased risk associated with it. Therefore, it can be concluded that with increase in height of the wall and reinforcement spacings, the risk factor or the probability of un-stabilisation increases. Hence, the length of the metallic strips required to cope up with the increased failure probability is boosted up.

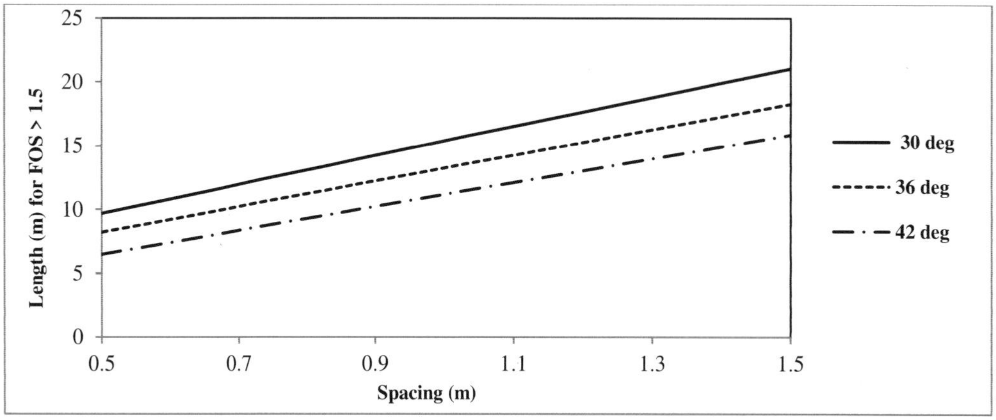

Fig 4. Variation of length w.r.t spacing for different friction angles

3.2 Analytical Method using PLAXIS 2D

3.2.1 Variation of FOS w.r.t Spacing for different soil

The lengths of the metallic strips obtained from the Lee et al. method have been used to develop the model in PLAXIS 2D simulating the field conditions. The outputs obtained after evaluating the models on PLAXIS 2D are in the form of Factor of Safety (FOS) or Σ_{msf}, which were obtained by analyzing the models of problem interest. The variations of FOS w.r.t spacing and friction angle of the soil have been graphically represented to understand their relationship in a better way, i.e., FOS increases with higher friction angle soil but decreases with increasing spacing when the lengths of the reinforcements are kept constant.

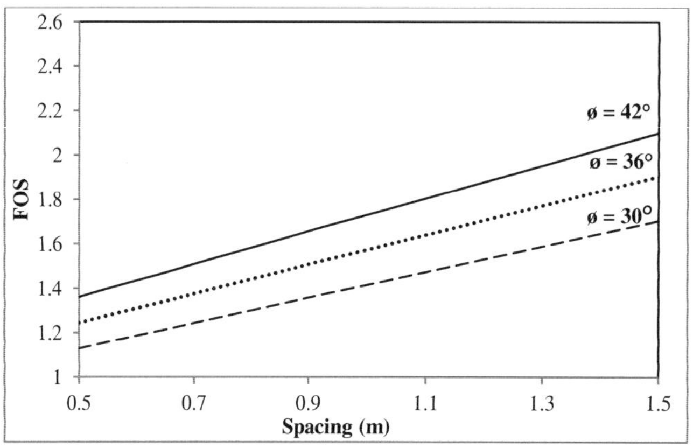

Fig 5 Graphical representation of Spacing v/s FOS for different friction angles.

3.3 Sensitivity Analysis

The most influential parameter affecting the performance of the MSE wall system can be determined by performing sensitivity analysis of the data considered. It is done by taking the averages of all the dependent and independent parameters of the wall system and considering 10% incremental and decremental parameters. It is being assumed that each input variable is an error independent variable for the analysis [21]. Further, the SPSS software has been used to predict the independent variable importance and normalised percentage for the parameters using Multilayer perceptron in ANN. The results obtained satisfy each other with friction angle of the soil having the most crucial influence on the wall system and length of the reinforcements the least role among the parameters considered.

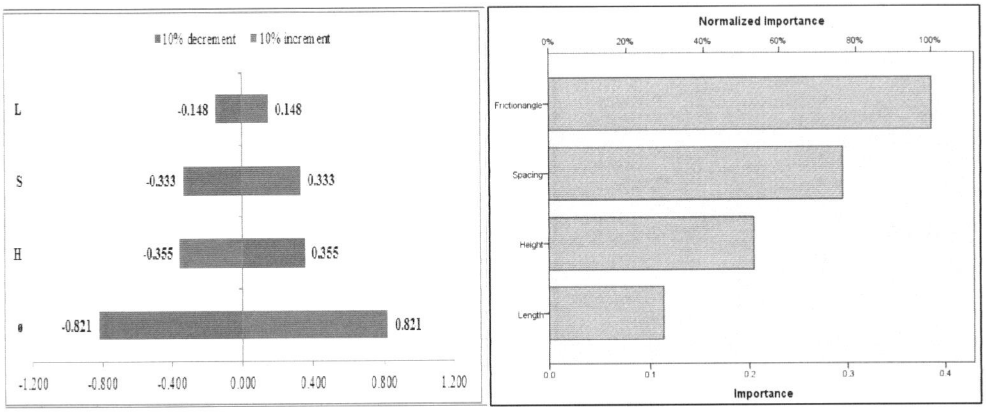

Fig 6. Sensitivity analysis and normalised importance from SPSS software respectively.

3.4 Linear Regression Analysis using SPSS

Out of 126 model data, around 112 model data (89%) have been used to perform linear regression analysis for developing the mathematical relationship between the input parameters and the output parameters of the wall system. The remaining data have been used to test the mathematical relationship obtained from the training data set and validated by comparing the RMSE and R^2 values.

The mathematical relationship obtained for the parameters of the MSE wall system by performing Multiple Linear Regression Analysis can be written as

$$\text{Predicted FOS} = 0.086 + 0.038\,.\,\phi - 0.058\,.\,\text{H} + 0.653\,.\,\text{S} + 0.028\,.\,\text{L} \qquad (5)$$

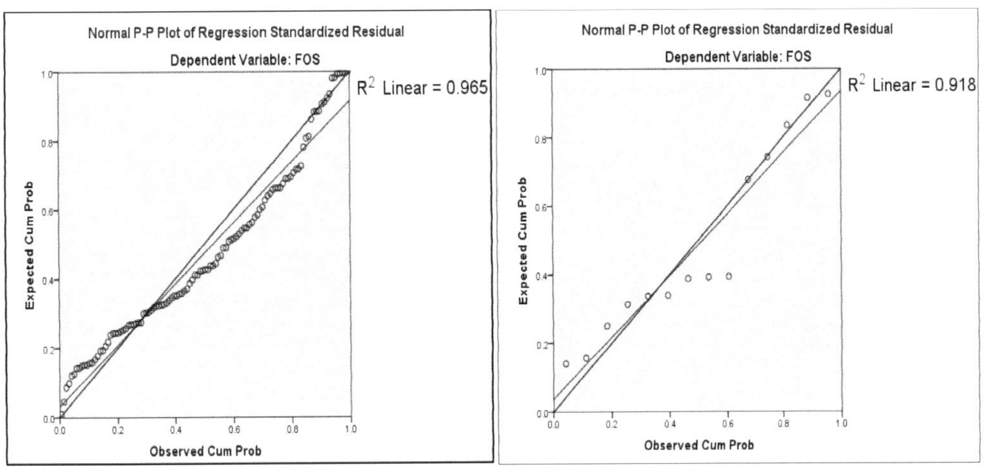

Fig 7. Regression analysis plots for training and testing data respectively.

4 Conclusions

In this paper, the parametric assessment of the RSW, i.e., MSE wall has been studied using a numerical method and analysed analytically by FEM using PLAXIS 2D program. Based on our study, the following are the main findings obtained, which are presented below:

1. The length of the metallic strips is directly dependent on the type of soil to be employed for construction of reinforced fill as well as the spacings between them. As it has been observed that as friction angle and length required for developing sufficient bond strength as well as to prevent pull-out failure follows inversely proportional relationship whereas the friction angle and spacing follows directly proportional relationship.
2. The FOS values obtained varies with length of the reinforcements. This indicates that the performance or design of the structure is directly dependent on the lengths of the reinforcements employed, indicating that higher the lengths employed, higher will be the FOS values. Hence, in many works involved in the past have focused on maintaining a proper proportionate of length to height (L/H) so that it can develop full bond strength as well as be economic enough to serve the purpose of method adoption.
3. The FOS values also varies with spacing between the reinforcements as well as friction angle of the soil employed. It has been observed that with the increase in friction angle and spacing, the FOS values also showed an increasing trend though with increased lengths to compensate the higher risks involved. This indicates that the performance or design of the structure is dependent on the soil type, spacing as well as on the position of the reinforcements employed.
4. Development of maximum tensile forces generally occurs at the bottom or lower part of the structure, which varies significantly with the area encompassed by the metallic strips. As the spacing increases, the encompassing area also increases and hence, the Tmax. Therefore, the variation of Tmax with respect to spacing have shown similar pattern with different types of soil involved, i.e., increasing trend line, but it shows a descending pattern when higher friction angle has been employed. It indicates that in cases of generation of higher tensile forces, increasing the friction angle of the soil can be an aid.
5. The friction angle of the soil is the most critical parameter that influences the performance of the MSE wall system followed by spacing, height of the wall and length of the reinforcements used. This compares well with the Multilayer Perceptron in SPSS software which also states that friction angle is the most influential parameter followed by wall height, spacing and length of the reinforcements.
6. The RMSE and R^2 values of the training and testing data obtained are 0.095, 0.094, 0.965 and 0.918 respectively which complements each other and fits really well.

References

[1] Duncan J M and Wright S G 2005. Soil Strength and Slope Stability. John Wiley & Sons, Inc., Hoboken, N.J
[2] Abramson L, Lee T, Sharma S, and Boyce G 2002 Slope Stability and Stabilization Methods. Second Edition, John Wiley & Sons, Inc., Hoboken, N.J.
[3] Cornforth D 2005 Landslides in Practice – Investigation, Analysis, and Remedial/Preventive Options in Soils. John Wiley & Sons, Inc., Hoboken, N.J.
[4] Collin J G 1986 Earth wall design, Ph.D. Dissertation, University of California, Berkeley, California.
[5] Damians I P, Bathurst R J, Josa A and Lioret A 2014 Numerical analysis of an instrumented steel-reinforced soil. *Int. Jr. of Geomechanics*, ASCE, pp. 01 to 15.
https://doi.org/10.1061/(ASCE)GM.1943-5622.0000394
[6] Singh H R 1988 Bearing capacity of reinforced soil beds, Ph.D thesis, Indian Institute of Science, Bangalore, India.
[7] Saran S and Khan I N 1988 Analysis of inclined retaining wall having reinforced cohesionless backfill with uniformly distributed surcharge. *Indian Geotechnical Journal*, Vol. **28**, No. 3, pp. 250-269.
[8] Swami S, Gerg K G and Bhandari R R 1992 Retaining wall with reinforced cohesionless backfill. *Journal of Geotechnical Engineering*, Vol. **118**, No. 12, ASCE, pp. 1869-1887.
https://doi.org/10.1061/(ASCE)0733-9410(1992)118:12(1869)
[9] Golam K, Sahadat M D and Sadik Khan H M 2014 Influence of Soil Reinforcement on Horizontal Displacement of MSE Wall. *Int. J. Geomechanics*, ASCE, Vol. **14**, pp. 130-141.
https://doi.org/10.1061/(ASCE)GM.1943-5622.0000297

[10] Mittal S, Gerg K G and Saran S 2006 Analysis and design of retaining wall having reinforced cohesive-frictional backfill. *Geotechnical and Geological Engineering*, Vol. **24**, pp. 499-522.
https://doi.org/10.1007/s10706-004-5153-9

[11] Hoe I N, Fumio T and Masaru T 1995 Simulating Performance of GRS-RW by Finite element procedure. *Journal of Geotechnical Engineering*, Div. ASCE, Vol. **121**, No. 04, pp. 330-340.
https://doi.org/10.1061/(ASCE)0733-9410(1995)121:4(330)

[12] Ho S K and Rowe R K 1996 Effect of wall geometry on the behaviour of reinforced soil walls. Geotextile and Geomembrane, Vol. 14, No. 10, pp. 521-541.
https://doi.org/10.1016/S0266-1144(97)83183-4

[13] Bilgin O and Mansour E 2014 Effect of reinforcement type on the design reinforcement length of mechanically stabilized earth walls. *Jr. of Engineering Structures*, Vol. **59**, pp. 663-673.
http://dx.doi.org/10.1016/j.engstruct.2013.11.013

[14] Yu H S and Sloan S W 1997 Finite element limit analysis of reinforced soils. *Computers and Structures*, Vol. **63**, No. 3, pp. 567-577.
https://doi.org/10.1016/S0045-7949(96)00353-7

[15] Yang G, Zhang B, Peng Lv and Zhou Q 2009 Behaviour of geogrid reinforced soil retaining wall with Concrete-rigid facing. *Geotextile and Geomembrane*, Vol. **27**, pp. 350-356.
https://doi.org/10.1016/j.geotexmem.2009.03.001

[16] Porbaha A and Goodings D J 1996 Centrifuge modelling of geotextile-reinforced cohesive soil retaining walls. *Journal of Geotechnical Engineering*, ASCE, Vol. **122**, No. 10, pp. 840-848.
https://doi.org/10.1061/(ASCE)0733-9410(1996)122:10(840)

[17] Sridevi J and Garg K G 1997 Finite element analysis of a reinforced earth wall. *Indian Geotechnical Journal*, Vol. **27**, No. 01, pp. 59-73.
https://doi.org/10.1061/AJGEB6.0000450

[18] Hatami K and Bathurst R J 2001 Investigation of seismic response of reinforced soil retaining walls. International conference on recent advances in geotechnical earthquake engineering and soil dynamics, ASCE, Paper 13, pp. 7-18.

[19] Dov L, Yuhui H and Jie H 2004 Limited reinforced space in segmental retaining walls. *Geotextile and Geomembranes*, Vol. **22**, pp. 543-553.
https://doi.org/10.1016/j.geotexmem.2004.04.002

[20] Ng C, Chew S, Karunaratne G, Tan S and Loh S 2000 Flexible and rigid faced MSE walls subject to blasting. Int. Conf: Advances in Transportation and Geoenvironmental Systems Using Geotextiles, GeoDenver 2000, Denver, Colorado, United states, ASCE, pp. 322-336.
https://doi.org/10.1061/40515(291)21

[21] Pandey M, Lam W H, Cui Y, Khan M A, Singh U K and Ahmad Z 2019 Scour around Spur Dike in Sand–Gravel Mixture Bed. Water. 2019; 11(7):1417.
https://doi.org/10.3390/w11071417

[22] Lee K L, Adams B D, and Vagneron J J 1972 Design of Reinforced Earth Walls. Report No. 7233, University of California at Los Angeles School of Engineering and. Applied Science, April, 1972, 219 p.

[23] Bell J R, Stilley A N, and Vandre B 1975 Fabric Retaining Earth Walls. Proceedings, Thirteenth Engineering Geology and Soils Engineering Symposium, Moscow, ID, pp. 271–287.

[24] Koerner R B 2005 Design with Geosynthetics, 5th edition, Pearson Education, Upper Saddle River, NJ.

Stability Analysis of Homogenous Embankment using Statistical Method

Pori Taipodia[1], Nabajyoti Kalita[2], Roshan Pradhan[3], Shubham Raj[4], Adarsh Pandey[5], Chau Kintana Chautang[6], Arunav Chakraborty[7]
[1,2,3,4,5,6] B.Tech, Department of Civil Engineering, Tezpur University, Tezpur, Assam, India
[7] Assistant Professor, Civil Engineering Department, Tezpur University, Tezpur, Assam, India
E-mail: taipodiapori@gmail.com

Abstract: Homogeneous earth embankments are man-made or artificial structure whose construction is an important part of highway, railway and flood structure projects. Soil exhibits different properties in different places. Reinforcing the soil with Geogrid can be effectively used for improving the durability and stability of embankment, this ensures uniform settlement of embankment and result in cost effective solution. This paper aims to develop two-dimensional stability charts and correlation for slope failure (slip circle) in homogeneous soil embankment. Sensitivity analysis is also done to determine the contribution of different soil parameter. The chart and correlation developed for stability analysis can be a convenient tool for preliminary estimation of factor of safety (FOS) and possibility of failure.

Keyword: Embankment, Geogrid, Regression Analysis, Sensitivity Analysis.

1. Introduction

Embankment is a structure built of earth/soil that is used to support roadways or railways above the ground surface. The cross section of embankment is trapezoidal, narrow at the top and wide at the bottom. In road construction embankment is provided when the water level rise above the ground level. Slope stability is one of the most important criteria for the design of embankment. The various factors affecting the slope stability viz., geometry of the slopes, ground water, cohesion, types of equipment used during the construction, angle of internal friction and so on. Translational, rotational, wedge and compound failure are the types of failure which occurs in an embankment. Under the rotational failure, Slope failure occurs when soil above the toe is weak. Condition under which the circular failure occurs, when particles do not get properly locked due to their shapes and compaction is not done up to maximum dry density. On the basis of area affected by the failure surface, circular failures may include, toe failure and base failure. Slope failure in road embankments occur when the load is in the form of a vehicle, which acts as a surcharge, increasing the soil's lateral earth pressure. Normally, we flatten the slope to make it more stable. As a result, it requires additional breadth, leads to the increase in construction costs.

A geosynthetic material made of polymer called geogrid can be used to reduce costs and increase stability. Geogrid are made of polyester, polypropylene etc. The function of Geogrid includes holding and capturing of aggregates together. Based on manufacturing, Geogrid can be extruded, woven and bonded. Also, on the basis of stretching direction done during manufacturing, geogrids are uniaxial and biaxial. Fellenius method of slices was the first to analyse the slope stability of embankment. Furthermore, various methods like Bishop Method developed over the time. Limit Equilibrium method

(LEM) is the most widely used method for analysing the slope stability.

Fellenius [1] published the first methodology of research which was related to slope stability analysis. However, due to ignorance of all interslices forces the fulfilment of equilibrium of individual slices was not met. Thus, effective stress at the slice base was incorrectly computed. Bishop [2] had given a method that fulfils the vertical force equilibrium and moment equilibrium around the centre of a circular slip surface. Still, this method cannot be used for the other slip surface. Janbu [3] provided a method that balances each slice's vertical force equilibrium but the net moment equilibrium was not met. So, he gave a correction factor for this inadequacy. Morgenstern and Price [4] provided a new method that included all considerations which were the drawbacks in the above three methods and can be used for any type of slip surface. Fredlund and Krahn [5] compared various 2D slope stability method and developed the FOS equation for solving various failure surfaces and loading conditions. A new method was introduced which satisfied both force equilibrium as well as moment equilibrium [6]. LEM, FEM, and FDM are also some of the different approaches used for slope stability analysis. Out of these methods, LEM is the most frequent method used to calculate the factor of safety based on the Mohr-Coulomb Criterion. Burman et al. [7] found that for the best consistent calculation of factor of safety value, the Ordinary slice method is the best. Majedi et al. [8] found that most suitable place to provide Geogrid as reinforcement is above the toe of the embankment. Also, the Horizontal displacement is reduced by the use of Geogrid. Raouf [9] after investigation founded that the ideal spacing between the Geogrid layers is 0.5m. Esmaeili et al. [10] found that with increase in tensile strength of Geogrid the bearing capacity increases. One of the initial FEM was placed forwarded by Smith and Hobbs [11] for $\varphi u=0°$ soil. Naseer and Evans [12] found that in FEM critical slip surface is near slope face whereas it is further away in LEM. Slope failure of an embankment is the most common failure type of an embankment. In order to provide correction for the slope failure, a good understanding of failure mechanism of embankment is required. Since then, various methods and techniques have evolved. Limit equilibrium methods are predominantly used for slope stability. Analysing stability of huge number of slopes using convectional LEM is difficult as it is very time consuming to develop slope model. Multiple regression method is better alternative which provide simple equation and statistical techniques to calculate factor of safety of slopes without using iterative procedure thereby decreasing time and complexity in slope stability evaluation [13]. Khan and Wang [14] developed correlations between Factor of Safety (FOS) and soil parameters using linear regression analysis to stabilize slope and achieve high shear strength. Azmoon et al. [15] gave deep learning-based methods to evaluate slope stability. To achieve this, a multiclass classification and regression models were employed.

The objectives of the present study are given in the following section.

1. To determine the FOS for slopes with varying soil parameter and Geogrid spacing from the toe of embankment using LEM and to develop stability charts for calculating the FOS.
2. To carry out a regression analysis to develop a correlation for determining the FOS of the slopes with Geogrid
3. To carry out sensitivity analysis to determine the most sensitive parameter.

2. Methodology

Three different slope angles of 30°, 45° and 60° for homogeneous earth embankment were considered. For each slope angle, three different embankment heights of 10m, 20m, and 30m were taken into consideration. For each slope and three different heights, random cohesion values were taken ranging from 10-40 kPa with random values of internal friction angle ranging from 10°-40° in reference with Maharashtra PWD Handbook on Earth and Composite Dam [16]. The Geogrid layers were placed at different spacing from toe of embankment for different embankment height and slope angle. Using

LEM, FOS were obtained for 2100 cases of embankment design for Fellenius method and Bishop's simplified method.

2.1 Development of Stability Charts

The FOS values were used to develop two stability charts based on data forecasting method for Ordinary and Bishop's method. The development of the new charts was prompted by the introduction of a legitimate tool for the quick evaluation of the slopes' safety, based on the stringent LEM. For a range of inclination angles, Bell [17] suggested that (F / tan φd) can be used as a function of (cd / γ H tan φd). He used the modified stability number, N*, denoted by (cd / γ H tan φd). Utilizing such a representation has the benefit of making parameter N* independent on safety factor F. Therefore, no iterative processes are needed for the estimate of the safety factor from charts that are provided as functions of N*. In order to take into account the spacing of the geogrid layer, (F m / tan φ) was used as a function of (c m / γ H tan φ) where (m = Gs / H). A typical example of embankment with geogrid is shown in Fig 1

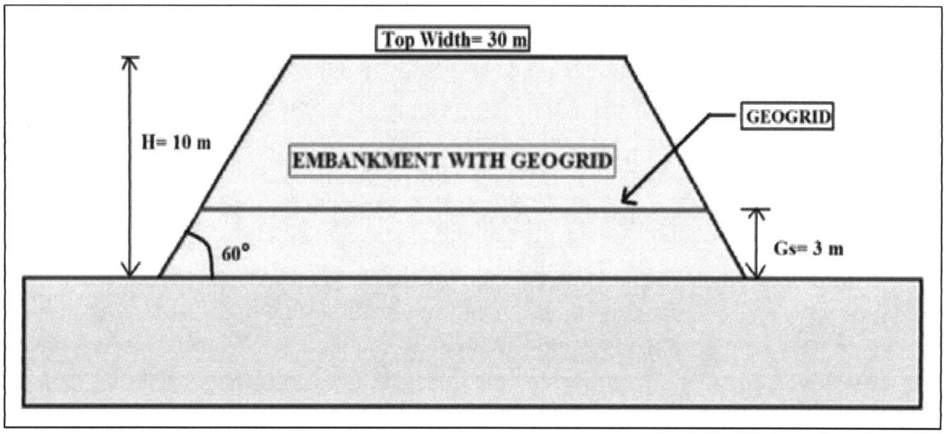

Fig 1. Embankment with Geogrid

2.2 Development of prediction model

The data used to develop the stability charts were also used to develop the prediction model. For the development of prediction model, regression analysis was used in MS Excel. The independent or input parameters include, slope angle (β), slope height (H), cohesion (c), angle of internal friction (φ) and spacing of geogrid layer (Gs). FOS is used as a dependent or output parameter.

2.3 Sensitivity Analysis

Sensitivity analysis was carried out on the pre-trained regression model to examine the impact of each predictor variables on the prediction of dependent variables. An approach based on data perturbation was employed for the sensitivity analysis [18]. Each input parameter was successively raised and lowered while all other input variables remained constant. Here, each input parameter was modified by ± 10%, and the percentage change in output was then calculated.The sensitivity analysis was done by taking the average value of all independent parameters. The dependent parameter is calculated by using the regression formula given in equations (1) or (2). If ΔX is the percentage increment or decrement and ΔY is the difference between the FOS considering 10% increment or decrement and the predicted FOS, then relative sensitivity (RS) is expressed as RS = (X . ΔY) / (Y' . ΔX).

The trained MLR models' prediction performance is assessed by determining the correlation percentage. Furthermore, the error analysis is carried out by calculating the root mean square error (RMSE) and mean absolute error (MAE).

3. **Results and discussions**

After carrying out the various analyses, the results are plotted in the following sections.

3.1. Stability charts

The computational results are used to develop two stability charts, firstly, for the Fellenius method and secondly, for the Bishop's method for different slope inclination having different values of geogrid spacing. The use of these charts is very straight forward. If the slope geometry, soil parameters and geogrid spacing are known, one can easily calculate the FOS thereby, eliminating the rigorous method of iteration. The stability charts for Fellenius method and Bishop's Method are shown in Fig 2 and Fig 3. Of course, these stability charts developed in terms of N* cannot be used for slopes with zero internal friction angle.

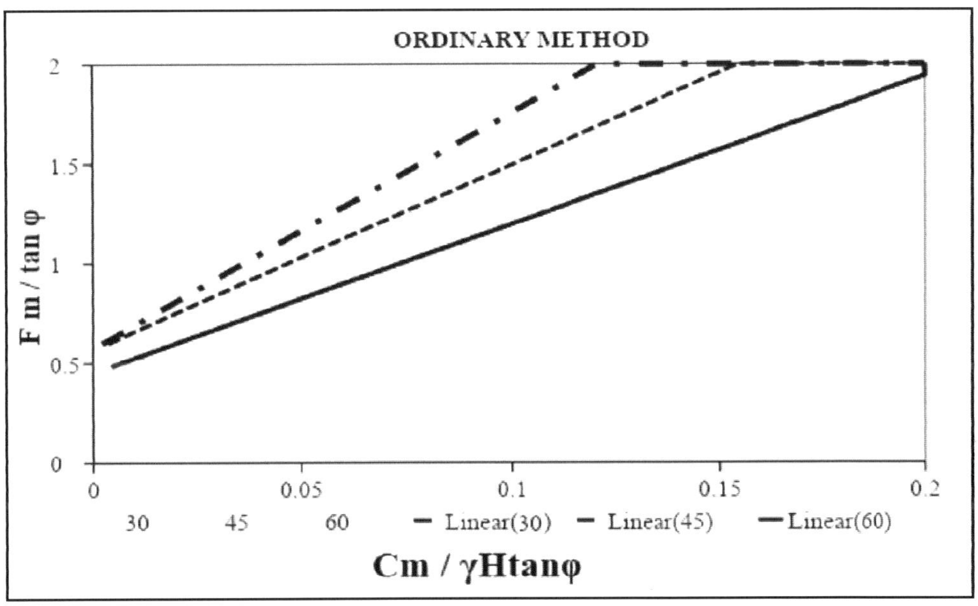

Fig 2. Stability chart for Fellenius method of slope stability

3.2 Prediction model

The equation of multiple linear regression (MLR) for Bishop's method for prediction of FOS is given in the following equation:

FOS=1.832-0.025 x β-0.042 x H+0.022 x c+0.037 x φ-0.001 x Gs

1

($R2 = 0.900$; MSE = 0.062)

The correlation matrix for FOS with other independent parameters for Bishop's method is shown in Table 1. This helps in identifying the presence or absence of a relationship between two variables clearly.

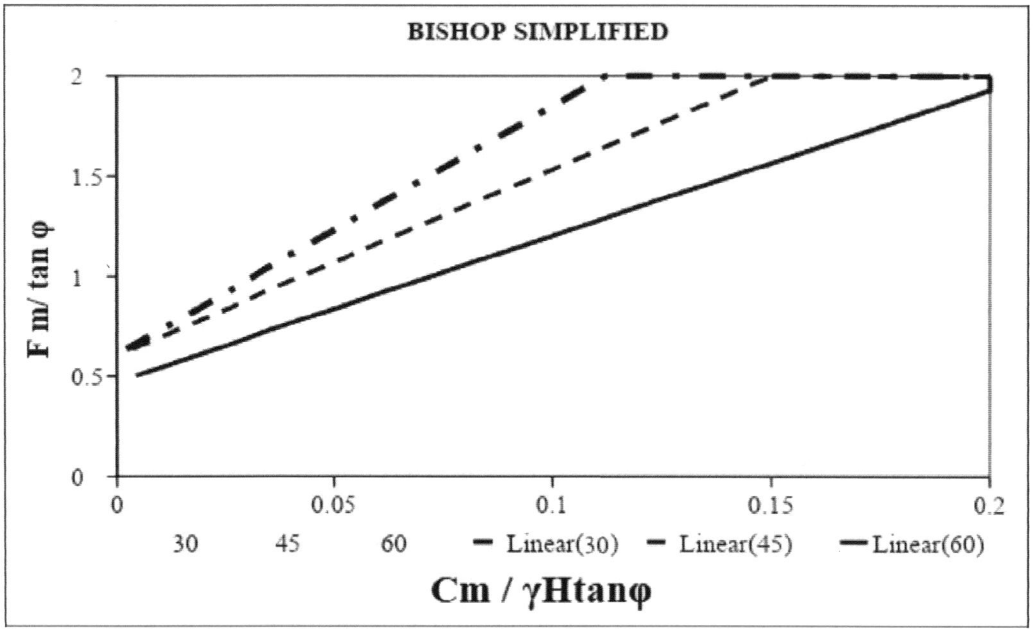

Fig 3. Stability chart for Bishop's method of slope stability.

Table 1. Correlation matrix for Bishop's method of slope stability

Parameters	β	H	c	φ	G_s	FOS (Bishop's Method)
β	**1**	-0.057	-0.009	-0.009	0.155	-0.378
H	-0.057	**1**	-0.018	0.019	0.498	-0.387
c	-0.009	-0.018	**1**	0.34	0.079	0.639
φ	-0.009	0.019	0.34	**1**	0.037	0.62
G_s	0.155	0.498	0.079	0.037	**1**	-0.218
FOS (Bishop's Method)	-0.378	-0.387	0.639	0.62	-0.218	**1**

The equation of multiple linear regression (MLR) for Fellenius method for prediction of FOS is given in the following equation:

FOS=1.967-0.029 x β-0.042 x H+0.023 x c+0.039 x φ-0.002 x Gs

2

(R2 = 0.896; MSE = 0.071)

The correlation matrix for FOS with other independent parameters for Fellenius method is shown in Table 2.

Table 2. Correlation matrix for Fellenius method of slope stability

Parameters	β	H	c	φ	G_s	FOS (Fellenius Method)
B	1	-0.057	-0.009	-0.009	0.155	-0.409
H	-0.057	1	-0.018	0.019	0.498	-0.369
C	-0.009	-0.018	1	0.34	0.079	0.627
φ	-0.009	0.019	0.34	1	0.037	0.621
G_s	0.155	0.498	0.079	0.037	1	-0.216
FOS (Fellenius Method)	-0.409	-0.369	0.627	0.621	-0.216	1

3.3 Sensitivity analysis

The data perturbation method, which involved altering the input values by ± 10% one at a time while keeping other parameters constant, was used to do sensitivity analysis on the created models. The percentage change in output due to input perturbation was determined in order to determine the parameters that were the most sensitive, and the findings are displayed in Tables 3 and Table 4.The tornado diagram [19] showing the most sensitive parameter for Bishop's method and Fellenius method is shown in Fig 4 and Fig 5.From Fig 4 and Fig 5, it can be seen that for both the slope stability methods, the slope inclination is found to be the most sensitive factor.

The well-trained MLR models are validated by considering 20 different data sets (shown in Table 5) not used during the model building process. In Fig. 6, model simulation results arepresented and statistical metrics, including correlation %, RMSE, and MAE are compared.

Table 3. Sensitivity analysis for Fellenius method of slope stability.

Inputs	**β**	**H**	**C**	**φ**	**G_s**	**FOS (Bishop's)**
Average (X)	44.742	19.723	29.789	27.82	8.97	1.536

10 % Increment	**INPUT**	**X**	**ΔX**	**X+ΔX**	**Y'**	**Y''**	**ΔY**	**RS**
	β	44.742	4.474	49.216	1.536	1.421	-0.113	-0.737
	H	19.723	1.972	21.695	1.536	1.452	-0.082	-0.540
	c	29.789	2.979	32.768	1.536	1.602	0.067	0.436
	φ	27.82	2.782	30.602	1.536	1.638	0.103	0.670
	G_s	8.97	0.897	9.867	1.536	1.534	-0.0008	-0.005

10 % Decrement	INPUT	X	ΔX	X-ΔX	Y'	Y''	ΔY	RS
	β	44.742	4.474	40.268	1.536	1.649	0.114	0.744
	H	19.723	1.972	17.751	1.536	1.619	0.084	0.547
	c	29.789	2.979	26.810	1.536	1.469	-0.066	-0.429
	φ	27.82	2.782	25.038	1.536	1.433	-0.101	-0.664
	G_s	8.97	0.897	8.073	1.536	1.536	0.001	0.012

Table 4. Sensitivity analysis for Bishop's method of slope stability

Inputs		β	H	C	φ	G_s	FOS (Fellenius)	
Average (X)		44.742	19.723	29.789	27.82	8.97	1.590	
10 % Increment	INPUT	X	ΔX	X+ΔX	Y'	Y''	ΔY	RS
	β	44.742	4.474	49.216	1.590	1.462	-0.127	-0.804
	H	19.723	1.972	21.695	1.590	1.506	-0.083	-0.525
	c	29.789	2.979	32.768	1.590	1.657	0.067	0.425
	φ	27.82	2.782	30.602	1.590	1.698	0.108	0.679
	G_s	8.97	0.897	9.867	1.590	1.588	-0.001	-0.007
10 % Decrement	INPUT	X	ΔX	X-ΔX	Y'	Y''	ΔY	RS
	β	44.742	4.474	40.268	1.590	1.7181	0.128	0.806
	H	19.723	1.972	17.751	1.590	1.6738	0.084	0.527
	c	29.789	2.979	26.810	1.590	1.5227	-0.067	-0.424
	φ	27.82	2.782	25.038	1.590	1.4822	-0.108	-0.678
	G_s	8.97	0.897	8.073	1.590	1.5915	0.002	0.009

Table 5. Data set for model validation

Bishop's Method			Fellenius Method		
LEM	**MLR**	**Stability Charts**	**LEM**	**MLR**	**Stability Charts**
2.398	2.301	2.367	2.533	2.504	2.598
1.929	2.005	1.911	2.038	2.177	2.098
1.903	1.976	2.02	2.006	2.153	2.122
1.838	1.946	1.816	1.942	2.112	1.994
1.676	1.821	1.761	1.765	1.983	1.878
2.2	2.161	2.331	2.316	2.363	2.424

1.29	1.471	1.293	1.371	1.636	1.389
1.4	1.567	1.406	1.487	1.741	1.513
1.766	1.877	1.732	1.875	2.08	1.893
1.665	1.812	1.67	1.771	2.016	1.773
1.733	1.962	1.607	1.799	2.172	1.677
1.232	1.424	1.3	1.282	1.589	1.51
1.139	1.312	1.119	1.185	1.456	1.175
1.29	1.496	1.27	1.352	1.65	1.339
1.027	1.172	1.164	1.067	1.315	1.237
1.09	1.24	1.113	1.13	1.38	1.157
0.898	1.009	0.953	0.931	1.122	0.968
0.638	0.285	0.467	0.662	0.351	0.5
0.7	0.41	0.489	0.734	0.48	0.56
0.689	0.381	0.559	0.716	0.455	0.541

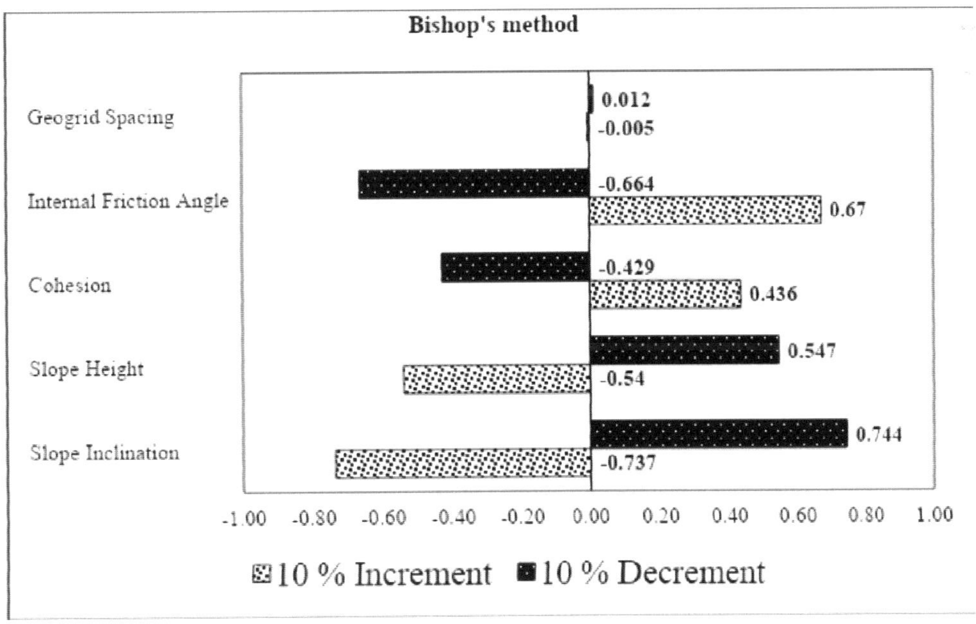

Fig 4. Tornado diagram for Bishop's method of slope stability

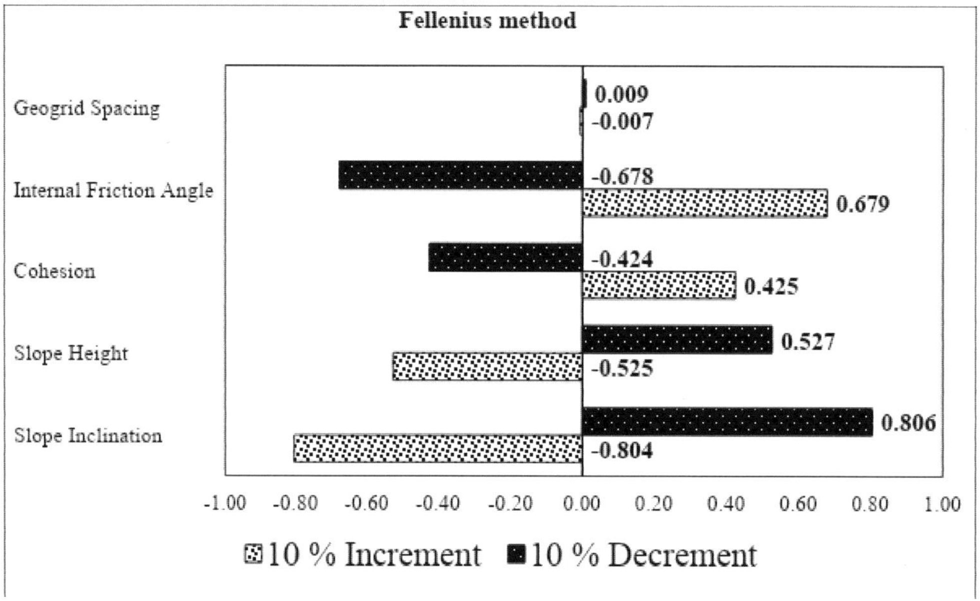

Fig 5. Tornado diagram for Fellenius method of slope stability

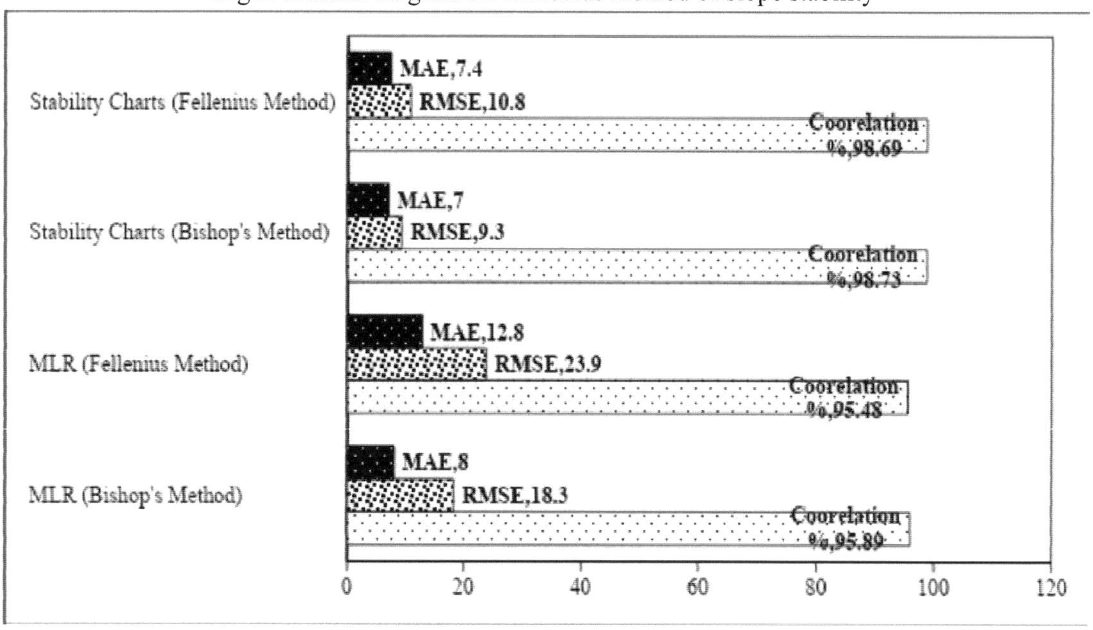

Fig 6. Model simulation results

3.4 Discussion

In this study, a distinctive method was employed to develop design charts and prediction models using 2100 artificial slope cases. The capacity of the created models and design charts to predict the FOS was implied by successful model validation with 20 different data sets not used during the model building process. The data perturbation method of sensitivity analysis shows that slope angle is the most influencing factor. Prediction of slope stability has been investigated previously over a 5-to-15-year period. Current study highlights the importance of using geogrid in the model. In developing countries like India, construction of houses, roads, tunnels, widening of roads, etc. are going on at a very fast rate. Hence, many slope cuttings are done which results in the increase in the stability issues. Geogrid can be an effective tool for enhancing the stability of the slopes. The presented methodology provides the site engineers a rapid and an effective way to arrive at a stable slope geometry.

4. Conclusion

For the slope stability analyses, stability charts and regression analysis were conducted. This design chart is convenient to use and it doesn't require any iterative process to determine the FOS. However, with zero value of angle of internal friction, the charts produced are not intended to be used. The following conclusions can be drawn from the present study.

1. Bishop's method gives higher FOS comparison to Fellenius method. Higher FOS is obtained when Geogrid touches slip circle.
2. Regression statistics for Bishop's method provides better FOS than Fellenius method. On comparing the MLR and stability charts, the FOS predicted using stability charts have higher correlation %, lower RMSE and lower MAE. Thus, the accuracy of stability charts over MLR is found to be high.
3. According to the sensitivity analysis tornado chart, slope angle is the most influencing parameter for the slope stability.

References

[1].Fellenius, W. (1936). Calculation of the stability of earth Dams. Proceedings of the 2nd Congress on Large Dam, Washington, D.C., 4:445-463.
[2].Bishop, A. W., (1955). The use of slip circle in the stability analysis of slope. Geotechnique, 5(1):7-17. https://doi.org/10.1680/geot.1955.5.1.7
[3].Janbu, N. (1957). Earth pressure and bearing capacity calculations by generalized procedure of slices. Proceeding of the 4th International Conference on Soil Mechanics, volume 2, pages 207-212, London.
[4].Morgenstern, N. R. and Price, V. W. (1965). The analysis of the stability of general slip surfaces. Geotechnique, 15(1):11-26.
https://doi.org/10.1680/geot.1965.15.1.79
[5].Fredlund, D. G., and Krahn, J. (1977). Comparison of slope stability method of analysis. Canadian Geotechnical Journal, 14:429-439.
https://doi.org/10.1139/t77-045
[6].Sarma, S. (1973). Stability analysis of embankments and slopes. Geotechnique, 23(3):423.
http://dx.doi.org/10.1061/AJGEB6.0000903
[7].Burman, A., Archarya, S. P., Sahay, R. R., and Maity, D. (2015). Comparative study of slope stability analysis using traditional limit equilibrium method and finite element method. Asian Journal of Civil Engineering (BHRC), 16(4):467-492.
[8].Majedi, P., Ghalehjough, B. K., Akbulut, S., and Çelik, S. (2017). Effect of Reinforcement on stability and Settlement of Embankment: A Finite Element analysis of Different Kinds of Reinforcing and Construction Conditions. European Journal of Advances in Engineering and Technology, 4(10):759-764.
[9].Raouf, M. A. E. (2020). Stability of Geogrid Reinforced Embankment on Soft Clay. Journal of Engineering Sciences, 48(5):830-844.
http://dx.doi.org/10.21608/JESAUN.2020.112941
[10].Esmaeili, M., Naderi, B., Neyestanaki, H. K., and Khodaverdian, A. (2018). Investigating the effect of Geogrid on stabilization of high railway embankments. Soils and Foundations, Tokyo, 58(2):319–332.
http://dx.doi.org/10.1016/j.sandf.2018.02.005
[11].Smith, I. M. and Hobbs, R. (1974). Finite element analysis of centrifuged and built-up slopes. Geotechnique, 24(4):531-559.
https://doi.org/10.1680/geot.1974.24.4.531
[12].Naseer, S., and Evans, E. (2020). Stability analysis of slopes using Limit Equilibrium and Finite Element Methods. In 2nd Conference on Sustainability in Civil Engineering, (CSCE'20), Saskatoon, Canada.
[13].Marrapu, B. M., and Jakka, R. S. (2017). Assessment of slope stability using multiple regression analysis. Geomechanics and Engineering, 13(2):237–254.

https://doi.org/10.12989/gae.2017.13.2.237
[14].Khan, M. I., and Wang, S. (2021). Slope Stability Analysis to Develop Correlations between Different Soil Parameters and Factor of Safety Using Regression Analysis. Polish Journal of Environmental Sciences, 30(5):4021-4030.
https://doi.org/10.15244/pjoes/131203
[15].Azmoon, B., Biniyaz, A. and Liu, Z. (2021). Evaluation of deep learning against conventional Limit equilibrium method for slope stability analysis. Applied Sciences (Basel, Switzerland), 11(13):6060.
https://doi.org/10.3390/app11136060
[16].Bhave, M. V. (1982). Maharashtra P.W.D. Handbook Chapter 21- Earth and Composite Dams. Maharashtra Engineering Research Institute, Government of Maharashtra.
[17].Bell, J. M. (1966). Dimensionless parameters for homogeneous earth slopes. J. Soil Mech. Found. Div., Am. Soc. Civ. Eng., 92(5), 51– 65.
[18].Pandey, M.; Lam, W.H.; Cui, Y.; Khan, M.A.; Singh, U.K.; Ahmad, Z. (2019). Scour around Spur Dike in Sand–Gravel Mixture Bed. Water, 11, 1417.
https://doi.org/10.3390/w11071417
[19].Bahreini, V., Pouraminian, M. & Tabaroei, A. (2022). Seismic sensitivity analysis of Musa Palas historic masonry arch bridge by Tornado diagram. J Build Rehabil 7, 71.
https://doi.org/10.1007/s41024-022-00215

Evaluation of Recycled Tyre Steel Fibres Effectiveness on the Properties of Concrete

Hussin A.M Yahia[1], Abdul-Aziz Al Maqbali[2], Said Saif Salem Al Maawali[3], Mohd Azizul Ladin[4], Sachin Kuckian[5] Kiran Kumar Poloju[6], C. Venkata Siva Rama Prasad[7]
[1,2,5,6] Department of Civil and Mechanical Engineering, Middle East College, Knowledge Oasis Muscat, Oman
[3] Department of Civil & Environmental Engineering, College of Engineering & Architecture University of Nizwa
[4] Faculty of Engineering, University Malaysia Sabah, UMS, Kota Kinabalu, Malaysia,
[7] Department of Civil Engineering, St. Peter's Engineering College, IndiaE-mail: hayhia@mec.edu.om

Abstract. Disposal of tire waste is a difficult task because tires are generally indestructible and have long lifetimes. Tire waste is managed via short-term solutions such as storing or dumping in illegal landfills. Tire recycling is a process where tires are recycled for reuse in vehicles; however, the recycled tires are not suitable for reuse because their recovery is difficult owing to faults such as holes. Thus, tires are the most problematic and largest source of waste owing to their high production and durability. Several tire recycling techniques that take advantage of various tire components have been implemented. This study focuses on the utilization of solid steel fibers are incorporated into a concrete tire frame. This is done to assess how well fibers work with reinforced concrete's design. Additionally, concrete's compressive strength was evaluated three times (0.1%, 0.2%, and 0.3%). The study's findings are based on an examination of how tire-recycled steel fibers affect concrete strength and their use to enhance the strength of concrete. Concrete reinforcement with steel fibers results was compared with the traditional concrete reinforcement and it was found that the iron fibers present in the tires enhanced durability and strength of concrete.

1. Introduction

The auto industry is one of the world's most significant economic sectors. The sector is still expanding, with a 30% increase. waste tire recycling has become increasingly popular in many nations across the world. Rubber, carbon, and metals make up the majority of waste tires. Although they have a high value for recycling resources, ineffective storage, regeneration, and disposal procedures can be harmful to people's health and the environment. As a result, figuring out how to recycle and dispose of waste tires in a reasonable and practical manner while avoiding pollution. Approximately one billion tires reach the end of their service life in the world[1]. The tires can be recycled or repurposed as cement kiln fuel, asphalt pavement building material, or concrete aggregates. Steel fibres in tire trash can be recovered using cryogenic, pyrolysis or shredding procedures when worn tires are recycled [2]. In developed countries, tires are recycled, and components are used in many industries [2]. As a result, it's critical to find a cost-effective and environmentally responsible way to handle the redevelopment.

Concrete structures are frequently reinforced with steel fibers produced in factories using tensile force across cracks and bridging them fiber-reinforced concrete reduces fracture propagation and enhances mechanical properties (such as the capacity and ductility of the material). This improves post-crack behavior The results show that recovered steel fiber from tire waste (RF) has the potential to replace concrete as a building material. The process of extracting materials from used tires is done by mechanical or thermal breakdown techniques.

1.1 Waste Management of Used Tyres

Previous studies have shown that Recycled Steel Fibers are used as a component in the production of fiber RC. [12][13][14]. Experiments on physical and chemical properties of RFRC, including as compressive strength, flexural strength, and pull-out behavior, have also been conducted [15][16]. Table 1 shows a summary of the experiments. [18] discovered that the pull-out behavior of RF is comparable to that of SF, and that varied fiber dose has no effect on compressive strength. However [18] discovered that increasing the dose of RF and SF while maintaining the same water content in the concrete mixtures resulted in a modest improvement in compressive strength.

Table 1: Testing of RFRC.

Authors	Length of Fibers	Fiber	Testing
Aiello et al.	15 to 25 mm 0.17 to 0.2 mm	0.13%–0.46% (Volume)	Flexural and Compressive strength
Hu et al.	15 to 40 mm Average 0.1 mm	10 kg/m3 –30 kg/m3	Flexural, pull-out test, strength test, Compressive strength
Leone et al	1mm to 13 mm 0.10 mm to 0.45 mm	0.46 % (Volume)	Flexural, pull-out test, strength test, Compressive strength

1.2 Problem Statement

Processing tire waste is a significant issue in waste management around the world. Only 4% of the 1.5 million tons of scrap tires that are produced annually are employed in civil engineering projects, according to estimates. These wastes are disposed of mostly as hazardous waste. Since concrete is the most popular building material owing to its simple preparation and low cost, reinforced concrete is being considered as a safe and effective method for the disposal of tyre waste where steel fibres from the tyre waste is recycled for this purpose. Bonding between the steel fibres and cement paste is weak and deteriorates quickly under stress owing the low modulus of elasticity in traditional concrete. Steel fibers from recycled tires were put to the test as industrial steel and showed promising results. To enhance the structural performance of rubberized concrete, improvements in toughness and strength. The potential use of tyre waste in the structural construction industry will be beneficial to the environment. Concrete is brittle in nature its tensile and bending strength are negligible. The tensile stress in concrete would be improved by strong conventional reinforcing in the form of steel rods of various grades. The Sultanate of Oman has experienced an unchecked increment in tyres over the past. Figure 1 shows scrap tyre waste.

Figure 1 waste tyre scrap

2. *Materials and Methods*

Cement, water, fine sand, and aggregates are the main ingredients of concrete. It is crucial for the development of infrastructure, including roads, buildings, and bridges. Although steel fibers can be applied to a wide range of applications, including foundations, runways, airports, ports, machinery floors, and heavy equipment [19], reinforced concrete appears to be the most logical and cost-effective way to use RTSFs in this study. The preparation of the materials and their distribution in relation to the various particle sizes is the first stage of the laboratory work, as seen in the following stages:

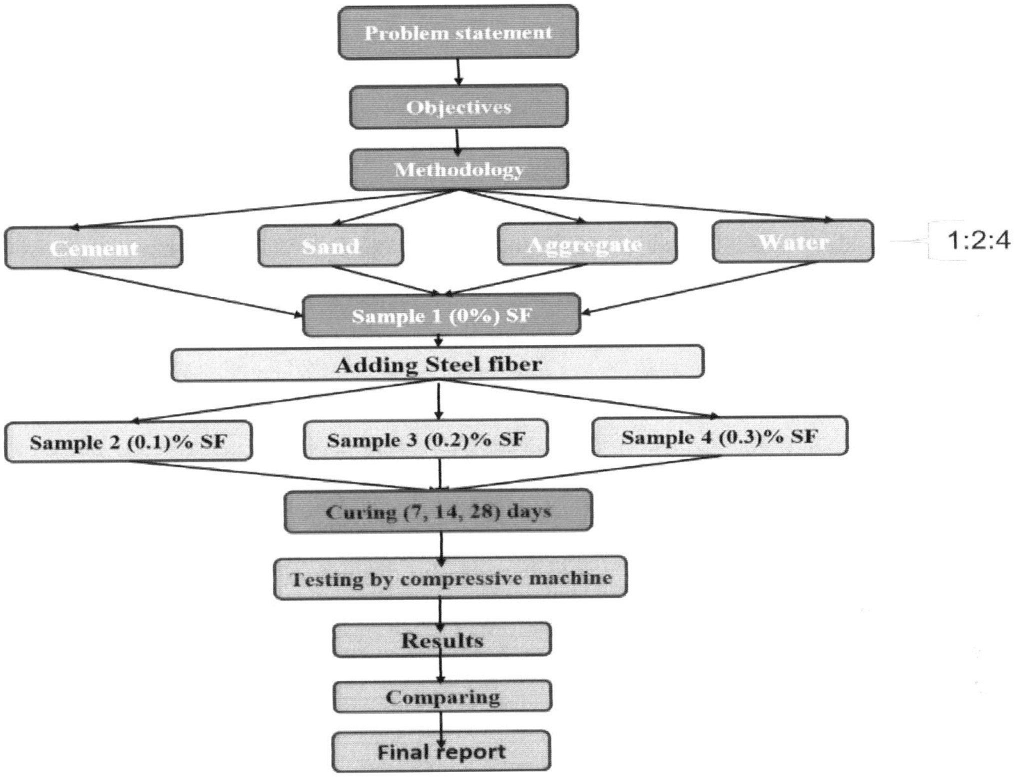

Figure 2 Research Design

2.1 *Raw Materials and Mix Proportion*

Cement: 80% lime, 20% clay, and processing agents make up cement, which is composed of limestone, calcite, silica, and bauxite (Cargill). Sand, Fe2O3, Al2O3, and Al2O3 combine to form cement (SiO 2). The delicate texture may solidify, turn opaque, and then glue the concrete components together when combined with water.

Sand : is a unique granular material that does not react with water or bulk cement. Currently, the most popular sand kind is calcium carbonate. Sand is a common non-renewable resource that is used more frequently in the production of concrete.

Aggregates: It might be sizable and yet be used as filler in a concrete mixture. A comprehensive examination reveals that the polymer plays a crucial role in the properties of moist concrete since it is feasible [7].

Water: which is a compound of (H2O), is crucial for the interaction of the total with the additives in the age of concrete and cement.

Steel Fibers: Figure 3 below illustrates the removal of steel strands from tire waste. These loop-shaped fibers have a diameter of around 0.025 meters.

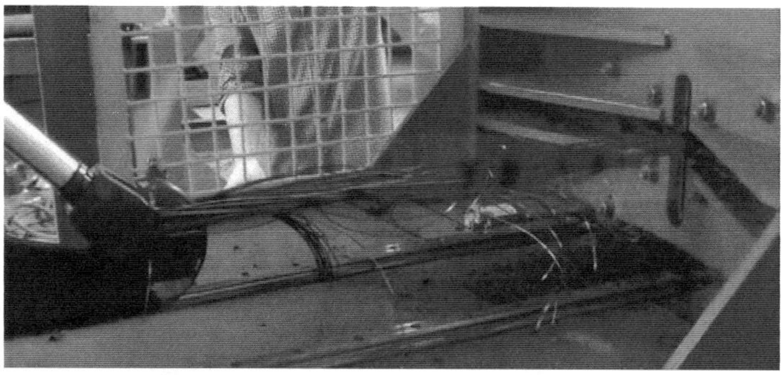

Figure 3 Extracting steel fibers from tires

Using an electric cutting machine, the steel fibers were extracted and then sliced into tiny pieces, each measuring about 3 inches, for the laboratory investigations as shown below in Figure 4.

Figure 4. Cutting steel fibres into small pieces

After cutting the steel fibres to about 3 inches, they were bent in the form of a U-shape as shown in the figure below.

2.2 Concrete mixing with 0.1% of Steel Fibre

The various volumes of concrete in each cubic meter were combined with steel fibers from tire leftovers (0.1, 0.2, and 0.3%) for each ratio of three cubes. The objective was to control the crushing of the surface shrinkage and to obtain the highest concrete compression strength (0.1%) using steel fibers. The purpose is to test the pressure strength of the concrete after 7, 14, and 28 days [4].

Figure 5 weight of steel fibre (0.1, 0.2, and 0.3%)

Concrete's resistance to corrosion and compressive strength can both be increased by adding steel fiber from tyre remnants. The weights used to calculate these percentages (0.11, 0.22 and 0.33 kg).

Figure 6: Addition of steel fiber

3. Results and discussion

Three different samples were used to obtain the results of the following tests: compressive strength testand slump test for cubes, and triple-mix experiments (0.1%, 0.2%, and 0.3%) For this study, a mixture of cement, sand, and aggregate in a weight-to-water ratio of 1: 2: 4 was used. Based on the mixing range, the material components of the mix were computed. For testing, twelve concrete cubes (15 cm ×15 cm×15 cm) were stacked. Nine cubes were arranged to utilize a variety of loose tire steel threads, while three trials were carried out without steel fibers from tire trash for comparison (0.1, 0.2, and 0.3).

Table 2 show quantities of samples

Experiments	**1**	**2**	**3**	**4**
Steel Fiber	0 %	0.1 % 0.11 kg	0.2 % 0.22 kg	0.3 % 0.33 kg
Cement	1.44 kg	1.44 kg	1.44 kg	1.44 kg
Sand	2.89 kg	2.89 kg	2.89 kg	2.89 kg

Aggregate	5.78 kg	5.78 kg	5.78 kg	5.78 kg
Water	0.72 L/1.44 kg	0.72 L/1.44 kg	0.72 L/1.44 kg	0.72 L/1.44 kg

3.1 Slump Test

A pilot test called "concrete slump" is used to evaluate the uniformity and completeness of concrete. Usually, the fall is measured, the appropriateness of the concrete (workability) is decided, and an exceptional cone is used to estimate the type and amount of water needed for the cement. The concreteused in different installations determines the proper surface (slump). 25 mm for highways, 50-60 mm for concrete, and 10–25 mm for homes.

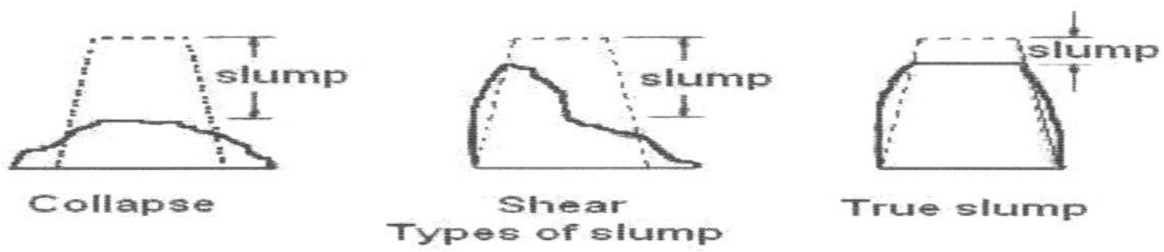

Figure 7. Concrete slump test measuring

The cubes inside surface were thoroughly cleaned to remove any water or concrete build-up. The cube was placed on a flat, smooth surface that was securely fastened and watertight. Three layers, each about a third thick, are used to fill the mold. It needs to be attached to the bottom surface and distributed uniformly over the top. After carefully and gradually filling the cone throughout the head, it is raised. The amount of drop in the measuring tape measured immediately after lifting the cone, as shown below in Figure 8.

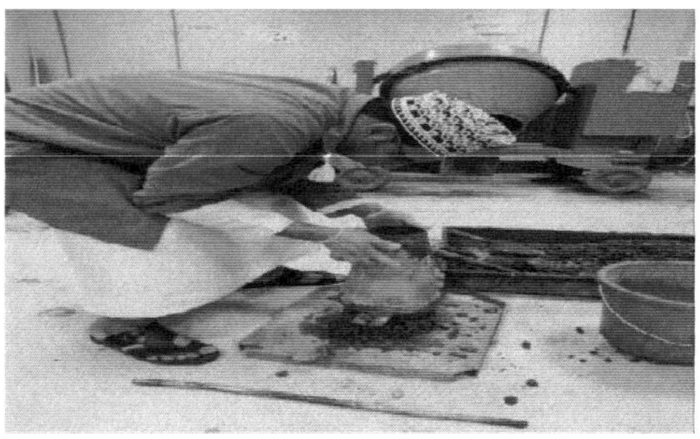

Figure 8 Slump Test

Slump	**0 to 20**	**10 to 40**	**30 to 120**	**100 to 120**	**180 to 220**
Consistency	**Dry**	**Stiff**	**Plastic**	**Wet**	**Sloppy**

3.2 Compressive strength test

This test is very important to ascertain the strength of concrete. The strength of the concrete is checked after 7 days, and to ensure the strength of the design and the durability of the concrete, it must be tested after 14 and 28 days. The code employed at the materials laboratory should be known prior to take the sample. Because the British code BS specifies a cube dimension of 15 x 15 x 15 cm, the size of the test cubes used in this investigation was in accordance with that standard. The cube test to be successful, the design mix must be approved, and the test must last for the required amount of time [5].

3.3 Steps and processes

The concrete underwent three steps of testing to determine its compressive quality and compressive strength. After seven days, the initial test started, and it had to achieve 70% accuracy. The instant test was carried out following a 14-day planning period. The strength achieved must range from 80% to 90%. A strength of between 90% and 100% must be attained after 28 days of curing.

Sample – 1. Original sample without adding's

Task	7 Days	14 Days	28 Days
Weight	8.4 kg	8.5 kg	8.7 kg
Maximum load (kN)	414	501	600
Strength of cubes	18 MPa	22 MPa	26 MPa

To ascertain the concrete cubes' strength, establish whether they meet the specifications, and determine whether they may be used with another component after the assessment of the previous concrete component has confirmed their validity, the concrete cubes must undergo a compressive strength test. After 7 days, the test should yield a result of 75% of the design compressive strength, after 14 days, 85%, and after 28 days, 100% of the planned concrete resistance. The outcomes for plain concrete devoid of any additives are shown in the table above.

Sample – 2 Adding Steel Fibers 0.1 %

Task	7 Days	14 Days	28 Days
Weight	8.01 kg	8.3 kg	8.41 kg
Maximum load (kN)	501	653	695
Strength of cubes	22 MPa	29 MPa	30 MPa

The results for the concrete with 0.1% steel fibers are displayed in the table above. The purpose of the test was to evaluate the strength of the identical concrete sample after 7, 14, and 28 days. The outcomes show that the cubes' strength continued to rise. This suggests that the amount of steel fibers in concrete affects how much weight it can support during construction.

Sample – 3 Adding Steel Fibers 0.2 %

Task	7 Days	14 Days	28 Days
Weight	8.35 kg	8.39 kg	8.5 kg
Maximum load (kN)	564	684	698
Strength of cubes	25 MPa	30.4 MPa	31 MPa

The findings show that the cubes' strength continued to rise. This shows that the load that concrete can support during construction increases with the quantity of steel fibers present.

Sample – 4 : Adding Steel Fibres (0.3%)

Task	**7 Days**	**14 Days**	**28 Days**
Weight	8.41 kg	8.43 kg	8.59 kg
Maximum load (kN)	571	677	710
Strength of cubes	25.4 MPa	30 MPa	31.5 MPa

The results for concrete with 0.3% steel fibers are depicted in the table above and the three graphs that follow on the seventh day following the sample's immersion in the machine. The test involved determining the strength of the same concrete sample after seven, fourteen, and 28 days. The outcomes show that the cubes' strength continued to rise. This suggests that the amount of steel fibers in concrete affects how much weight it can support during construction. Typically, concrete cubes are shattered after 7, 14, and 28 days to assess its durability. The average compressive strength of all selected samples made from the same sample was calculated and tested by the following equation: Compressive strength= P/A.

To compressive force test, samples of three blocks of each kind (0%, 0.1%, 0.2%, and 0.3%) were gathered. Three samples were initially collected without the presence of steel fibers. After 28 days, the maximum results were attained (26 MPa). Second, 3 of the analyzed samples had 0.1% steel. After 28 days, the greatest result was 30 MPa. After 28 days, the third sample with 0.2% solid steel fibers produced 31 MPa, and the fourth sample with 0.3% steel fibers produced 31.5 MPa. The results showed that the concrete's strength increased as more steel fibers were added to the cubes.

4. Conclusion

The utilization of steel fibers from tyre waste is summarized in this study since these wastes are significant barriers to waste management on a global scale; as a result, we show the value of these wastes in the field of civil engineering. In the sphere of engineering, the government authorities ought to assist tire waste recycling or reuse. The goal of this investigation is to recover steel fibers from car tires for use in concrete. The goal is to make conventional concrete stronger when fibers are added at various percentages (0.1%, 0.2%, and 0.3%). For samples of typical concrete, the results were 18 MPa after 7 days, 22 MPa after 14 days, and 26 MPa after 28 days. However, the samples were submerged in water after the concrete and steel fibers had been combined. Three samples were taken with a concentration of 0.1%, and the findings were 22.2 MPa after 7 days, 29 MPa after 14 days, and 30 MPa after 28 days. Like the 0.2% sample, 3 samples were taken, and the findings were 25 MPa after 7 days, 30.4 MPa after 14 days, and 31 MPa after 28 days. Once more, samples were taken for 0.3%, and the findings were 25.4 MPa after 7 days, 30 MPa after 14 days, and 31.5 MPa after 28 days. As a result, a high-strength test research was carried out after adding 0.3% steel fibers within 28 days. This study's findings support the notion that the presence of steel fibers makes concrete less brittle and less prone to breaking. Additionally, although the improvement was not significant, it improved the concrete compressor's strength.

Acknowledgement

The authors extend their appreciation to the Center for Research and Consultancy at Middle East College, Oman for funding this work through the Undergraduate Research Grant (URG) under grant number [BFP / URG/ EI / 19 / 045] Research Council, Oman.

References:

[1] B.S. Thomas, R.C. Gupta, V.J. Panicker. Recycling of waste tire rubber as aggregate in concrete:durability-related performance, J. Clean. Prod. 112 (Part 1) (2016) 504–513.

[2] Badorul Hisham et al. Cumulative Effect of Crumb Rubber and Steel Fiber on the Flexural Toughness ofConcrete (2017),

[3] C.G. Papakonstantinou, M.J. Tobolski, Use of waste tire steel beads in Portland cement concrete, Cement Concrete Res. 36 (September (9)) (2006) 1686–1691.

[4] D. Bjegovic, A. Baricevic, S. Lakusic, D. Damjanovic, I. Duvnjak, Positive interaction of industrial and recycled steel fibers in fiber reinforced concrete, J. Civ. Eng. Manage. 19 (2013) 50–60.

[5] H. Hu, P. Papastergiou, H. Angelakopoulos, M. Guadanini1, K. Pilakoutas, Flexural performance of steel fiber reinforced concrete with manufactured and recycled tyre steel fibers, Construction Materials for Sustainable Future (2019) University of Zagreb, Faculty of Civil Engineering, Zagreb, Croatia, pp. 904– 911. ISBN 978-953-8168-04-8
https://core.ac.uk/download/pdf/55626748.pdf

[6] Huang CH, Lin SK, Chang CS, Chen HJ. Mix proportions and mechanical properties of concrete containing very high-volume of Class F fly ash. Constr Build Mater. 2013; 46:71-78.

[7] J.A.O. Barros (2013), Assessment of the potentialities of recycled steel fibres for the reinforcement of cement-based materials [online] available from

[8] K. Pilakoutas, K. Neocleous, H. Tlemat, Reuse of tyre steel fibres as concrete reinforcement, Proc. ICE –Eng. Sustain. 157 (3) (2004) 131–138 ISSN 1478-4629.

[9] K. Pilakoutas, K. Neocleous, H. Tlemat, Reuse of tyre steel fibres as concrete reinforcement, Proc. ICE –Eng. Sustain. 157 (3) (2004) 131–138 ISSN 1478-4629.

[10] M. Leone, G. Centonze, D. Colonna, F. Micelli, M.A. Aiello, Fiber-reinforced concrete with low content of recycled steel fiber: Shear behaviour, Constr. Build. Mater. 161 (2018) 141–155, doi: http://dx.doi.org/10.1016/j.conbuildmat.2017.11.1011

[11] M.A. Aiello, F. Leuzzi, G. Centonze, A. Maffezzoli, Use of steel fibers recovered from waste tyres as reinforcement in concrete: pull-out behaviour, compressive and flexural strength, Waste Manage. 29 (6) (2009) 1960–1970 2009/06/01/.

[12] M. Mastali, A. Dalvand, Use of silica fume and recycled steel fibers in self-compacting concrete (SCC), Constr. Build. Mater. 125 (2016) 196–209, https://doi. org/10.1016/j.conbuildmat.2016.08.046.

[13] Mehta A, Siddique R. An overview of geopolymer derived from industrial by-products. Constr Build Mater. 2016; 127:183-198.

[14] Mohd. Gulfam Pathan and Ajay Swarup. A Review on Steel Fiber Reinforced Concrete (2017).

[15] Muhammad Akram et al., Utilization of Soft Drink Tins as Fiber Reinforcement in concrete, 2017

[16] Rahul Chaudhary, Tapeshwar Kalra. To Study the Strength Properties of Concrete by Using Tire Waste. (2017).

[17] Y. Wang, Utilization of Recycled Carpet Waste Fibers for Reinforcement of Concrete and Soil.2016.

[18] Yuanxun Zheng, Mechanical Properties of Steel Fiber-Reinforced Concrete by Vibration Mixing Technology. 2018

[19] Sachin Kuckian and Hussin Yahia, Economically Prudent Design (EPC) of Outrigger-Based Structural Systems, 2022, The Asian Review of Civil Engineering, ISSN: 2249-6203 Vol.11 No.2, 2022,pp.1-9.

COMPARATIVE STUDY OF LINEAR STATIC ANALYSIS OF G+10 STOREY BUILDING WITH AND WITHOUT SHEAR WALL IN ETABS

P Lavanya[1], P Sridhar[2], and V Manasa[3]
[1]Assistant professor, Department of Civil Engineering, Shri Vishnu Engineering College For Women(A),
E-mail: plavanyacivil@svecw.edu.in,
[2]Associate Professor, Department of Civil Engineering, Shri Vishnu Engineering College For Women(A),
E-mail: parupallysridhar@gmail.com
[3]Assistant professor, Department of Civil Engineering, Shri Vishnu Engineering College For Women(A), E-mail: vmanasace@svecw.edu.in

Abstract: Planning, designing, and execution of high-raise building while taking into account the impact of lateral loads such as wind and seismic loads computation with or without shear wall technology in conventional mode is a crucial task for civil engineers. The lateral loads develop high lateral stresses and sway movement in structural members and need adequate stiffness to resist lateral forces along with sufficient strength against vertical gravity loads. In this case, civil engineers establish an effective system for construction of high raise building with the usage of shear wall system is high in-plane stiffness and torsional resistance. Present the authors are considered a G+10 storey commercial shopping mall with a central cut-out at Bhimavaram location which lies in seismic zone III as, per IS 1893(part 1): 2016. The load cases from IS 875 Part I (dead loads), Part II (imposed loads), and Part III (wind loads) are used to create the load combinations in accordance with IS 456: 2000. Two models, with and without shear wall system are taken for comparing the parameters like storey drift, storey displacement, stiffness, shear force, base shear, bending moment and center of rigidity by performing linear static seismic analysis and wind analysis for both models in ETABS software. The simulations concluded that Shear wall system is the efficient and economical way of resisting lateral loads in high rise or multistory buildings.

Keywords: ETABS, Seismic Load, Wind Load, Live Load, Shear wall

1. Introduction:

In recent years, civil engineers have been under pressure to build high-rise buildings due to land scarcity, increased land costs in several cities and towns, and to facilitate the increased population of megacities. The construction of high-rise buildings for residential, commercial, and celluloid uses and among others in an area must be designed efficiently to ensure people's safety. The Civil engineers planning and design of high-rise building structures need to consider which are impacted by lateral loads like wind and earthquakes. In designing high-rise buildings, civil engineers consider the seismic load component, which is a fundamental concept load in earthquake engineering. Seismic load occurs sudden movement or disturbance in the earth due to sudden break in underground rockand motion in along the fault, which releases lot of energy rapidly. The unpredictable releases of energy damages any building structures. As a civil engineer, while designing the building structure, it is to be withstanding the lateral forces to be achieved through stiffness improvement. One good wayto resist these loads is by adopting a shear wall system in a building. Shear wall is a structural component in a building with high in-plane stiffness and transfers the wind and earthquake loads tothe foundation more efficiently than a typical reinforced-concrete (RC) framed building.

Structural engineers must aim for the most precise, effective, and affordable design possible in order to guarantee that the structure's final design and the building itself must be functional for their anticipated to use lifetime on the basis of it designed. For designing and analysis of building structures several software's are available at globally such as ETABS, MIDAS, RISA, RAM, SAP, STAADPRO, STRUDL. However, the current work utilises ETABS software to explore linear static analysis of a multi-story (G+10) building with and without shear walls. To achieve the frame work the following objectives are framed for the study area.

2. OBJECTIVES

- To Investigate about Shear force, Bending Moment, Story drifts, story displacement, Storystiffness, story shear, centers of mass rigidity of buildings with and without shear walls.
- To safeguard the structure against all loading conditions
- To analyze seismic and wind loads using code of part 11893:2016 and IS 875:2015 (Part III).

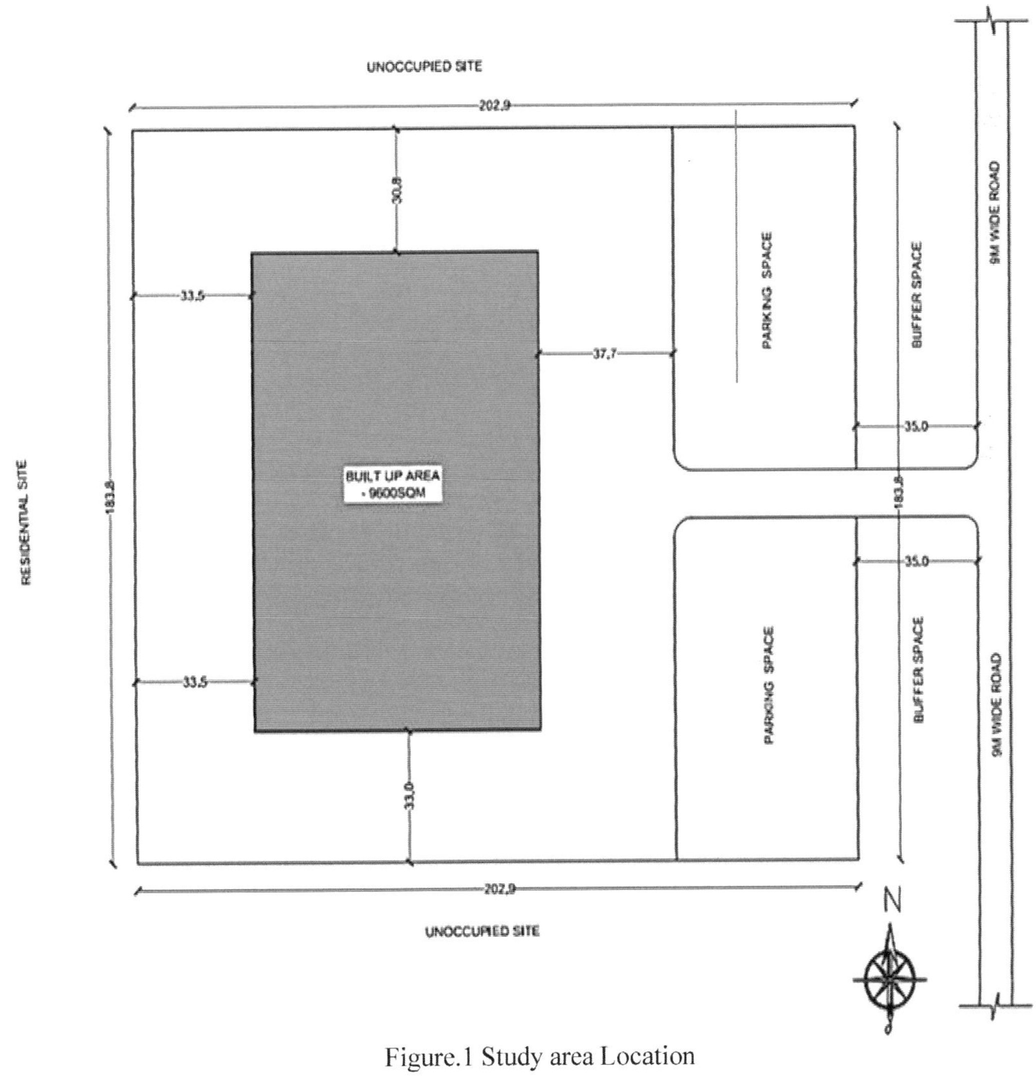

Figure.1 Study area Location

3. METHODOLOGY

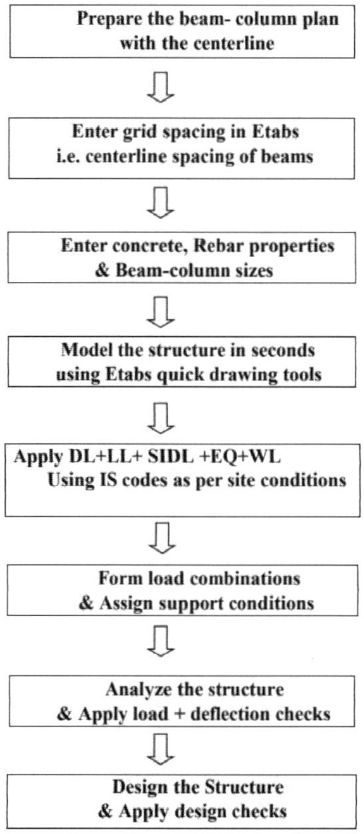

Figure.2 Methodology Flow Chart

According to Indian standard standards, the ETBAS programme considers loads such dead loads, wind loads, imposed loads, and seismic loads. Part-1 of IS: 875-1987 is used to determine dead load, while Part 2 is used to determine imposed load. According to IS:875-1987of part-1, the combined weight of a building's walls, the number of partitioned levels, false floors,celling, and colour mounts on are considered to represent the building's dead load. The part-2 is imposed load considers the distribution and concentrated loads, total movable members weight, dust and vibration loads. Seismic, wind and snow loads, shrinkage and creep occur in the structure due to the influence of temperature that does not include imposed loads. The horizontal wind load isconsidered as a wind load compared to the vertical load on the building member as per standard of IS: 875-2015 as part 3. Anemometers or anemographs fitted with varied heights of 10 to 30 meters above the ground are used to measure wind speed and are set up by the meteorological department to determine the wind speed in the area. The expression of wind speed is

$$V_b = k_1 * k_2 * k_3 * k_4$$

V_b – Basic wind speed at any Height Z in $\frac{m}{s}$, k_1 – Probability factor, k_2 - Terrian, k_3 – topography, k_4 – factor for cyclonic region

The Seismic Loads utilised in ETBAS software are computed initially as lateral force delivered to the building in accordance with Indian Standards (IS:1893-2002) PART1. Depending on the floor diaphragm action, the total design seismic force that is subsequently generated at each floor level must be transferred to specific lateral load resisting devices.The equation of A_h = Spectrum of horizontal acceleration

Z = seismic zone factor; I
= importance factor is given in IS 1893 (Parts 1 to 5)
a) 1.5 for critical and lifeline structures;
b) 1.2 *for business continuity structures*; *and c*) 1.0 *for the rest. R*
= *response reduction factor given in IS* 1893 (*Parts* 1 *to* 5) *for the corresponding structures.*

$$[Sa/g] = design\ acceleration\ coefficient\ for\ different\ soil\ types.$$

$$A_h = ([Z/2][I/R])/[S_a/g]$$

Force Distribution in the Design Base Shear Vertical Distribution to various Floor Levels.

The following guidelines must be followed while distributing the design base force (VB) along the building's height:

Expression

$$Q_i = \left(\frac{W_i\, h_i}{\sum_{i=1}^{n} W_j h_j^2}\right) V_B$$

- Design lateral force at the floor is Q_i.
- H_i is the height of the floor measured from the base, where W_i is the seismic weight of the floor.
- The number of levels where the masses are situated is determined by the number of stories in the building, or n.

4. RESULTS

4.1 Center of Mass And Center of Rigidity

The centre of mass, also known as the centre of rigidity, is where the entire storey's mass acts, whereas the centre of stiffness, also known as the centre of rigidity, is where the entire stiffness acts. Vertical resisting systems like columns and shear walls—the latter being stiffer—contribute significantly to the stiffness. However, beams also add to mass. The eccentricity, or distance between the centres of mass and rigidity, must be at a minimum for a structure to be resistant to a torsional moment. The comparative plot between the centre of mass and the centre of rigidity in both the X and Y directions with and without shear wall situations is shown in Figure 5 and was derived using ETABS after analysis.

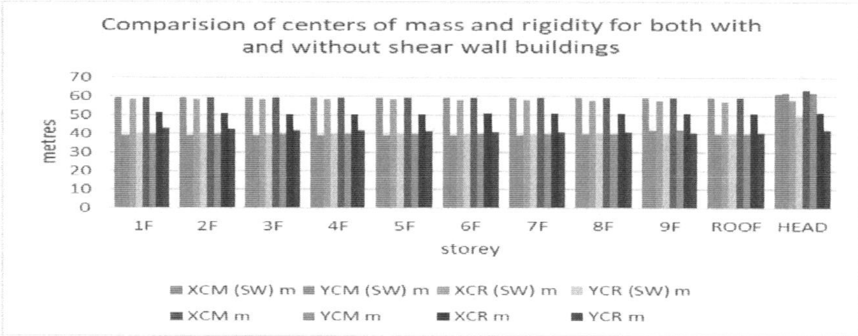

FIGURE 3. Comparative plot of center of mass and center of rigidity for buildings with and without shear walls in X and Y directions

4.2 Story Stiffness

The ratio of storey shear to storey drift is known as storey stiffness. The comparative storey stiffness plot for structures with and without shear walls in the X and Y directions is shown in Figure 6.

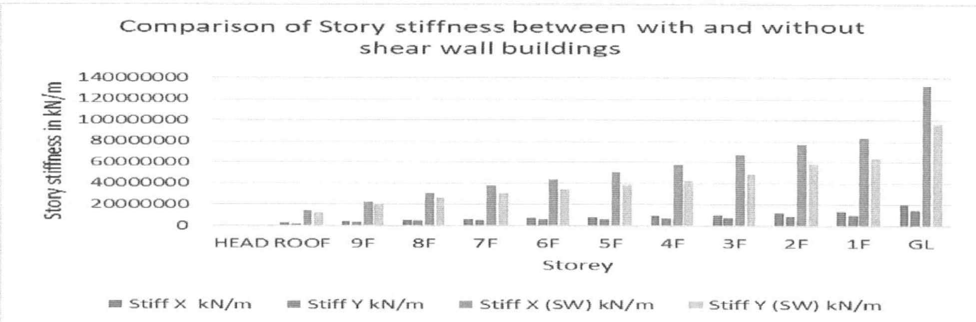

FIGURE 4. Comparative plot of Storey stiffness for buildings with and without shear walls in X and Y directions

4.3 Story Displacement

The displacement of a storey with regard to the ground is known as storey displacement. The comparison plot of storey displacement for structures with and without shear walls for EQ-X and EQ-Y loadings is shown in Figure 7.

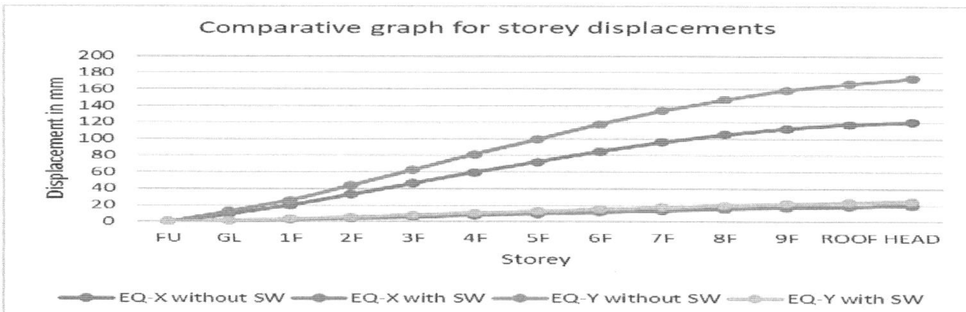

FIGURE 5. Comparative graph of storey displacement for buildings with and without shear wall for EQ-X and EQ-Y loadings Storey Drift

Storey drift is the difference between the height of a floor and the displacement of two adjacent levels. It is a parameter with no dimensions. The charts in Fig. 8 compare the storey drift for earthquake and wind loads in the X and Y axes. Storey drift shouldn't be more than 0.004 times the height of the storey.

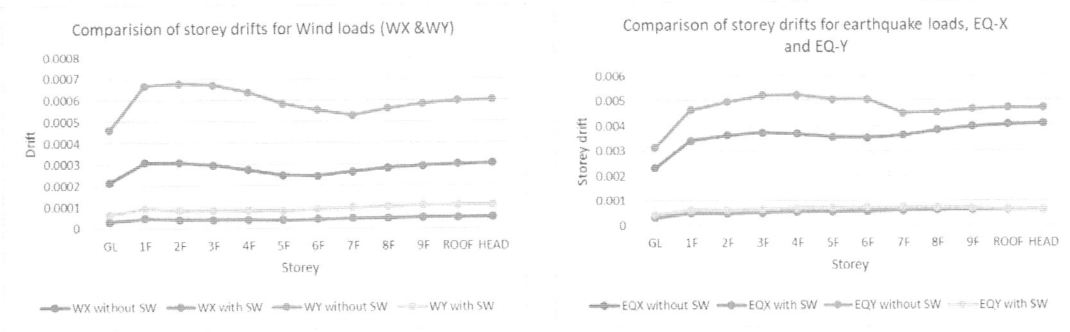

FIGURE 6. Comparative plot of Storey drifts for building with and without shear wall for (a) WX and WY loads (b) EQ-X and EQ-Y loads

4.4 Bending Moment And Shear Force of Beams And Columns

As there are large number of beams and columns, we have considered beam series '108' and column series '42', which are nearer to perimeter, for comparison. Figure 9 is the elevation view of frame 5- A-B-C-D-E-F-G-H-I-J-K-L-M-N-O-P. The results of bending moment and shear force for load combination, (1.2 D.L+1.2 L.L+1.2 EQ-X) are considered for comparison.

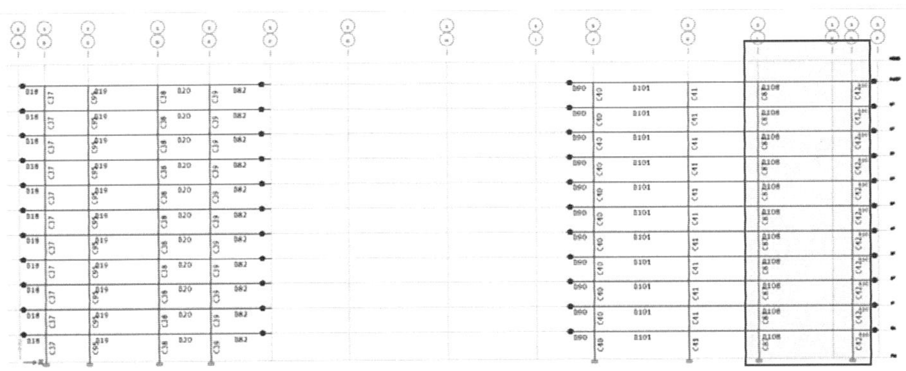

FIGURE 7. Elevation frame 5-A-B-C-D-E-F-G-H-I-J-K-L-M-N-O-P

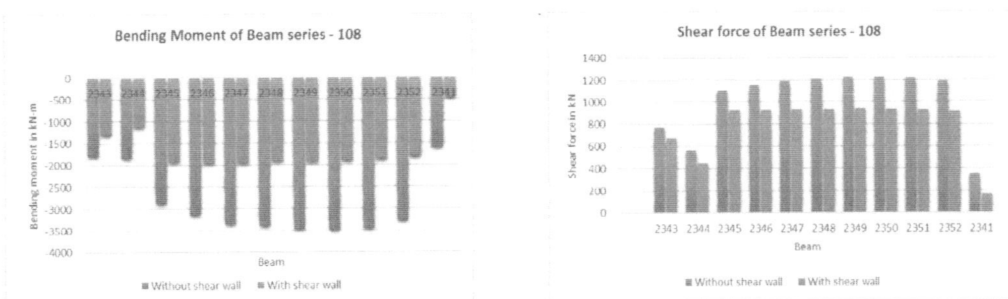

FIGURE 8. Comparison plot for building with and without shear wall for (a) bending moment (b) shear force of beams

Table 1: shows the bending and shearforce of columns for building shear wall system. Figure 11 shows the comparative plots for the same.

TABLE 1. Bending moment and shear force of columns

Storey	Column unique name	B.M (Without shear wall) (kN-m)	B.M (With shear wall) (kN-m)	S.F (Without shear wall) (kN)	S.F (With shear wall) (kN)
GL	517	5046.669	709.103	1351.691	184.4546
First Floor	528	1817.045	406.4446	894.2432	336.1058
Second floor	527	2798.252	1204.461	1378.235	627.901
Third Floor	526	2364.064	1131.788	1300.474	623.7127
Fourth Floor	525	2280.662	1193.261	1301.182	648.3045
Fifth Floor	524	2159.657	1279.093	1312.627	755.2702
Sixth Floor	523	1853.677	966.0368	1139.552	579.9347
Seventh Floor	522	1840.727	1201.372	1115.759	667.2865
Eight Floor	521	1572.813	1171.935	1010.855	668.9713
Ninth floor	520	1212.879	1098.37	697.2312	519.6755
Roof	519	599.0481	673.1739	656.2858	605.7955

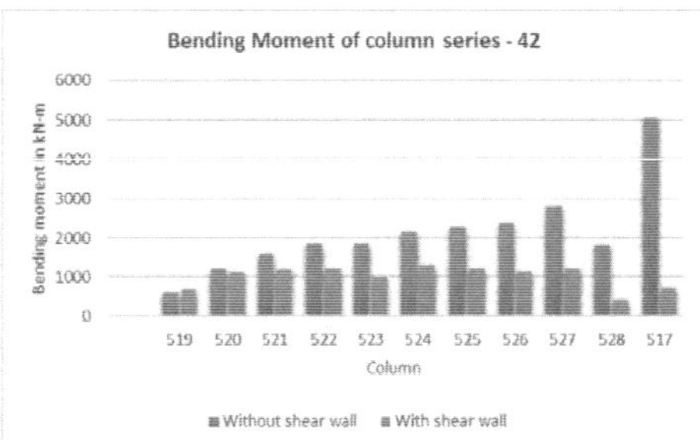

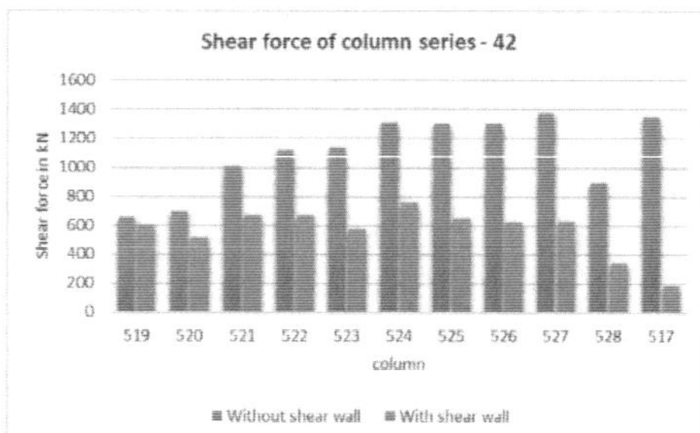

FIGURE 9 . Comparative plot for buildings with and without shear walls of (a) bending moments (b) Shear forces for columns

5. CONCLUSIONS:

The study is revealed that the models have been safe against all external loads like lateral loads and internal loads like live loads, dead loads and superimposed dead loads according to IS 875 (parts I & II) for commercial shopping complex. The positioning of shear wall symmetrically in the building has brought considerable change in parameters like bending moment, shear force, storey displacements, drifts, center of mass, rigidity, storey shear, and storey stiffness. When compared to buildings without shear walls, the bending moment for beams and columns is reduced by up to 43%, while the shear force is reduced by up to 23.56 to 45.8%. Maximum storey displacement for building without shear wall was observed as 175 mm for earthquake load in Y direction which got reduced to 25mm when shear wall system is adopted which is about 85% reduction of storey displacement. Storey drifts for building with shear wall has been reduced by 86.9% as compared to that of building with shear wall. For earthquakes and with wind loads in both X and Y directions, storey drifts were more frequently seen in the second and third stories. The experiment shows that using the Shear Well method in a building as opposed to one without a Shear Wall greatly increases the storey stiffness.

REFERENCES

[1] P Nagendra Babu, S J (May, 2017). Comparative study of shear wall in multi-stored rc building. international journal & magazine of engineering, Technology, *Management and Research,* Volume 4 (Issue 5), 780-788.

[2] Harne, V. R. (November, 2014). Comparative study of strength of rc shear wall at different location on multi-storied residential building, *International Journal of Civil Engineering Research*, Volume-5, 391-400.

[3] IS 1893 (Part 1): 2016 Criteria for earthquake resistant design of structures, (December, 2016). *Bureau of Indian Standards.*

[4] IS 456: 2000 - Code of practice for plain and reinforced concrete. (2000), *Bureau of Indian Standards.*

[5] IS 875 (Part 1): 1987 - Dead loads - unit weights of building materials and stored materials. (1987), *Bureau of Indian Standards.*

[6] IS 875 (Part 2): Code of practice for design loads (other than earthquake loads) for buildings and structures, part 2: Imposed loads. (n.d.).

[7] IS 875 (Part 3): 2015 Wind loads on buildings and structures (2015), *Bureau of Indian Standards.*

[8] P Kalpana, R D (December, 2016), Analysis of Building with and Without Shear Wall at Various Heights and Variation of Zone III and Zone V, *International Journal of Engineering Research and Application*, Vol. **6**(Issue-12), 05-11.

[9] Provisions For Construction and Regulation of Multiplex Complexes (2012), *In Andhra Pradesh Building Rules,* G.O.Ms.No.119 (pp. 157 - 167), Municipal Administration and Urban Development Department.

[10] Sonali Pandey, Dr. Krishna Murari, & Ashish Pathak (December, 2017), A Review on Shear wallin High Rise Buildings, *International Journal of Engineering Inventions*, Volume-**6**(Issue-12), 19-21.

[11] Tarun Magendra, Abhyuday Titiksh, & A A Qureshi (June, 2016), Optimum Positioning of Shear Walls in Multistory Buildings, *International Journal of Trend in Research and Development*, Volume-3 (Issue-3).

Study of the Effect of Installation of Stone Columns on the Stability of Soil Slopes using Finite Element Method

A Goswami [1] and S Deka [2]
[1] Research Scholar, Tezpur University, Tezpur, Assam, India
[2] Assistant Professor, Tezpur University, Tezpur, Assam, India
E-mail: anasuya.goswami27@gmail.com, shailendeka@gmail.com

Abstract. There are several ways to mitigate the failure of a soil slope. The installation of stone columns for this purpose, is easy as well as economical. As their strength is greater than that of soil, they disrupt the slip surfaces. Also being more porous, they dissipate the pore pressure generated in the soil in the vicinity of the stone columns to a certain extent. This paper illustrates 2D finite element analyses carried out to simulate the behavior of stone columns in cohesive soil using the computer program PLAXIS. Parametric study was conducted to analyze the effect of number, length and diameter of the stone columns on the Factor of Safety of the soil slope. The F.O.S. of the soil slope was predicted using prediction models created using artificial neural networks (ANN) considering slope geometry like slope angle and height, material properties like cohesion and angle and internal friction of soil as well as granular material used in the columns, and the stone column parameters like the stone column's number, length, diameter and the centre to centre (c/c) distance between two adjacent columns as input parameters. It was found that introduction of stone columns significantly improved the stability of the soil slopes. Further, it was seen that all the parameters considered in the study influenced the Factor of Safety of the soil slope.

Keywords: Slope stability, Stone columns, Finite element analysis, Artificial Neural Network.

1. Introduction

There are several ways to mitigate the failure of a soil slope. The installation of stone columns for this purpose, is easy as well as economical. As their strength is greater than that of soil, they disrupt the slip surfaces. Also being more porous, they dissipate the pore pressure generated in the soil in the vicinity of the stone columns to a certain extent. The main objectives served by the installation of stone columns are reduction of settlement and improvement of bearing capacity and stability of soil slopes [1-4]. Various numerical methods such as Finite Element Method (FEM) and Finite Difference Method (FDM) have been used successfully to estimate the stability of slopes by calculating the Factor of Safety (F.O.S.) [4-7]. The F.O.S. has been seen to increase with an increase in the diameter of the stone column [5,8]. This paper describes the influence of number, length as well as diameter of the stone columns on the F.O.S. of the soil slope. The F.O.S. of the soil slope was predicted using prediction models created using artificial neural networks (ANN) considering slope geometry like slope angle and height, material properties like cohesion and angle and internal friction of soil as well as granular material used in the columns, and parameters of the stone columns like the stone column's number, length, diameter and their distance from one another at the centre, as input parameters. ANN is a nonlinear projection method and because of its propensity to overparameterize, ANN tends to be relatively insensitive to problems of multicollinearity, that is encountered often in regression analysis. When training is done with adequate data, ANN was found to be have considerably lesser predictive error than other basic machine learning tools like linear regression [9].

2. Methodology

A geometry model is a two-dimensional representation of a three-dimensional problem, consisting of points, lines, and clusters. The geometry is created in the global coordinate system's X-Y plane, with the z-axis pointing away from the plane. The default element chosen for the study is a 15-node triangular element. The 15- node element is more accurate and yields more realistic results. It should have a representative subsoil division into separate soil layers, building stages, and loading. The model must be large enough that boundary conditions do not affect the conclusions of the problem under investigation. The Mohr-coulomb failure criteria and plane strain condition is used for developing the model. For the soil slope, material type is selected as undrained while for the stone columns it is slected as drained. The fine element distribution is used for generation of mesh.

The unsaturated unit weight (γ_{unsat}) is taken as 19 kN/m^3 and 15 kN/m^3 for the soil slope and stone columns respectively. The saturated unit weight (γ_{sat}) is taken as 20 kN/m^3 for both soil slope and stone columns. Both k_x and k_y (coeffecients of permeability in x and y direction) are 1x10^{-3} m/day for soil and 10.368 m/day for stone. The Young's modulus (E) is taken as 1x10^4 kN/m^2 for soil and 5x10^4 kN/m^2 for stone. The Poisson's ratios for soil and stone are taken as 0.35 and 0.3 respectively.

The parameters considered are slope angle, β (°), slope height, H (m), cohesion of soil, c_{soil} (kN/m^2), cohesion of stone, c_{stone} (kN/m^2), friction angle of soil, φ_{soil} (°), friction angle of stone, φ_{stone} (°), stone column number, N, stone column length, L (m), stone column diameter, d (m) and c/c spacing between two stone columns, S (m). The methodology is divided into deterministic and probabilistic studies. In the deterministic approach, a soil slope of height 15 m and slope angle 45° is modelled in PLAXIS 2D with c_{soil} = 35 kN/m^2 , c_{stone} = 42 kN/m^2, φ_{soil} = 10° φ_{stone} = 0.1° and stone columns are installed on the slope (figure 1).

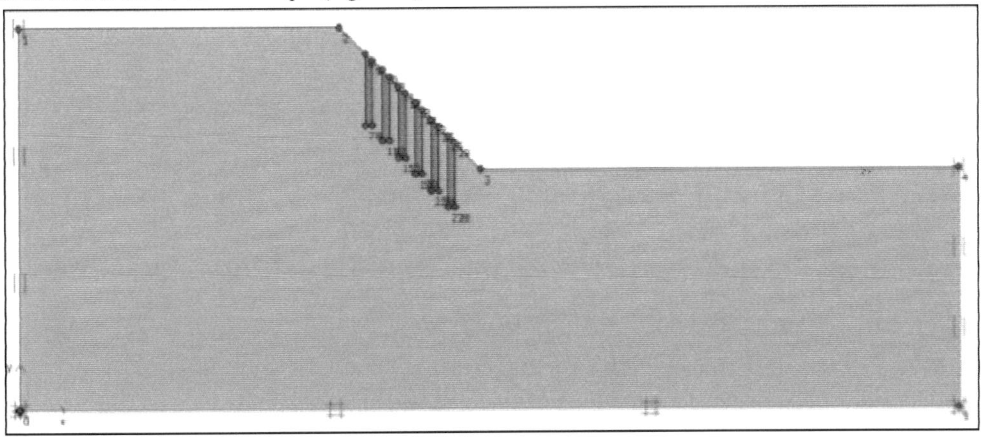

Fig. 1 Soil slope with stone columns modelled in PLAXIS 2D

The effects of N, L, d and S on the F.O.S. of the soil slope are studied by keeping the other parameters constant. The probabilistic study includes development of prediction models for predicting the F.O.S. of the soil slope using ANN considering all the parameters as input parameters. A total of 150 datasets are considered for development of the models are another 20 datasets are used for validation of the models. The input parameters and their corresponding ranges used are shown in Table 1. A trial-and-error approach is used because there is no established algorithm in the literature for finding the figure of hidden layer neurons. The figure required in a model is calculated using Jeff Heaton's approach and various neural networks are then built between the highest and lowest values [10]. In this study, the network training function (TRAINLM) based on the Levenberg-Marquardt back- propagation training technique is utilised, which has a faster convergence than other methods and is frequently recommended [11-12]. The flow chart for the development of ANN is shown in figure 2.

Table 1. Input parameters and their corresponding ranges used in probablistic study

PARAMETER	**RANGE**
Slope Angle (°)	20-60
Slope Height (m)	10-15
Cohesion of soil (kN/m^2)	30-36
Friction angle of soil (°)	5-10
Cohesion of stone (kN/m^2)	0-5
Friction angle of stone (°)	40-45
Stone column number	2-6
Stone column length (m)	5-10
Stone column diameter (m)	0.6-1.0
C/c spacing of stone columns (m)	1.4-12.5

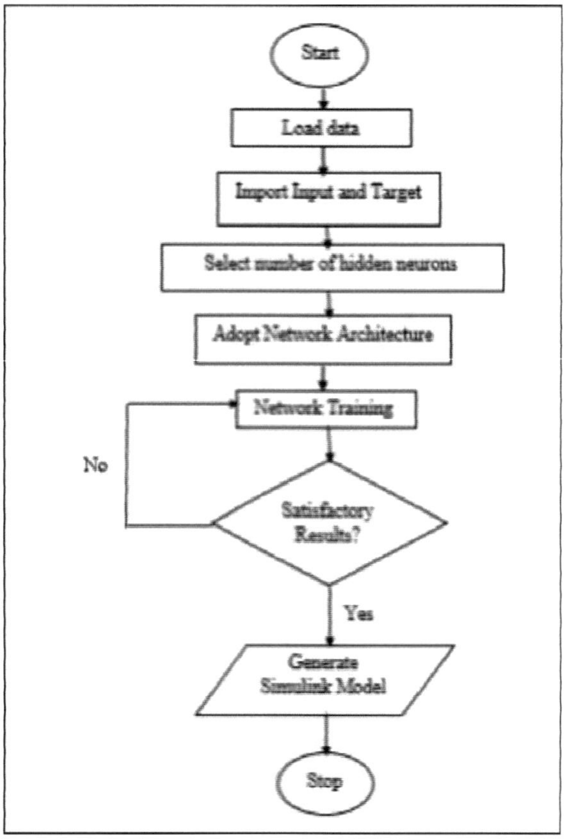

Fig 2. Flow Chart for adopted ANN

3. Results and Discussions

In the deterministic approach, a soil slope of height 15 m and slope angle 45° is modelled in PLAXIS 2D with csoil = 35 kN/m2 , cstone = 42 kN/m2, φsoil = 10° φstone = 0.1° and the effects of N, L, d and S on the Factor of Safety (F.O.S.) of the soil slope are studied by keeping the other parameters constant.

3.1. Effect of stone column number

F.O.S. were calculated for soil slopes installed with stone columns of diameter, d = 0.75 m and c/c spacing between two adjacent stone columns, S =1.75 m with stone column number varying from 0 to 6 and length of stones columns (L) varying from 5 m to 10 m. It was found that introduction of stone columns significantly improved the stability of the soil slopes, and also the F.O.S. increased with the increase in the stone column number. Figure 3 shows the plot between stone column number and F.O.S. for different lengths of stone columns.

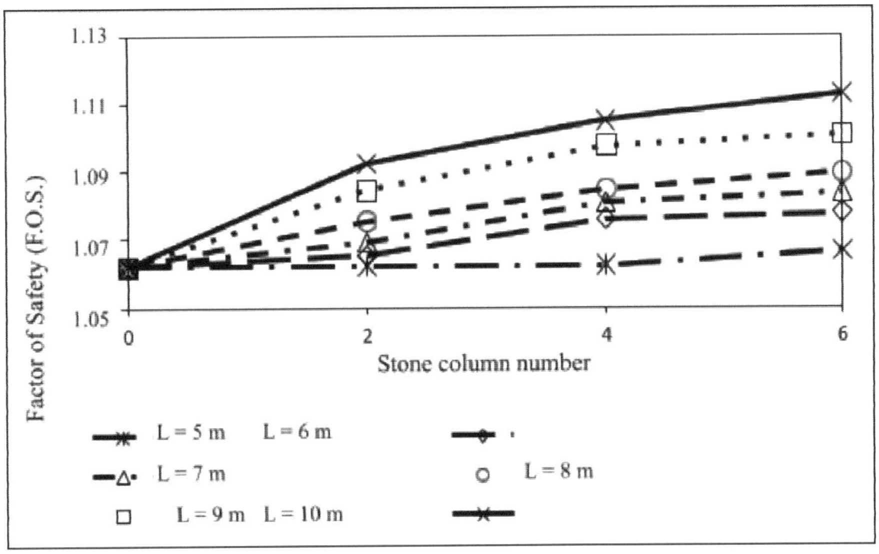

Fig 3. Effect of stone column number on F.O.S.

3.2. Effect of stone column length

From the plot shown in figure 3, it can also be seen that for different numbers of stone columns, F.O.S. increased as the stone column length increased.

3.3. Effect of stone column diameter

Considering 6 stone column number of length 10 m, factors of safety were calculated for stone column diameters varying from 0.6 m to 0.9 m. It was observed that the F.O.S. increased with the increase in the diameter of the stone columns (figure 4).

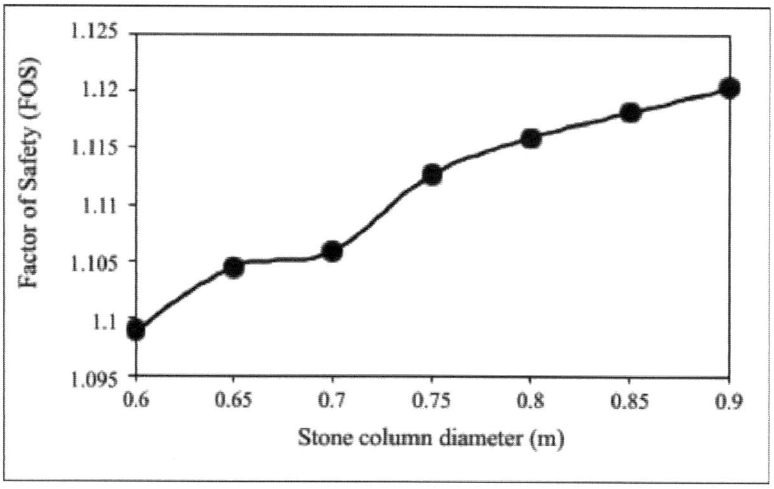

Fig 4. Effect of diameter of stone columns on F.O.S.

3.4. Prediction Model

The development of prediction model is done using ANN. A hidden layer with 7 neurons is found to be the most effective for F.O.S. prediction in an ANN architecture after each network is assessed based on its R and MSE values. The adopted network architecture is shown in figure 5.

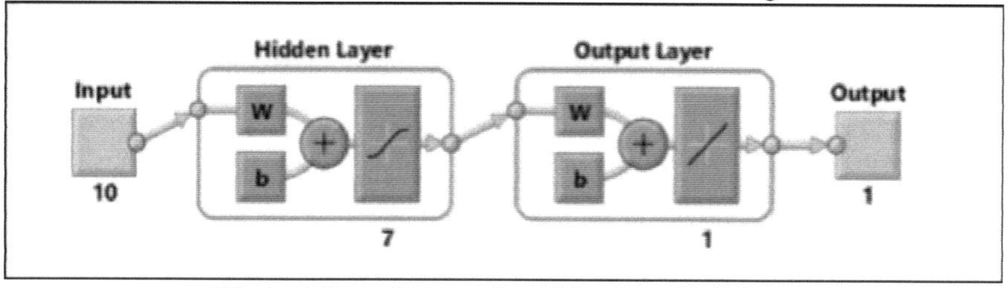

Fig 5. Adopted network architecture 10-7-1

The overall regression plot of the adopted network model is shown in figure 6. Very high correlation coefficient is observed for training and testing data in ANN topology.

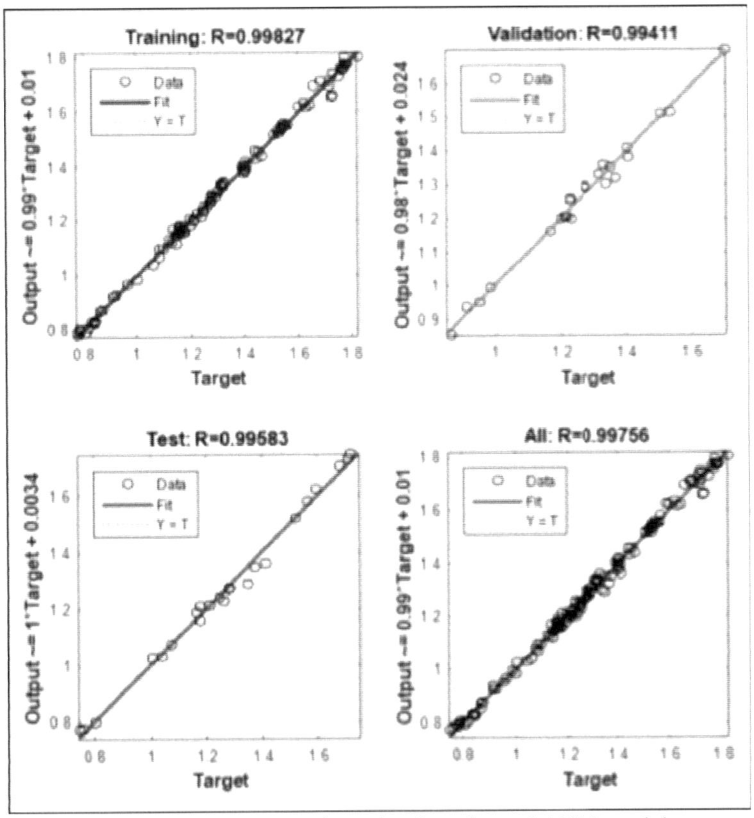

Fig 6. Regression plot for adopted ANN model

20 datasets are used for validation of the models. The error analysis is carried out by calculating the RMSE and MAE. The RMSE and MAE values come out to be 0.060 and 0.047 respectively. A comparison is done between the calculated value and predicted value and it is shown in figure 7. It can be seen that the forecasted results were found to be very similar to the calculated results.

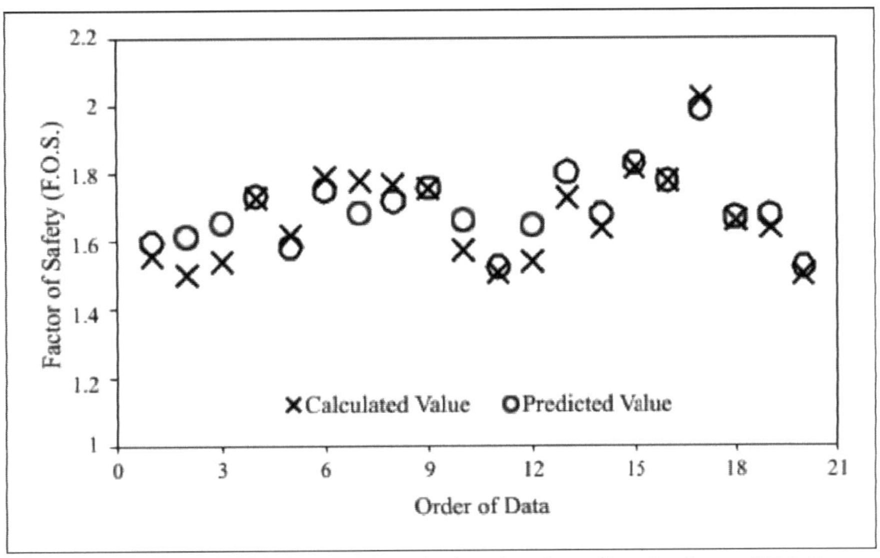

Fig. 7 Comparison between calculated and predicted values of F.O.S.

4. Conclusions

The study's findings lead to the following conclusions:

- Introduction of stone columns significantly improved the stability of the soil slopes.
- The F.O.S. of soil slopes increased as the stone column number increased.
- The F.O.S. of soil slopes increased as the stone column length increased.
- The F.O.S. of soil slopes increased as the stone column diameter increased.
- Predictive model can be successfully developed using ANN for the prediction of F.O.S. of soil slopes considering slope geometry like slope angle and height, material properties like cohesion and angle and internal friction of soil as well as granular material used in the columns, and parameters of the stone columns like number, length, diameter of the stone columns and c/c spacing between two adjacent stone columns, as input parameters. The RMSE and MAE are found to be 0.060 and 0.047 respectively. The forecasted results were found to be very similar to the calculated results.

5. References

[1] Ambily AP and Gandhi S; 2007; Behavior of Stone Columns Based on Experimental and FEM Analysis; *Journal of Geotechnical and Geoenvironmental Engineering*; **133**(4), pp. 405-415.

[2] Han J; 2012; Recent Advances in Column Technologies to Improve Soft Foundations; *Proceedings of International Conference on Ground Improvement and Ground Control;* **Vol. 1**; pp. 99-113.

[3] Lo S, Zhang R anf Mak J; 2010; Geosynthetic-encased stone columns in soft clay: a numerical study; *Geotextiles and Geomembranes*; **28**(3); pp. 292-302.

[4] Zhang Z, Han J and Ye G; 2014; Numerical investigation on factors for deepseated slope stability of stone column-supported embankments over soft clay, *Engineering Geology*; **168**; pp. 104-113.

[5] Abubashar SW and Han J; 2011; Two-dimensional deep-seated slope stability analysis of embankments over stone column improved soft clay; *Engineering Geology*; **120**(1); pp. 103-110.

[6] Han , Chai J, Leshchinsky D and Shen SL; 2004; Evaluation of deepseated slope stability of embankments over deep mixed foundations; *GeoSupport 2004: Innovation and Cooperation in the Geo-Industry*; pp. 945-954.

[7] Han J, Chen J, Hong Z, and Shen S; 2010; Mitigation of levee failures using deep mixed columns and geosynthetics; *Geomechanics and Geoengineering*; **5**(1); pp. 49-55.

[8] Kadhim S, Parsons, RL, Han J; 2015; Stability Analysis of Embankments Supported by Geosynthetic Encased Stone Columns; *IFCEE 2015*; pp. 2318-2327.

[9] Veaux, RDD and Ungar LH; 1994; Multicollinearity: A tale of two nonparametric regressions; *Lecture Notes in Statistics*; **Vol**. 89; pp. 393-402.

[10] Heaton J; 2008; Introduction to Neural Networks with JAVA; *Heaton Research, Inc.*; **2**.

[11] Demuth HB, Beale MH, Jesus O and Hagan MT; 2014; Neural network design. Martin Hagan; Stillwater; **2**.

[12] Ayaz M; 2022; Estimation of release history of groundwater pollution source using ANN model; *Modeling Earth Systems and Environment*; **8**(3); pp. 925–937.

An experimental study on the effect of particle size and non-homogeneity on the shear strength of soil

R A Shabong[1], A J Sonowal[2], A Pukhrambam[3], R Das[4], A Anand[5], S Bobing[6] andS Deka[7]*

Department of Civil Engineering, Tezpur University, Tezpur, Assam-784028, India

deka.s@tezu.ernet.in

Abstract. Soil deposits available in nature are often present as homogeneous mixtures of particles of different sizes and base minerals or as a combination of soils in different layers forming a non-homogeneous configuration. This variability in soil has great influence on the shear strength behaviour. In this study, different naturally occurring soil particle sizes and shapes (angular and flaky) have been utilized as reinforcements in the form of layers to determine their influence on the soil shear strength. Therefore, triaxial tests have been conducted on homogeneous sand (where particles are mixed proportionately), and on non-homogeneous sand (different layer combinations of different particle sizes as well as the inclusion of angular and flaky particles). It has been observed that there was a substantial increase in the shear strength for non-homogeneous arrangement compared to homogeneous sand sample. Furthermore, the addition of flaky particles from 10% to 30% by weight of fine grained sand increased the angle of internal friction (ϕ) and cohesion (c) values thereby increasing the shear strength. However, for angular-fine sand layer combination, the addition of 10- 20% of angular particles by weight shows a significant increase in shear strength in comparison to 30-100% of angular particles.

Keywords: Non-homogeneous layered soil, Shear strength, Consolidated undrained triaxial test

1. Introduction

1.1. General overview

Natural soil deposits largely exist as mixtures of soils which consists of particles of various sizes, shapes and minerals due to which it cannot be classified as a particular soil type i.e., clay, silt or sand but rather into a combination of soils i.e., they are available in non-homogeneous forms. It is very difficult to select a relatively homogeneous soil sample which specifically represents the three classes of soil. Generally even within a certain depth of 60cm, the soil make-up could change considerably which ranges from silty-clay to silts or sandy silts [5]. Soils are found in composite form due to their formation process. Mechanical behaviour of soil such as shear strength, compressibility, erodibility, permeability are dependent on composition of soil [2]. Many investigations were conducted to acquire a comprehensive insight on the behaviour of soil mixtures [1][2][3][4][11][14]. Normally, studies were divided into two primary combinations such as gravel-sand mixtures and sand-clay mixtures. Several studies were also madeto understand the shear behaviour of silt-clay or sand-clay soil mixtures [1][11]. A study investigated the influences on the dilation and shear strength of sand due to addition of gravel content. [8]. The findings of the study indicated that when gravel is added to sand, the dilation and shear strength increases due to the resistance offered by gravel particles towards rotation. [12] found that the fine content in coarse-fine sand mixture greatly influences the load bearing capacity. When fine sand is added, an increase in the angle of friction and ultimately the shear strength of coarse-fine sand mixture is observed due to strong interlocking between the particles [2]. When subgrade soil is mixed with gravel, it was seen that the stiffnessand internal friction angle were enhanced by irregular gravel but less impact on cohesion was observed[16]. [15] found that the large size gravels have a good inter-locking and re-arrangement capacity, which increases the shear strength. Direct shear and triaxial test on soil mixture of granular material with oversized particles were conducted over the years [9]; [6]; [10]; [7];[13] found that shear strength of the soil mixture rely on the relative concentration of oversized particles. In a mixture containing 48% to 78% larger particles (by weight), friction between larger particles controlled the shear strength. Smaller particles dominates the shear strength if the concentration of larger particles is less than 48% by weight. Investigation into the impact of gravel content on the shear behaviour of soil-rock mixtures revealed that the gravel content is a vital element altering the mechanical traits of soil-rock mixture in the course of sample preparation [17]. When less than 25% of gravel is present, the shear parameters of the pure soil will mostly define the overall strength of soil-rock mixture as the gravel content will not be in close contact with each other and are in a state of suspension. Above 70% of gravel content, the interlocking action of gravel particles is what is responsible for the change in shear strength as they are in close contact with each other. Therefore, the gravel content of the tested soil-rock mixture samples should range from 25% to 70%. It was observed that most of the conducted researches and studies only considered the homogeneous behaviour of soil i.e., behaviour of only soil mixtures but not the non-homogeneous behaviourthat is when soil is a combination of layers. Very few studies have been made regarding non-homogeneous soil [11] but these studies are limited.Thus, it becomes necessary to understand the stress-strain behaviour ofnon-homogeneous soil in geotechnical engineering.

1.2. Objective of the study

This study aims to understand the behaviour of non-homogeneous sand sample. The following objectives are outlined for the study:

(a) To find the effect on the shear strength when soil is arranged as a non-homogeneous layered soil sample.
(b) To find the variation of shear strength with different particle size.
(c) To understand the stress-strain behaviour of fine sand sample when reinforced with angular and flaky particles

1.3. Necessity of the study

For earth construction projects, it is crucial to assess the soil's shear strength and other properties. Soil of a particular type may not have adequate geotechnical or mechanical properties which have to be used for road embankment, foundation and construction material, etc. Innovative techniques have been used to stabilize and modify deficient soils by means of local accessible industrial or environmental waste materials, so as to minimize the construction costs. In the present study, utilization of different soil particle sizes and shape (angular and flaky) as reinforcements in the form of layers and studying the behaviour of these type of soils in finding outdesired shear strength.

2. Methodology

This section includes the detailed discussion of the materials used, sample preparation and test procedure of the Consolidated Undrained triaxial test. A flowchart of the study is given in Fig.1.

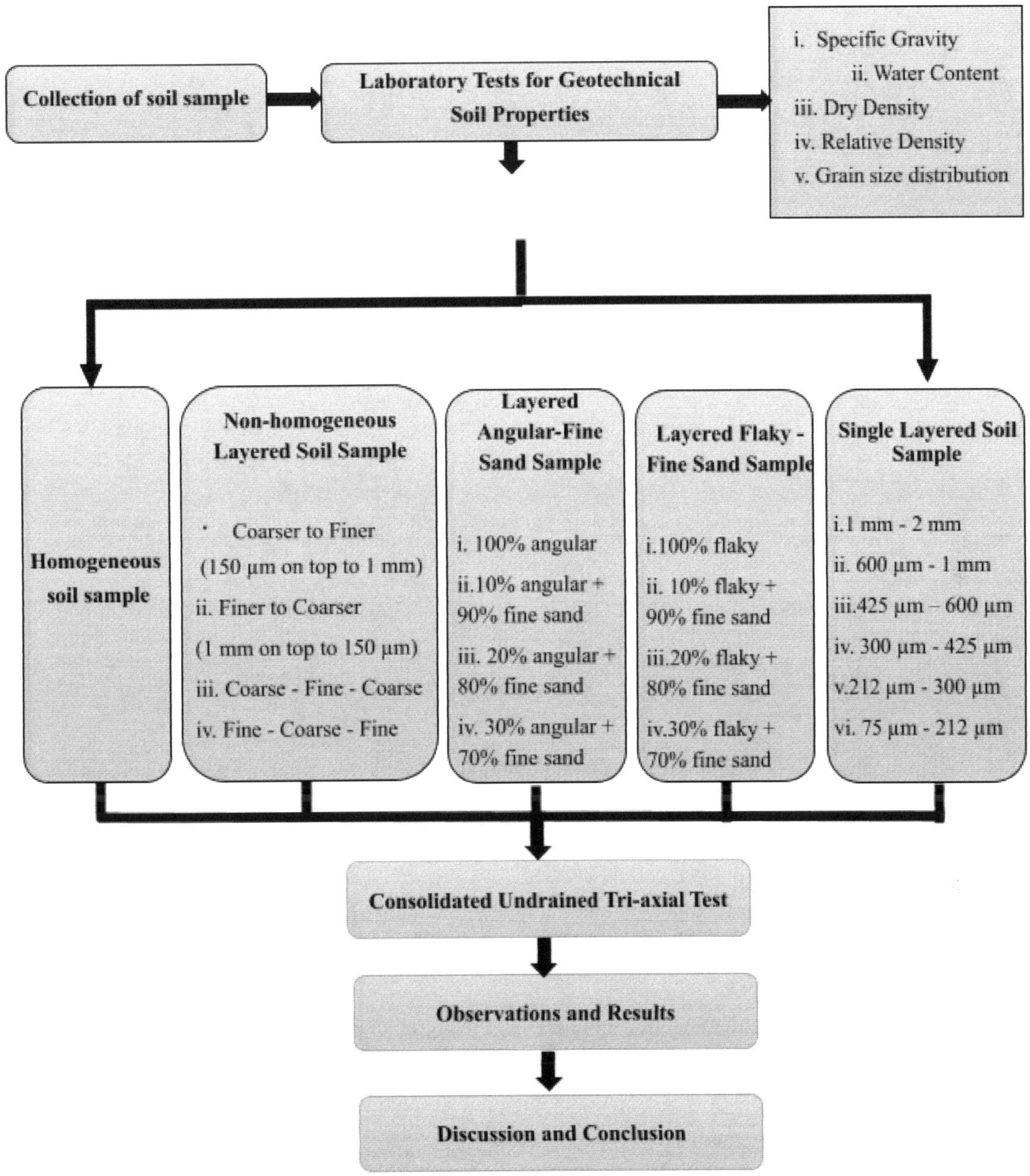

Fig.1. Flow Chart of the methodology

2.1. Materials Used

In the current study, the soil sample is collected near Jiabharali River (26°45'11.5" N 92°51'57.1" E) of Tezpur, Assam. The general properties of the soil were studied in the laboratory. The soil is tested for moisture content, specific gravity, relative density, etc. The results obtained from the preliminary tests on the sand sample are tabulated in Table 1. The triaxial test is then conducted to understand the shear behaviour of layered sand samples.

Table 1. General properties of Soil

Properties	Sample data
Soil classification	SP (Poorly graded sand)
Water content	0.72
Specific gravity, G_s	2.645
Dry Density, γ_d	16.08 kN/m^3
Relative dry density, I_d	69 %

The sieve analysis of the soil is performed. From the gradation curve shown in Fig.2, the following are obtained:

D_{10}= 0.308 Uniformity coefficient, Cu= 1.88
D_{30}= 0.441 Coefficient of curvature, Cc=1.087
D_{60}= 0.581 Since Cu<6 and 1<Cc<3, the sample is SP

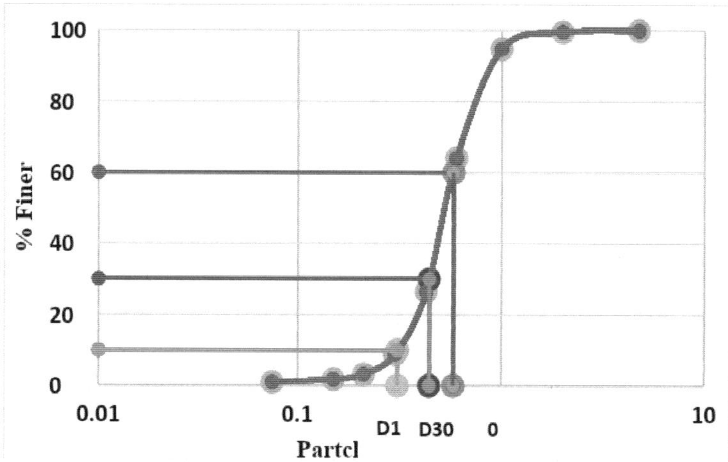

Fig.2. Particle size distribution curve

2.2. Sample Preparation

The sample is firstly sieved using sieve sizes ranging from 0.075 mm to 2 mm. Each sample retained in these sieves is arranged in layers according to the required combinations as shown in Table 2.The sample retained in each sieve is collected as shown in Fig.3 with angular and flaky samples also sorted out from samples retaining in 2mm sieve. The sample is then compacted layer by layer in different combinations according to the particle sizes by mild tamping. The required mass of the sample is determined by knowing the dry density of sand and the volume of the mould and is found out to be 145 gm. Therefore the mass of soil for each layer is taken as 23.5 gm. Coarse sand (1 mm <= d < 2 mm) having flaky and angular particles was used to make another phase of soil combinations along with fine sand (0.075 mm <= d < 0.425 mm). The coarse sand is added in proportions of 10 % -30 % by weight of fine sand. The whole sample is divided as per required fractions and a layer- by- layer sample is made. The first layer is filled by pouring the first fraction and compacting it with a tamping rod. The remaining layers are then filled in the same manner.

Table 2. Different combination of non-homogeneous soil sample

Combination	Particle size
1	Finer to Coarser (0.150 mm to 1 mm)
2	Coarser to Finer (1 mm to 0.150 mm)
3	Coarse-Fine-Coarse*
4	Fine-Coarse-Fine
5	Fine- Angular particles
6	Fine- Flaky Particles

**Coarse: (1 mm-0.600 mm)Fine: (0.075 mm - 0.425 mm)*

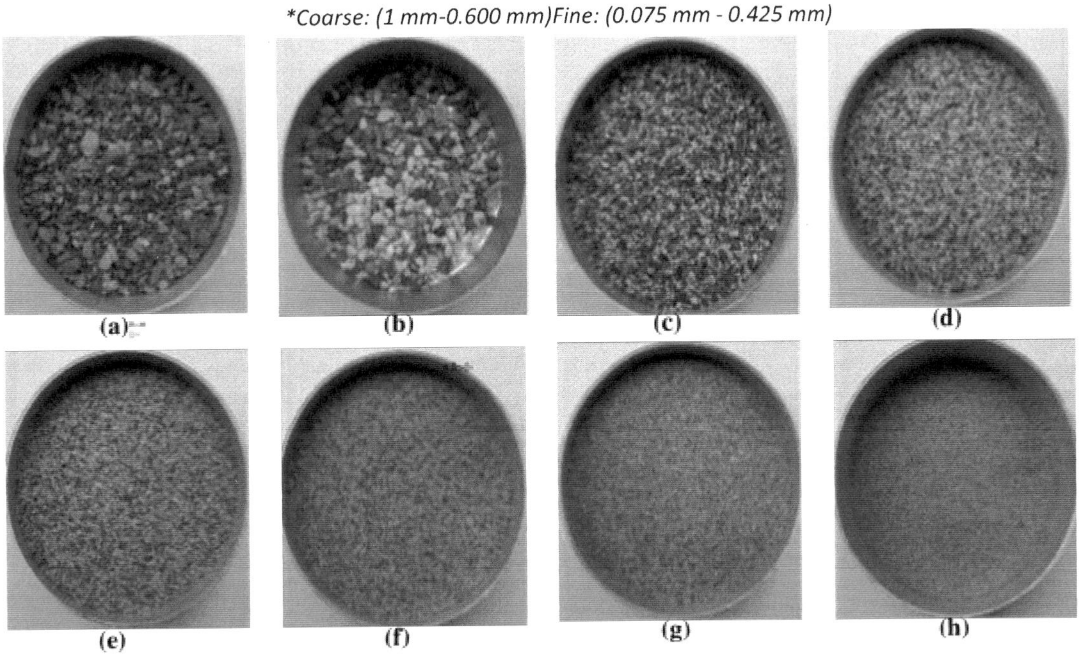

Fig.3. Different particle sizes (a) Flaky particles (b) Angular particles (c) 1mm-2 mm (d) 0.6mm - 1 mm (e) 0.425 mm – 0.6mm (f) 0.3 mm- 0.425 mm (g) 0.212 mm – 0.3 mm (h) 0.075 mm – 0.212 mm

2.3. *Consolidated undrained (CU) Triaxial Test*

The CU triaxial test is carried out on the test sample. As the drainage is not allowed in the shearing stage, so pore pressure develops in this stage. The triaxial test gives the measure of both the total and effective stresses under different confining pressures. In this study, all the soil specimens are tested under three confining pressures i.e., 50kN/m^2, 100kN/m^2 and 150kN/m^2. The consolidated drained triaxial test is conducted as per IS 2720 Part XII (1981). In this test, the failure plane's stress distribution is relatively uniform and the specimen can fail on any weak plane or can simply bulge.

The shear strength at failure of the sample is calculated using the Mohr's- Coulomb equationas shown in the equation below:

$\tau_f = c + \tan\phi$ (1)

where,

τ_f is the shear stress at failure plane of the soil sample.

C is the cohesion.

σ is the normal stress.

ϕ is the angle of internal friction.

The CU triaxial test was performed for the following samples with different arrangements:

2.3.1. Homogeneous and non-homogeneous sand sample

The arrangement of soil including both homogeneous and non-homogeneous sand in the CU triaxial apparatus is shown in Fig.4. For homogeneous sand sample the soil particles ranging from sizes passing 2mm and retaining 1mm to those retaining in 75 micron are homogeneously mixed as shown in Fig.4(a). Non-homogeneous sand sample is arranged in layers in the following combinations:

(i) Coarser to finer- The sand specimen is arranged in layers with particles ranging from those retaining in 1mm sieve to those passing 150 micron sieve as shown in Fig.4(b) ;

(ii) Finer to coarser- The sandspecimen is arranged in layers from particles ranging from those passing 150 micron to those retaining in 1mm sieve as shown in Fig.4(c);

(iii) Coarse-Fine-Coarse- The sand specimen is arranged in layers such that finer layers are sandwiched between coarser layers. The sample is prepared in this arrangement is as shown in Fig.4(d);

(iv) Fine-Coarse-Fine- The sand specimen is arranged in layers such that finer layers are sandwiched between coarser layers as shown in Fig.4(e).

Fig.4. CU Triaxial Test samples: (a) Homogeneous sand sample (b) Coarser to finer (c) Finer to Coarser (d) Coarse-Fine-Coarse (e) Fine -Coarse-Fine

2.3.2. Flaky -fine sand arrangement

The flaky and fine sand particles are arranged in the following combinations:

(i) 10% Flaky and 90% Fine Soil Sample : 10 % by weight of flaky particles (retaining on 2mm) and 90 % by weight of fine sand (passing 0.425mm sieve).

(ii) 20% Flaky and 80% Fine Soil Sample: 20 % by weight of flaky particles and 80 % byweight of fine sand.

(iii) 30% Flaky and 70% Fine Soil Sample: 30 % by weight of flaky particles 70 % byweight of fine sand.

(iv) 100% Flaky: Flaky particles without addition of finer sand layer

Each of the above combinations are taken and arranged in two layers, the flaky particles in the bottom layer and then fine soil sample is poured on the top and compacted by mild tamping. Fig.5 shows the arrangement of flaky particles before fine sand is poured.

Fig.5. Flaky-fine particle arrangement before fine sand is poured

2.3.3. Angular-fine sand arrangement

The angular and fine sand particles are arranged in the following combinations:

(i) 10% Angular and 90% Fine Soil Sample : 10 % by weight of angular particles (retaining on 2mm) and 90 % by weight of fine sand (passing 0.425mm sieve).

(ii) 20% Angular and 80% Fine Soil Sample: 20 % by weight of angular particles and 80 % by weight of fine sand.

(iii) 30% Angular and 70% Fine Soil Sample: 30 % by weight of angular particles and 70 % by weight of fine sand.

(iv) 100% Angular: Angular particles without addition of finer sand layer.

Each of the above combinations are taken and arranged in two layers, the angular particles in the bottom layer and then fine soil sample is poured on the top and compacted by mild tamping.

2.3.4. Different particle sizes The different particle particles sizes used are as follows:

(i) 1 mm-2 mm: Soil particles of sizes passing through 2 mm and retaining on 1mm

(ii) 0.6mm-1 mm: Soil particles of sizes passing through 1mm and retaining on 0.6mm

(iii) 0.425 mm-0.6mm: Soil particles of sizes passing through 0.6mm and retaining on 0.425 mm

(iv) 0.3mm-0.425 mm: Soil particles of sizes passing through 0.425 mm and retaining on 0.3 mm

(v) 0.212mm-0.3mm: Soil particles of sizes passing through 0.3 mm and retaining on 0.212 mm

(vi) 0.075-0.212mm: Soil particles of sizes passing through 0.212mm

3. Results

3.1. Effect of Homogeneous and Non-homogeneous arrangement on the soil shear strength

It is observed that when the soil sample is arranged in non-homogeneous layer the shear strength improved in comparison to homogeneous sand sample. The c and ϕ values increased for non-homogeneous samples in comparison to non-homogeneous sample. It was found that the c and ϕ values increased the most in 'fine-coarse-fine' layer arrangement The maximum deviator stresses, c and ϕ values obtained are tabulated in Table 3. The Mohr circles are drawn as shown in Fig.6

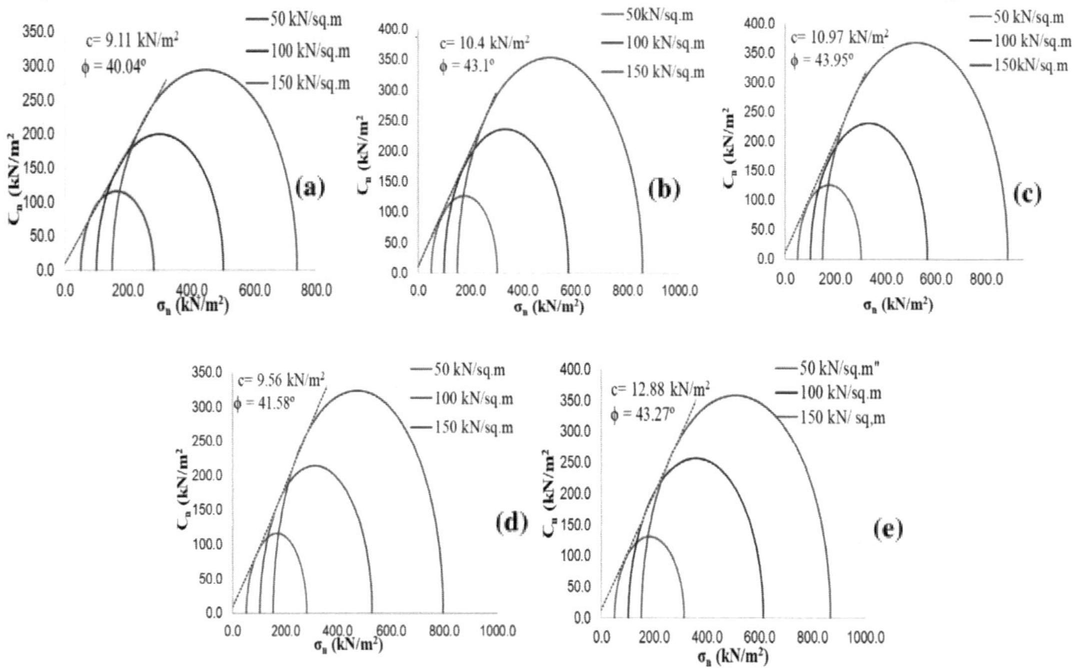

Fig.6.Mohr Circles for (a) Homogeneous sand sample (b) Coarser to finer (c) Finer to Coarser (d) Coarse-Fine-Coarse (e) Fine -Coarse-Fine

Table 3. Deviator stresses at failure, c and ϕ for homogeneous and non-homogeneous soil sample

Soil	Peak Deviator stresses(kN/m²)			c(kN/m²)	ϕ º
	σ_3=50kN/m²	100 kN/m²	150 kN/m²		
Homogeneous	231.7	399.95	588.82	9.11	40.04
Non-homogeneous(layered: Coarser to finer)	253.8	446.01	707.10	10.4	43.1
Non-homogeneous(layered: Finer to coarser)	252.0	463.37	734.49	10.17	43.95
Non-homogeneous(layered: coarse-fine-coarse)	232.2	427.27	646.38	9.56	41.58
Non-homogeneous(layered: fine-coarse- fine)	262.9	514.9	717.9	12.88	43.27

3.2. *Effect of flaky-fine sand arrangement on the soil shear strength*

It is observed that when flaky particles are added to fine sand the shear strength of soil is enhanced. The variation of the strength with addition of flaky particles in different percentages by weight is shown in Fig. 7(a). The deviator stresses and c and ϕ values obtained are given in Table 4. The Mohr circles are drawn as shown in Fig.8.

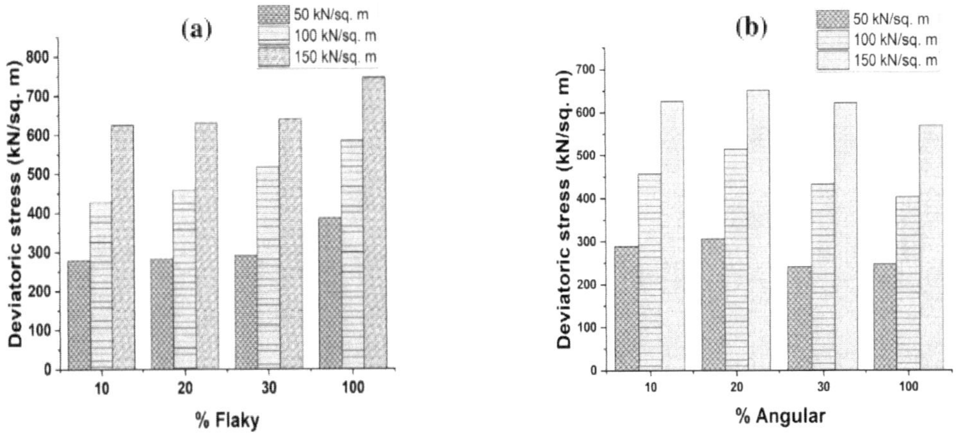

Fig.7 Effect of: (a)% flaky particles (b) % angular particles on the deviator stress

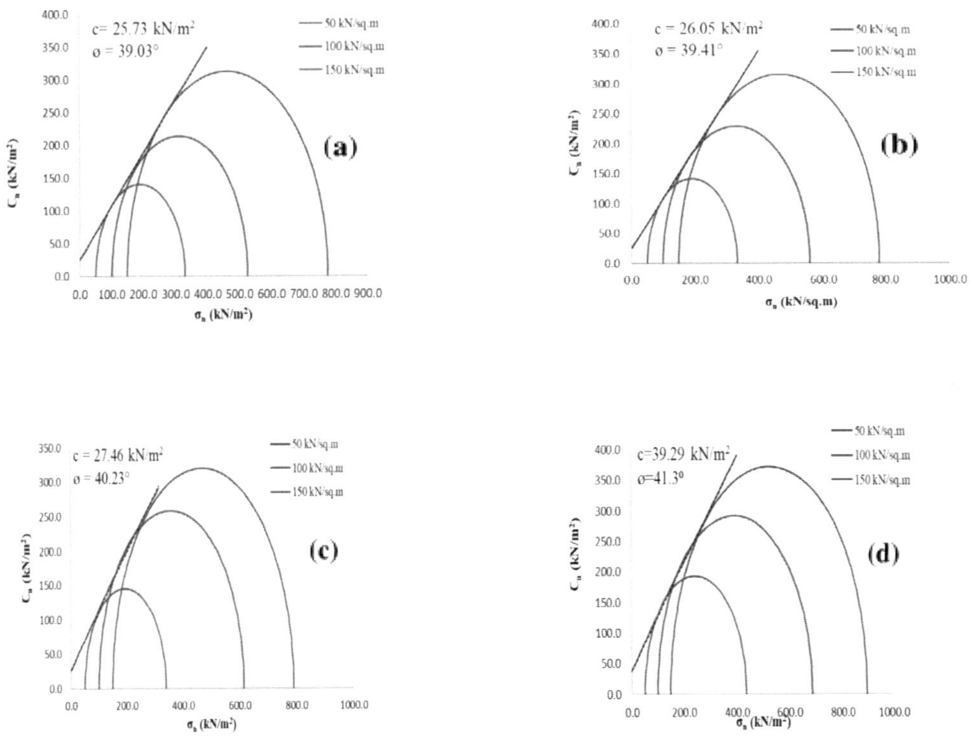

Fig.8. Mohr circle for (a) 10% Flaky-90% Fine sand (b) 20% Flaky and 80% Fine sand (c) 30%Flaky-70% Fine sand (d) 100% Flaky

Table 4. Deviator stresses at failure, c and ϕ values for Flaky-Fine particles arrangement

Soil	Peak Deviator Stresses (kN/m^2)			c(kN/m^2)	ϕ º
	σ_3=50 kN/m^2	100kN/m^2	150 kN/m^2		
10 % Flaky-90 % Fine	279.36	426.65	625.51	25.73	39.03
20 % Flaky-80 % Fine	282.84	457.82	630.95	26.05	39.41
30 % Flaky-70 % Fine	290.32	516.73	639.9	27.46	40.23
100 % Flaky-0 % Fine	386.10	584.47	745.33	39.29	41.3

3.3. *Effect of angular particles on the shear strength*

It is observed that upto 20% addition of angular particles, the soil shear strength improved after which the strength decreases. The variation of the strength with addition of angular particles in different percentages by weight is shown in Fig.7(b). The deviator stresses at failure, and c and ϕ values obtained are tabulated in Table 5. The Mohr circles are drawn as shown in Fig.9.

Table 5. Deviator stresses at failure, c and ϕ values for Angular-Fine particles arrangement

Soil	Peak Deviator Stresses (kN/m^2)			c(kN/m^2)	ϕ º
	σ_3=50 kN/m^2	100kN/m^2	150 kN/m^2		
10 % Flaky-90 % Fine	288.9	456.89	625.51	29	38.87
20 % Flaky-80 % Fine	306.75	514.91	650.93	24.85	40.59
30 % Flaky-70 % Fine	241.00	432.05	622.01	23.55	39.37
100 % Flaky-0 % Fine	246.6	402.5	569.69	14.69	38.37

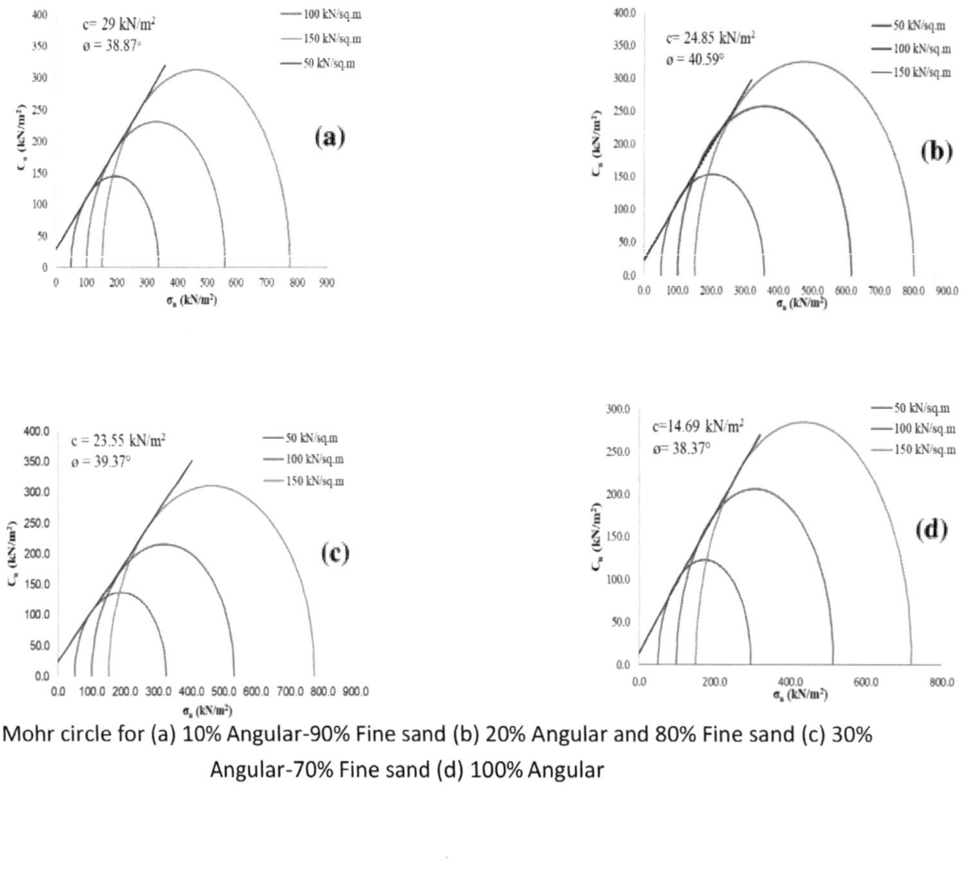

Fig.9. Mohr circle for (a) 10% Angular-90% Fine sand (b) 20% Angular and 80% Fine sand (c) 30% Angular-70% Fine sand (d) 100% Angular

IOP Conf. Series: Materials Science and Engineering 1282 (2023) 012021 doi:10.1088/1757-899X/1282/1/012021

3.4. *Effect of particle size on the shear strength*

It is observed that the soil's shear strength varies with the particle size. This variation is as shown in Fig.10. It is observed that with decreasing particle sizes there is a gradual decrease in c and ϕ values. The impact of the size of particles on the friction angle and cohesion are also shown in Fig.11. The Mohr circles for different particle sizes that are tested are also drawn as shown in Fig.12 and c and ϕ values are obtained. The deviator stresses and c and ϕ values obtained are tabulated in Table 6 respectively.

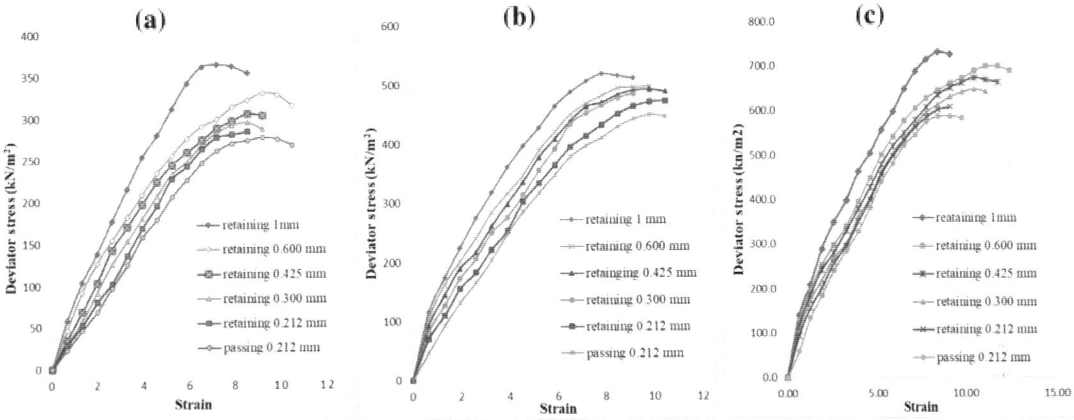

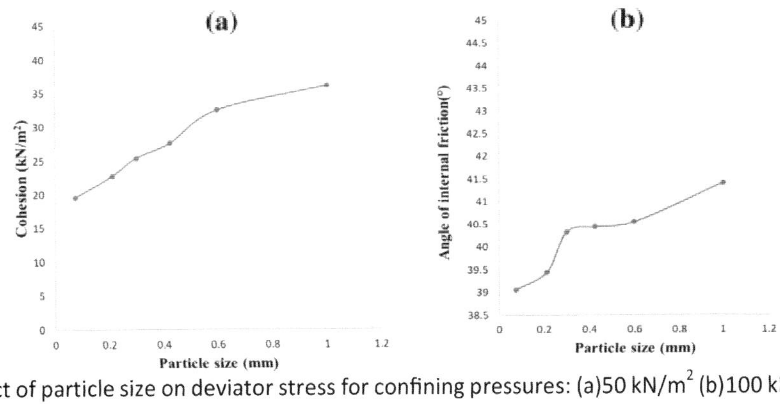

Fig.10 Effect of particle size on deviator stress for confining pressures: (a)50 kN/m^2 (b)100 kN/m^2 (c)150 kN/m^2

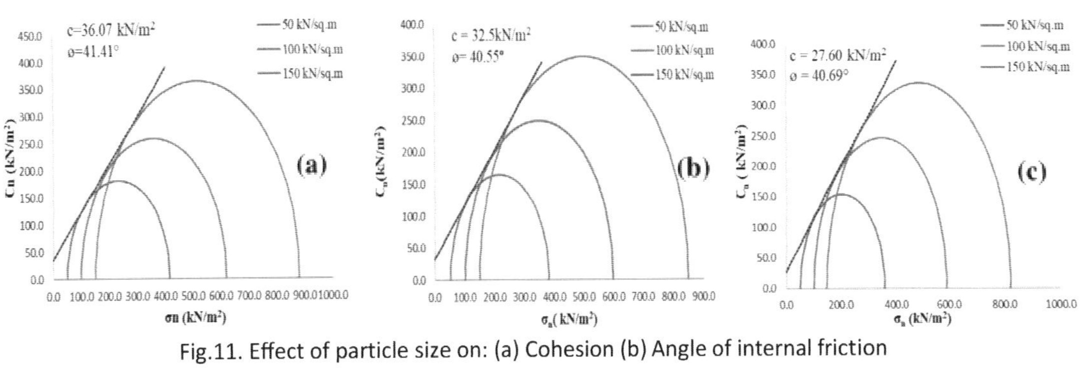

Fig.11. Effect of particle size on: (a) Cohesion (b) Angle of internal friction

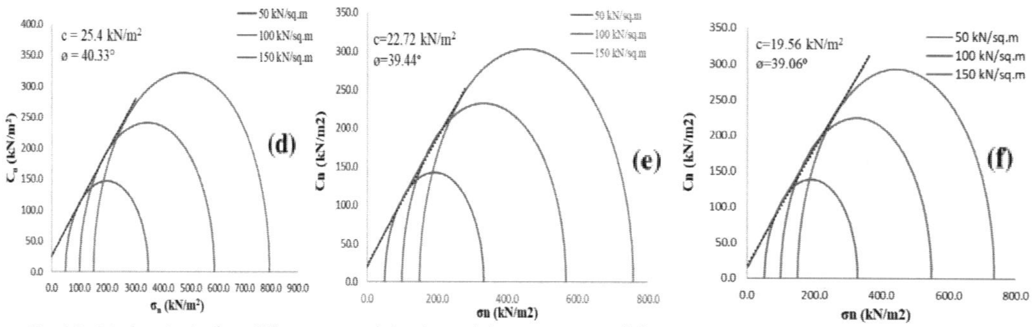

Fig.12. Mohr circle for different particle sizes: (a) 1mm-2 mm (b) 0.6mm-1mm (c) 0.425mm-0.6mm (d) 0.3mm-0.425mm (e) 0.212mm-0.3mm (f) 0.075mm-0.212mm

Table 6. Deviator stresses at failure, c and ϕ values for different particle sizes

Soil	Peak Deviator Stresses (kN/m^2)			c(kN/m^2)	ϕ º
	σ_3=50 kN/m^2	100kN/m^2	150 kN/m^2		
1 mm-2 mm	364.43	518.46	729.24	36.07	41.41
0.6mm-1 mm	380.05	596.78	847.84	32.5	40.55
0.425mm-0.6mm	305.64	491.5	671.53	27.6	40.69
0.3mm-0.425mm	294.91	484.44	645.53	25.4	40.33
0.212mm-0.300mm	284.19	466.92	607.77	22.72	39.44
0.075mm-0.212mm	276.82	449.22	585.58	19.56	39.06

4. Discussion

The following discussions can be made from the study:

- In the study, it was found that the c and ϕ values increased for non-homogeneous soil sample in comparison to homogeneous sample and ultimately the shear strength.
- However, when the soil was arranged such that the finer particles are placed in between two coarser layers the c and ϕ values came out lesser as compared to other arrangements. This can be because in triaxial test, the shear planes go through the middle and on the top and bottom there will be restraints from the caps. The fine layer is at the middle so the shear zone will mainly go to the portion having lesser ϕ value. When the coarse layer is in the middle the shear plane will go to the middle and a higher ϕ value is obtained.
- The shear strength tends to increase when flaky particles are included. An increase in the ϕ values was observed when the flaky particles was increased from 10% to 30% by weight. The maximum angle of internal friction was observed when 100% flaky particles was taken. The increase in cohesion is also observed as flaky content increases. The maximum cohesion is shown when flaky content is 100%.The increase in ϕ values as flaky content increases could be due to the shape and arrangement of flaky particles. Flaky particles have lesser void percentage as compared to that of the angular particles which has more void percent, thus providing a denser arrangement in the flaky sample.
- In case of angular and fine particle arrangement, the cohesion was observed to decrease with the increase in angular particle content from 10% to 20%. On the other hand, the internal friction angle increased by about 4% from the previous value containing 10% angular content. With the further increase in angular content to 30% by weight, both c and ϕ values decreased to some amount. Fine sand particles with concentration of angular particles between 10% -20% dominate the shear strength. As the % of angular particles is increased the density becomes lesser as the

angular particles concentration by weight increases.

- On comparing following combinations:
  (a) 10% flaky- 90% fines and 10% angular- 90% fines
  (b) 20% flaky- 80% fines and 20% angular- 80% fines and
  (c) 30% flaky- 90% fines and 30% angular- 70% fines, it was observed that the angular arrangement upto 20% showed a higher value of shear stress as compared to that of 20% flaky arrangement.

But with further increase in the angular content with upto 30% by weight the shear stress decreased for all the confining pressures, compared to flaky arrangement.

- Out of all the flaky-fine sand combination and angular-fine sand combinations, the highest deviator stress was observed in the sample having 100% flaky soil content, whereas in the sample having 100% angular content the lowest deviator stress was observed.
- The shear behaviour of the soil as well as its contact surface and rolling and sliding resistance are influenced by the corresponding voids among the grain mass as a result of the change in theparticle size . The maximum shear strength as well as angle of internal friction increases when the particle size increases. A gradual decrease in c and ϕ value was observed with decreasing particle sizes. This could probably be due to the reason that with more fine content although the soil seems to be densely packed, the volume of voids is more in the fine sand content than that of the coarse sand. This could probably have resulted a decrease in the friction angle.

5. Conclusion

Based on the data observed from the results and discussions, these conclusions can be made as follows:

- The shear strength of the soil improved when the soil was arranged in layers as compared to when the soil was homogeneous. It was found that the c and ϕ values increased the most in fine-coarse-fine layer arrangement. The c value increased by 3 kN/m^2 and ϕ increased from 40.04° to 43.27° respectively.
- Sometimes layers found in the soil may not be very thick but can only be few centimeters thin. However, this might affect the overall shear strength of the soil. Thus, this study is an effort to find out the shear behaviour of such layered soils.
- In the field, samples are not cylindrical and there are no restraints, therefore we cannot explainthe behaviour yet. If there are alternate layers, the combined shear strength of the layers will be probably be governed by the layer which is weaker. The non-homogeneity of the soil therefore affects the overall shear strength of the soil
- Particle shape and size plays a significant part in affectiong the mechanical behaviour of soil. In this study it is seen that the change in particle size influences the angle of internal friction and in turn the shear strength. The shear strength was seen to improve in case of angular-fine arrangement when compared with flaky-fine soil arrangement. But this increase was only upto when the angular content was about 20% by weight. With angular content of 30% or more the shear strength decreases probably due increase in void content. Thus, angular particles could serve as a good reinforcement to soils. But if the angular content is in excess, as compared to the fines then a decrease in strength could be observed.
- In this study the soil particles were limited to certain sizes and they can further be increased to obtain test results. The constraints imposed by the smaller size of the sample provided some limitations for preparation of a large sized samples . A further research using a larger size triaxial test can be carried out for increased particle sizes.
- The angular and flaky particles were prepared in layers with the fine sand. A further researchon mixing them homogeneously in different proportions could help in understanding the strength behaviour.
- Rough or smooth textured angular and flaky sand can also be used for understanding the shear parameters.

References

[1] Anantanasakul P, Yamamuro J A and Kaliakin V N 2012 Stress-strain and strength characteristics of silt-clay transition soils *Journal Of Geotechnical and Geoenvironmental Engineering* **138(10)**, 1257-1265.

[2] Balaban E, Smejda A and Onur M I 2019 An Experimental Study on Shear Strength Behavior of Soils under Low Confining Pressure *In Proceedings of the 4th World Congress on Civil, Structural,and Environmental Engineering,* 7-9.

[3] Banupriya S, Soundarya M and Chakravarthy P R K 2015 Stress-strain and strength characteristics of sand-silt mixtures *International Journal of Latest Trends in Engineering and Technology* **6(1)**.

[4] Chen Y and Meehan C L 2011 Undrained strength characteristics of compacted bentonite/sand mixtures *In Geo-Frontiers 2011: Advances in Geotechnical Engineering*, 2699-2708.

[5] Cola S and Simonini P 2002 Mechanical behavior of silty soils of the Venice lagoon as a function of their grading characteristics *Canadian Geotechnical Journal* **39(4)**, 879-893.

[6] Doddiah D, Bhat H S, Somasekhar P V, Sosalegowda H B and Ranganath K N 1969 Shear characteristics of soil-gravel mixtures *Journal of the Indian National Society of Soil Mechanics and Foundations Engineering* **8(1)**, 57-66.

[7] Fragaszy R J, Su J, Siddiqi F H and Ho C L 1992 Modeling strength of sandy gravel *Journal of Geotechnical Engineering* **118(6)**, 920-935.

[8] Hamidi A, Yazdanjou V and Salimi N 2009 Shear strength characteristics of sand-gravel mixtures *International Journal of Geotechnical Engineering* **3(1)**, 29-38.

[9] Holtz W G and Gibbs H J 1956 Triaxial shear tests on pervious gravelly soils *Journal of the Soil Mechanics and Foundations Division* **82(1)**, 867-1.

[10] Marsal R J and Fuentes de la Rosa A 1976 Mechanical properties of rockfill soil mixtures *In Transactions of the 12th International Congress on Large Dams, Mexico City* **1**, 179-209.

[11] Ronad H 2014 An Experimental Study on Variation of Shear Strength for Layered Soils *International Journal of Research in Advent Technology* **2(5)**.

[12] Vahidi-Nia F, Lashkari A and Binesh S M 2015 An insight into the mechanical behavior of binary granular soils *Particuology* **21**, 82-89.

[13] Vallejo L and Lobo-Guerrero S 2012 The shear strength of granular materials containing dispersed oversized particles:DEM analysis *International Journal of Geotechnical Engineering* **6(3)**, 371-380.

[14] Wang C and Zhan C N 2011 Gravel content effect on the shear strength of clay-gravel mixtures. *In Applied Mechanics and Materials* **71**, 4685-4688.

[15] Wei H Z, Xu W J, Xu X F, Meng Q S and Wei C F 2018 Mechanical properties of strongly weathered rock-soil mixtures with different rock block contents *International Journal of Geomechanics*, **18(5)**, 04018026.

[16] Xu D S, Tang J Y, Zou Y, Rui R and Liu H B 2019 Macro and micro investigation of gravel content on simple shear behavior of sand-gravel mixture *Construction and Building Materials* **221**, 730-744.

[17] Yao Y, Li J, Ni J, Liang C and Zhang A 2022 Effects of gravel content and shape on shear behaviour of soil-rock mixture: Experiment and DEM modelling *Computers and Geotechnics* **141**, 104476.

AUTHOR INDEX

Institute of Physics
Dirac House, Temple Back
Bristol BS1 6BE UK

ISSN: 1757-8981
ISBN 978-1-7138-8184-1

9 781713 881841

CURRAN ASSOCIATES INC.
proceedings.com